Pipefitting
Level Two

Trainee Guide
Third Edition

PEARSON

Upper Saddle River,
New Jersey
Columbus, Ohio

NCCER

President: Don Whyte
Director of Curriculum Revision and Development: Daniele Dixon
Pipefitting Project Manager: Daniele Dixon
Production Manager: Tim Davis
Quality Assurance Coordinator: Debie Ness
Editors: Bethany Harvey, Brendan Coote
Desktop Publisher: Jennifer Jacobs, James McKay

NCCER would like to acknowledge the contract service provider for this curriculum:
Topaz Publications, Liverpool, New York.

This information is general in nature and intended for training purposes only. Actual performance of activities described in this manual requires compliance with all applicable operating, service, maintenance, and safety procedures under the direction of qualified personnel. References in this manual to patented or proprietary devices do not constitute a recommendation of their use.

ISBN 10:   0-13-227314-4
ISBN 13: 978-0-13-227314-5

PREFACE

TO THE TRAINEE

There are some who may consider pipefitting synonymous with plumbing, but these are really two very distinct trades. Plumbers install and repair the water, waste disposal, drainage, and gas systems in homes and in commercial and industrial buildings. Pipefitters, on the other hand, install and repair both high- and low-pressure pipe systems used in manufacturing, in the generation of electricity, and in the heating and cooling of buildings.

If you're trying to imagine a setting involving pipefitters, think of large power plants that create and distribute energy throughout the nation; think of manufacturing plants, chemical plants, and piping systems that carry all kinds of liquids, gaseous, and solid materials.

If you're trying to imagine a job in pipefitting, picture yourself in a job that won't go away for a long time. As the US government reports, the demand for skilled pipefitters continues to outpace the supply of workers trained in this craft. And high demand typically means higher pay, making pipefitters among the highest-paid construction workers in the nation.

While pipefitters and plumbers perform different tasks, the aptitudes involved in these crafts are comparable. Attention to detail, spatial and mechanical abilities, and the ability to work efficiently with the tools of their trade are key. If you think you might have what it takes to work in this high-demand occupation, contact your local NCCER Training Sponsor to see if they offer a training program in this craft, or contact your local union or non-union training programs. You might make the perfect "fit".

We wish you success as you embark on your second year of training in the pipefitting craft and hope that you'll continue your training beyond this textbook. There are more than a half-million people employed in this work in the United States, and as most of them can tell you, there are many opportunities awaiting those with the skills and desire to move forward in the construction industry.

We invite you to visit the NCCER website at **www.nccer.org** for the latest releases, training information, *Cornerstone* magazine, and much more. You can also reference the Pearson product catalog online at **www.crafttraining.com**. Your feedback is welcome. You may email your comments to **curriculum@nccer.org** or send general comments and inquiries to **info@nccer.org**.

NCCER STANDARDIZED CURRICULA

NCCER is a not-for-profit 501(c)(3) education foundation established in 1995 by the world's largest and most progressive construction companies and national construction associations. It was founded to address the severe workforce shortage facing the industry and to develop a standardized training process and curricula. Today, NCCER is supported by hundreds of leading construction and maintenance companies, manufacturers, and national associations. The NCCER Standardized Curricula was developed by NCCER in partnership with Pearson, the world's largest educational publisher.

Some features of NCCER's Standardized Curricula are as follows:

- An industry-proven record of success
- Curricula developed by the industry for the industry
- National standardization, providing portability of learned job skills and educational credits
- Compliance with the Office of Apprenticeship requirements for related classroom training (*CFR 29:29*)
- Well-illustrated, up-to-date, and practical information

NCCER also maintains a Registry that provides transcripts, certificates, and wallet cards to individuals who have successfully completed a level of training within a craft in NCCER's Curricula. *Training programs must be delivered by an NCCER Accredited Training Sponsor in order to receive these credentials.*

Contents

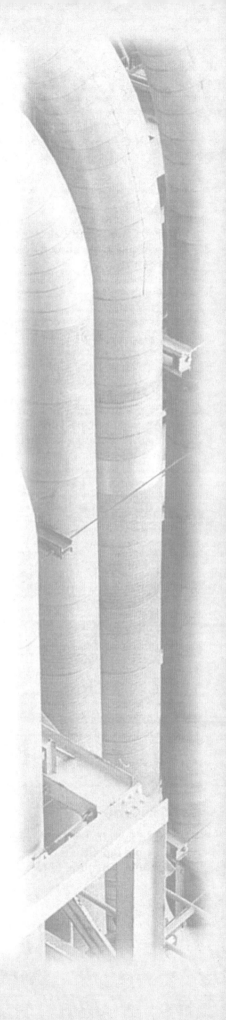

NCCER Standardized Curricula

NCCER's training programs comprise more than 80 construction, maintenance, pipeline, and utility areas and include skills assessments, safety training, and management education.

Boilermaking
Cabinetmaking
Carpentry
Concrete Finishing
Construction Craft Laborer
Construction Technology
Core Curriculum:
    Introductory Craft Skills
Drywall
Electrical
Electronic Systems Technician
Heating, Ventilating, and
    Air Conditioning
Heavy Equipment Operations
Highway/Heavy Construction
Hydroblasting
Industrial Coating and Lining
    Application Specialist
Industrial Maintenance Electrical
    and Instrumentation Technician
Industrial Maintenance
    Mechanic
Instrumentation
Insulating
Ironworking
Masonry
Millwright
Mobile Crane Operations
Painting
Painting, Industrial
Pipefitting
Pipelayer
Plumbing
Reinforcing Ironwork
Rigging
Scaffolding
Sheet Metal
Signal Person
Site Layout
Sprinkler Fitting
Tower Crane Operator
Welding

Maritime

Maritime Industry Fundamentals
Maritime Pipefitting
Maritime Structural Fitter

Green/Sustainable Construction

Building Auditor
Fundamentals of Weatherization
Introduction to Weatherization
Sustainable Construction Supervisor
Weatherization Crew Chief
Weatherization Technician
Your Role in the Green Environment

Energy

Alternative Energy
Introduction to the Power Industry
Introduction to Solar Photovoltaics
Introduction to Wind Energy
Power Industry Fundamentals
Power Generation Maintenance
    Electrician
Power Generation I&C Maintenance
    Technician
Power Generation Maintenance
    Mechanic
Power Line Worker
Power Line Worker: Distribution
Power Line Worker: Substation
Power Line Worker: Transmission
Solar Photovoltaic Systems Installer
Wind Turbine Maintenance
    Technician

Pipeline

Control Center Operations, Liquid
Corrosion Control
Electrical and Instrumentation
Field Operations, Liquid
Field Operations, Gas
Maintenance
Mechanical

Safety

Field Safety
Safety Orientation
Safety Technology

Supplemental Titles

Applied Construction Math
Careers in Construction
Tools for Success

Management

Fundamentals of Crew Leadership
Project Management
Project Supervision

Spanish Titles

Acabado de concreto: nivel uno,
    nivel dos
Aislamiento: nivel uno, nivel dos
Albañilería: nivel uno
Andamios
Aparejamiento básico
Aparajamiento intermedio
Aparajamiento avanzado
Carpintería:
    Introducción a la carpintería,
    nivel uno; Formas para
    carpintería, nivel tres
Currículo básico: habilidades
    introductorias del oficio
Electricidad: nivel uno, nivel dos,
    nivel tres, nivel cuatro
Encargado de señales
Especialista en aplicación de
    revestimientos industriales: nivel
    uno, nivel dos
Herrería: nivel uno, nivel dos, nivel
    tres
Herrería de refuerzo: nivel uno
Instalación de rociadores: nivel uno
Instalación de tuberías: nivel uno,
    nivel dos, nivel tres, nivel cuatro
Instrumentación: nivel uno, nivel
    dos, nivel tres, nivel cuatro
Mecánico industrial: nivel uno, nivel
    dos, nivel tres, nivel cuatro, nivel
    cinco
Paneles de yeso: nivel uno
Seguridad de campo
Soldadura: nivel uno, nivel dos,
    nivel tres

Portuguese Titles

Currículo essencial: Habilidades
    básicas para o trabalho
Instalação de encanamento
    industrial: nível um, nível dois,
    nível três, nível quatro

Acknowledgments

This curriculum was revised as a result of the farsightedness and leadership of the following sponsors:

Becon
Cianbro
Flint Hills Resources / Koch Industries
Fluor Global Craft Services
Kellogg Brown & Root
Lee College
Zachry Construction Corporation

This curriculum would not exist were it not for the dedication and unselfish energy of those volunteers who served on the Authoring Team. A sincere thanks is extended to the following:

Glynn Allbritton
Tom Atkinson
Ned Bush
Adrian Etie
Daniele Gomez
Tina Goode
Ron Harper
Ed LePage
Toby Linden

NCCER PARTNERING ASSOCIATIONS

American Fire Sprinkler Association
Associated Builders and Contractors, Inc.
Associated General Contractors of America
Association for Career and Technical Education
Association for Skilled and Technical Sciences
Carolinas AGC, Inc.
Carolinas Electrical Contractors Association
Center for the Improvement of Construction Management and Processes
Construction Industry Institute
Construction Users Roundtable
Construction Workforce Development Center
Design Build Institute of America
GSSC – Gulf States Shipbuilders Consortium
Manufacturing Institute
Mason Contractors Association of America
Merit Contractors Association of Canada

NACE International
National Association of Minority Contractors
National Association of Women in Construction
National Insulation Association
National Ready Mixed Concrete Association
National Technical Honor Society
National Utility Contractors Association
NAWIC Education Foundation
North American Technician Excellence
Painting & Decorating Contractors of America
Portland Cement Association
SkillsUSA®
Steel Erectors Association of America
U.S. Army Corps of Engineers
University of Florida, M. E. Rinker School of Building Construction
Women Construction Owners & Executives, USA

08201-06

Piping Systems

08201-06
Piping Systems

Topics to be presented in this module include:

Overview

Piping systems vary widely in materials, components, and procedures. The systems are subject to specific standards, depending on the materials being transported and the context of the system. Color codes warn personnel of the safety requirements for piping systems. Expansion of pipe materials due to heating and cooling is a factor in the design of pipe systems. Insulation serves several purposes in piping systems, including preventing flow interruption by freezing or liquefaction, and protecting personnel from injury.

Objectives

When you have completed this module, you will be able to do the following:

1. Identify and explain the types of piping systems.
2. Identify piping systems according to color-coding.
3. Explain the effects and corrective measures for thermal expansion in piping systems.
4. Explain types and applications of pipe insulation.

Trade Terms

Acid
Ambient temperature
Caustic
Condensate
Haunch
Heat-tracing
Material safety data sheet (MSDS)
Pounds per square inch gauge (psig)
Saturated steam
Slurry
Superheated steam
Water hammer

Required Trainee Materials

1. Pencil and paper
2. Appropriate personal protective equipment

Prerequisites

Before you begin this module, it is recommended that you successfully complete *Core Curriculum*; and *Pipefitting Level One*.

This course map shows all of the modules in the second level of the *Pipefitting* curriculum. The suggested training order begins at the bottom and proceeds up. Skill levels increase as you advance on the course map. The local Training Program Sponsor may adjust the training order.

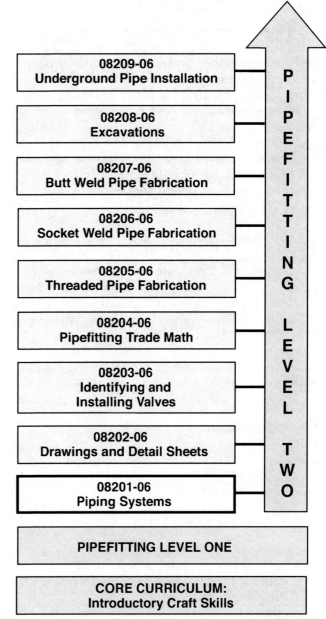

08209-06
Underground Pipe Installation

08208-06
Excavations

08207-06
Butt Weld Pipe Fabrication

08206-06
Socket Weld Pipe Fabrication

08205-06
Threaded Pipe Fabrication

08204-06
Pipefitting Trade Math

08203-06
Identifying and Installing Valves

08202-06
Drawings and Detail Sheets

08201-06
Piping Systems

PIPEFITTING LEVEL ONE

CORE CURRICULUM:
Introductory Craft Skills

PIPEFITTING LEVEL TWO

202CMAP.EPS

1.0.0 ◆ INTRODUCTION

A piping system is a complete network of pipes, fittings, valves, and other components that are designed to work together to convey a specific material. Piping systems must be designed and fabricated according to local building codes, plans, specifications, and component installation instructions. Where conflicts between these sources are found, always use the one that is the most conservative and provides the greatest margin of safety. This module explains the different types of piping systems, thermal expansion of pipes, and pipe insulation.

2.0.0 ◆ TYPES OF PIPING SYSTEMS

Piping systems are designed to convey a wide variety of gases and fluids. Each system must meet special requirements depending on the type of fluid to be conveyed and the temperature, pressure, flow rate, and working atmosphere. The following sections explain the specific design guidelines and hazards of the following types of piping systems:

- Chemical
- Compressed air
- Fuel oil
- Steam
- Water

2.1.0 Chemical Piping Systems

The correct selection of piping materials is critical for chemical piping systems because of the corrosive nature of many chemicals. Manufacturing processes involving chemicals use acids, caustics, and other chemical substances that can attack the piping, fittings, and valves through which they pass. The atmosphere surrounding these systems is also highly corrosive, so the piping systems are attacked from the outside as well as the inside. Chemical piping systems can be either vapor systems or liquid systems. Follow these safety guidelines when working around chemical piping systems:

- Always know exactly what chemicals are contained in the piping systems and know the location of the material safety data sheet (MSDS) for all chemicals in your work area.
- Wear personal protective equipment as specified by your employer and the MSDS, including a chemical resistance suit.
- Always know the location of the nearest eyewash fountain and quick-drench shower.

- If someone inhales chemicals, move the person from the exposure area into fresh air, keep the person warm and at rest, and seek medical attention immediately.
- If a person is unconscious in a hazardous area, do not enter the hazardous area to retrieve that person. Contact the emergency response team immediately.

 WARNING!
If you notice a gas or vapor cloud, notify operations immediately!

- Never walk through a vapor cloud in a hazardous area. Go upwind or walk around the cloud.

2.1.1 Acid Piping Systems

Sulfuric acid and hydrochloric acid are commonly used in waste water neutralization processes at chemical plants. These acids are typically used at 20 to 50 percent concentrations. Solutions of 30 to 50 percent concentrations have a relatively low freezing point of –20°F. At 20 percent concentration, the freezing point is 8°F, and the system may require some type of freeze protection. When freeze protection is required, heat-tracing and insulation must be used.

The recommended piping materials for sulfuric acid and hydrochloric acid are either Teflon®-lined or polypropylene-lined flanged steel pipe. Corrosion-resistant thermoplastic piping and fiberglass-reinforced pipe are also becoming more widely used for conveying acids. Be sure to obtain clear information on chemical resistance before using any material in acidic or caustic systems. All valves in these systems should be installed with the stem pointing upward and be equipped with a splash shield to protect the operator and the equipment. Many times, slurry from the acid in the system will clog valves, preventing tanks and pipelines from draining completely. When working with acid systems, be aware of this possibility. Protect yourself from exposure to the acid by wearing the proper personal protective equipment.

2.1.2 Caustic Piping Systems

The most widely used of the commercial caustic materials is caustic soda, or sodium hydroxide, which is used in chemical manufacturing, petroleum refining, and pulp and paper manufacturing. Caustic soda is generally used at 20 to 50 percent concentrations. The higher the concentration, the higher the freezing point will be. Caustic soda at 50 percent concentration freezes at 54°F. At 30 percent concentration, the freezing point drops to 32°F, and at 20 percent concentration, the freezing point is at about –18°F.

The operating pressure of a caustic system is about 50 pounds per square inch gauge (psig) at ambient temperature. Caustic piping systems fall into the American National Standards Institute (ANSI) 150-pound class, a standard for the manufacture of pipe, valves, and flanges. Caustic systems require carbon steel materials that can be flanged, screwed, or welded. Caustics corrode some piping materials, including copper and cast iron. Lead and tin must not be used with caustics, because flammable hydrogen gas is released when caustics react with these metals. If threaded pipe is used, Teflon® tape must be used on all threads. Copper, brass, zinc, or aluminum accessories should not be used, because these metals are not resistant to the caustic. Many piping systems now use new plastic linings, including fusion-bonded epoxy or liquid epoxy, to protect metal piping from corrosion or other chemical damage.

2.2.0 Compressed Air Piping Systems

Compressed air piping systems are classified as either instrument air or utility air. Air that operates many of the plant instruments is instrument air. Instrument air is compressed and dried to remove the moisture, because moisture in the air fouls or damages the equipment. The air is routed to all of the plant pneumatic instruments through piping systems. Instrument air piping systems are usually galvanized steel, stainless steel, or copper, which protects against internal corrosion and rust.

Utility air is air that is compressed and then used for driving air motors in power tools, for propelling fluids, for operating various cleaning services, and for emptying and drying piping. Most utility air systems are carbon steel welded pipe. Air pressure for each system varies, depending on the application, but usually ranges between 50 and 125 psig. *Figure 1* shows a typical compressed air piping system.

Follow these safety guidelines when working around compressed air systems:

- Because compressed air is under pressure, precautions must be taken in the event of fire. Excessive heat may cause vessels to rupture explosively.

- Never direct the airflow from an air nozzle directly on your skin because an air bubble could enter the skin and the bloodstream.
- Never stop a leak in a compressed air piping system with your finger or any other part of your body.
- Do not inhale compressed air because this may cause decompression sickness, which has a variety of adverse effects and can lead to long-term illness and even death.
- In case of inhalation, move the person to fresh air immediately. If breathing has stopped, seek medical attention immediately, and perform artificial respiration while waiting; keep the person warm and at rest.
- Do not strike vessels or fittings, as the seals could break and release high pressure air and pieces of the fittings.

2.3.0 Fuel Oil Piping Systems

Most fuel oil piping systems convey fuel oil from a tank or vessel, usually located underground, to some type of fuel-dispensing island or to a heating unit. Pipefitters install, connect, and replace thousands of fuel oil tanks each year. Even after the tank is emptied, the vapors remaining in fuel tanks and piping can be dangerously flammable; be very cautious in working around them. Leaking tanks and pipes present a threat to the environment because the fuel oil seeps into the ground and into underground water supplies. Fuel oil piping systems must be made of suitable materials that can withstand the operating pressures, structural stress, and chemicals they will be subjected to. Piping, fittings, and valves must also be compatible with the products being stored and must be resistant to underground corrosion. Where practical, piping should run in a single trench from the tank along the shortest possible route to the dispensers. Piping across the tanks should be avoided whenever possible. All piping should be sloped at least 1/8 inch per foot back to the tanks. When selecting materials and the type of connections to use in fuel oil piping systems, always refer to the job specifications. *Figure 2* shows a schematic diagram of a typical fuel oil piping system.

2.4.0 Steam Piping Systems

Steam is a gas that is generated by adding heat energy to water in a boiler. It is a very efficient and easily controlled heat-transfer medium. The steam piping system is used to transfer energy from a

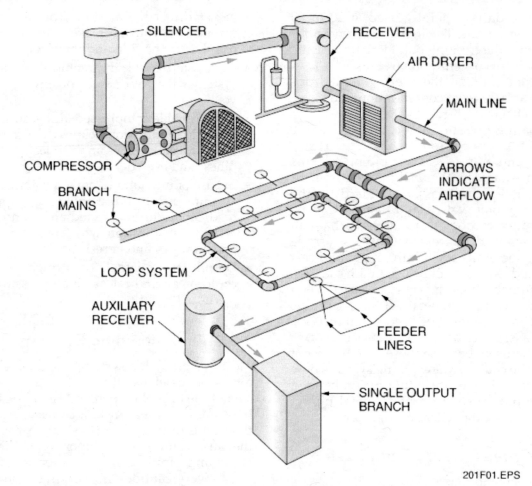

201F01.EPS

Figure 1 ◆ Compressed air piping system.

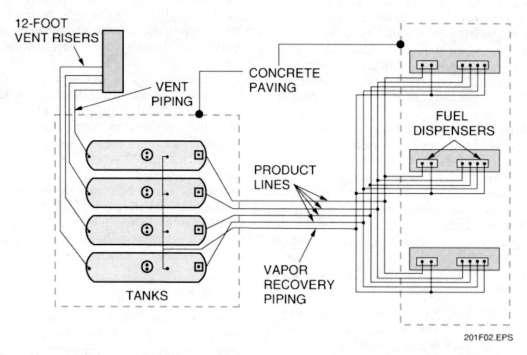

201F02.EPS

Figure 2 ◆ Fuel oil piping system.

central area of the plant, or the boiler, to other areas of the plant. Steam is commonly classified according to three ranges of pressure. The steam used in household heating systems is generally at a pressure of less than 15 psi. Saturated steam is used in a number of industrial applications and has a pressure range from 50 to 150 psi. Superheated steam is used in heavy industrial applications, such as for driving large steam turbines or heating reboilers, and is rated at 200 psi and up. Superheated steam has two distinct advantages over saturated steam: it is more efficient than saturated steam, and since it is dryer, it is less likely to condense while moving through the turbine.

Steam flows naturally from a point of higher pressure, as is generated in the boiler, to a point of lower pressure in the steam mains. Because of this, the natural flow of steam does not require a pump, such as that needed for hot water heating. Steam also circulates through a heating system much faster than other types of fluids.

Condensate is the by-product of a steam system and is formed in the distribution lines when the steam condenses to water. Once the steam condenses, the hot condensate must be removed from the system immediately and returned to the boiler. Hot steam that comes in contact with condensate that has cooled below the temperature of the steam may cause water hammer, which makes an annoying banging sound and can damage the pipes and fittings. You must therefore ensure that there is no water in a system when the steam is introduced, because the water will expand at a high rate of speed and will damage the system. *Figure 3* shows a schematic diagram of a typical steam piping system.

2.5.0 Water Piping Systems

Among the types of water carried by piping systems are utility water, potable water, demineralized water, distilled water, and cooling water. Utility water is used primarily for cleaning, washing down, and other industrial purposes. It is untreated water that comes directly from a stream, reservoir, or deep well. Potable water is drinkable water that comes from the city water main and must be installed according to city and county codes. Demineralized water is treated to remove minerals that may corrode piping or damage equipment in the plant. Distilled water is a specially purified type of water. Cooling water is used as a medium for displacing heat and is also untreated water that comes directly from a stream, reservoir, or deep well.

Cooling water systems commonly run underground because pipe racks cannot support the weight of the pipes that are used. Cooling water

systems typically use 78- and 84-inch pipe in large industrial settings. A wide range of piping materials can be used to convey water. The engineering specifications govern the type of materials used based on the type of water, operating pressure, surrounding conditions, and application. Utility water systems and cooling water systems usually use carbon steel or PVC pipe; potable water systems use threaded galvanized steel, copper, stainless steel, ductile iron, or PVC pipe. These systems must be sanitized, usually by being flushed and treated with chlorine, before use.

> **NOTE**
> In some states, a third-party contractor is required by code to perform the sanitizing procedures on potable water systems.

Both demineralized water systems and distilled water systems use stainless steel pipe. *Figure 4* shows a typical potable water piping system.

Figure 5 shows a typical cooling water piping system.

3.0.0 ◆ IDENTIFYING PIPING SYSTEMS

In many industries, the piping systems are very complex, transmitting not only steam and water, but also dangerous and valuable fluids. The American National Standards Institute (ANSI) has approved a Standard Identification Scheme for Piping Systems. This standard has been adapted for all industries and is designed to be recognized by all workers and by city fire departments. The primary purpose of this identification scheme is to reduce hazards to life and property in mills and plants.

This identification scheme places piping systems into five classes, each with its own color-code. Color bands are painted on the pipes at regular intervals, preferably close to valves and fittings, to ensure ready recognition during operation, repairs, or in times of emergency. Sometimes, the entire length of the piping system is painted the classification color. In PVC pipe, color is usually inserted during the pipe extrusion process. When the pipes are located above the operator's line of vision, the color band should be placed below the horizontal center line of the pipe so it can be easily recognized.

> **NOTE**
> Color coding systems may vary from one plant to another. In some facilities, written labels are used instead of color codes.

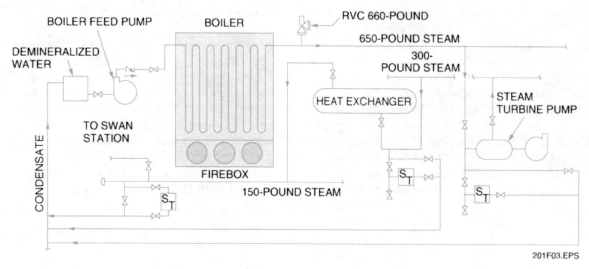

Figure 3 ♦ Steam piping system.

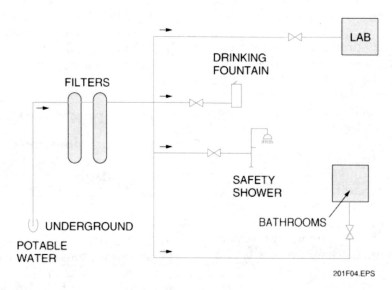

Figure 4 ♦ Potable water piping system.

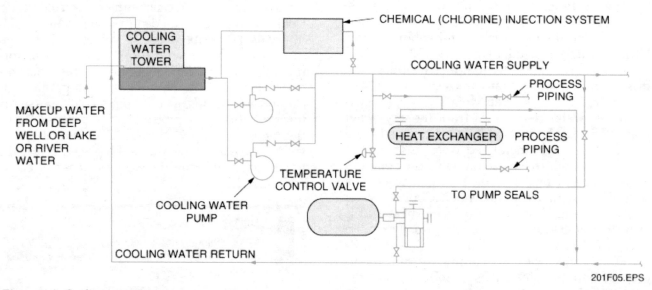

Figure 5 ♦ Cooling water piping system.

The ANSI recommendations apply to all piping systems in industrial and power plants, with the exception of pipes buried in the ground and electrical conduit lines. Fittings, valves, and pipe coverings are also covered by the standard. *Table 1* shows the Standard Identification Scheme for Piping Systems.

Class F refers to fire-protection materials and equipment. These piping systems include sprinkler systems and other fire-fighting equipment and are designated by the color red.

Class D refers to dangerous materials. These piping systems contain materials that are hazardous to life and property because they are poisonous, flammable, or produce poisonous gases. Fire-producers and explosives are also included in this category. These piping systems are designated by the color yellow. Yellow is also used for steam systems.

Class S refers to safe materials, which are products that pose no hazard. A pipefitter making repairs to a Class S system will run no risks in breaking into this system, even if the material has not been completely emptied from the system. These piping systems are designated by the color green. Other colors that can also be used with Class S systems are white, black, and gray. Potable water falls into this category.

Class P refers to protective materials, which include materials that a plant has available to prevent or minimize the hazards of other dangerous materials in the plant. Many plants have piping systems that contain these protective materials. The protective materials are released and serve as an antidote for hazardous fumes from another system in the event of an emergency. An example is the fire retardant foams used in restaurant kitchens. Class P piping systems are designated by the color bright blue.

Class V refers to valuable materials. Examples of Class V materials might include fluid solutions of silver or platinum. Silver nitrate is used in some photochemical processes. Platinum compounds are used in catalytic reactions, such as the processing of

natural gas. These materials (coatings) are safe, but have either a very high monetary value or a high value to a certain process in the plant. These piping systems are designated by the color deep purple. Purple pipe also is used for recycled water from treatment plants, used for cooling or irrigation.

Another way to identify the content of a piping system is to label the pipe with the name of the substance that flows through the pipe. The name of the contents is stenciled onto a background color that complies with the classes above.

4.0.0 ◆ THERMAL EXPANSION

All materials change in size, to some extent, when the temperature changes. This is also true with piping materials. A pipefitter must be able to account for thermal expansion and contraction while working out the details for the fabrication of a pipeline.

Under normal conditions, metallic piping changes several inches in a 100-foot length for a change of 300°F. If only one point in a pipeline were kept fixed during a change in length, movement would take place in perfect freedom, and no stress would be imposed on the pipe or the connection. However, piping systems have more than one fixed point. They are nearly always restrained at terminal points by anchors, guides, stops, rigid hangers, or sway braces. These restraining points resist expansion and put the line under stress, which causes it to deform when the temperature changes.

Stress imposed by expansion must be kept within the strength capability of the pipe and also of the points to which the pipe is anchored. Equipment that is fastened solidly, such as a pump or a turbine, will be put under excessive stress if the pipe fastened to it expands toward it. This must be prevented by allowing expansion away from the equipment. Failure to do so can result in ruptured pipe or damage to the structure or the equipment to which the pipe is anchored.

All materials do not expand and contract at the same rate. This can cause problems when different materials are used in the same piping system. *Table 2* shows the expansion of different types of pipe.

Table 1 Standard Identification Scheme for Piping Systems

Classification	Application	Predominant Color of System
Class F	Fire-protection materials and equipment	Red
Class D	Dangerous materials	Yellow
Class S	Safe materials	Green
Class P	Protective materials	Bright blue
Class V	Valuable materials	Deep purple

201T01.EPS

Table 2 Expansion of Different Types of Pipe

Coefficient of Expansion	
Material	**inches or mm**
Carbon steel	0.0000063
Aluminum	0.0000124
Cast iron	0.0000059
Nickel steel	0.0000073
Stainless steel	0.0000095
Concrete	0.0000070

201T02.EPS

To find out how much a length of pipe would expand for a given change in temperature, multiply the length of the pipe in inches, times the temperature change in °F, times the coefficient of expansion for that material from *Table 2*. For example, a hundred feet of carbon steel pipe (1,200 inches) that is heated from 100°F to 500°F would expand by 1,200 (length) × 400 (temperature change) × 0.0000063 (coefficient of expansion) = 3.024 inches of expansion.

There are three accepted methods used to account for expansion and to avoid excessive stresses on the line: flexibility in the layout, installing expansion loops, and cold springing.

4.1.0 Flexibility in Layout

Flexibility can often be built into the layout of a piping system. If a pipeline is built with small diameter pipe, and if it will not have great temperature changes, the system can be designed to account for expansion. Flexibility in layout consists of making many changes in direction between anchor points. These changes in direction allow for some thermal expansion of the line without damaging the joints and fittings. *Figure 6* shows an example of flexibility in layout of a pipeline constructed of 4-inch Schedule 40 steel pipe carrying steam indoors at 400°F.

4.2.0 Installing Expansion Loops

The second method used to account for expansion is expansion loops that are installed in the pipe run to provide extra flexibility between anchors. Expansion loops accommodate thermal expansion and contraction and sideways bowing. These loops absorb the movement of the line caused by expansion and contraction. The pipe is anchored along its length and sometimes in the loop itself. Guides are placed at designated intervals along the pipe run to ensure that the pipe moves toward the expansion loop as it expands.

Expansion loops are made with pipe bends or welded pipe and fittings. They should not be fabricated with flanges or threaded fittings, because these types of joints cannot withstand the stress and will leak. Fabricated expansion U-loops made of pipe welded to long radius elbows are usually very simple in shape. They can be fabricated in the shop according to the specifications on the layout drawing. The number and types of expansion loops used differ with each installation. *Figure 7* shows an expansion loop.

4.3.0 Cold Springing

The third way to deal with expansion is to use cold springing, along with one of the other two methods used to account for expansion. Cold springing is a method of shortening the pipe run when it is installed so that when the pipe heats up, it expands to its intended length. Cold springing also relieves the stress that thermal expansion exerts on the pipe.

The cold spring gap refers to the difference between the length of the installed pipe and the length that the pipe should be. The location and the size of the cold spring gap must be calculated accurately and the piping must be installed according to those calculations. Cold springing pipe at the wrong location or by the wrong amount can damage the pipe joints, anchors, valves, or attached equipment, so it is not advisable to guess at cold spring dimensions and locations. There are several ways to implement a cold spring gap. The engineers will calculate the size of the gap. *Figure 8* shows examples of cold springing.

5.0.0 ◆ PIPE INSULATION

Before proceeding with any work in the area of pipe insulation, be sure to review MSDSs to identify any hazardous materials requirements. Piping is often covered by layers of insulation to maintain the temperature of the fluid in the pipe. Failure to insulate pipelines that convey steam and other hot fluids results in a tremendous loss of heat. Piping used to transport very cold materials, such as liquefied natural gas, should also be insulated. Insulated pipelines also provide safer and more comfortable working conditions. Pipefitters must be able to recognize three types of insulation: personal protection, heat conservation, and cold conservation. Personal protection insulation provides protection to workers who may come in contact with hot pipes. Any pipe that is hotter than 110°F must be insulated between the working level and a height of 7 feet. The working level can be considered the ground or a suspended work platform that the pipe is routed past. If a hot pipe passes by a work platform, the personal protection insulation is placed on the pipe at the working level up to 7 feet above the platform.

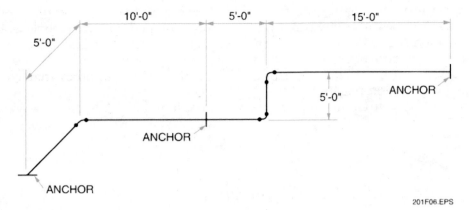

201F06.EPS

Figure 6 ◆ Flexibility in layout.

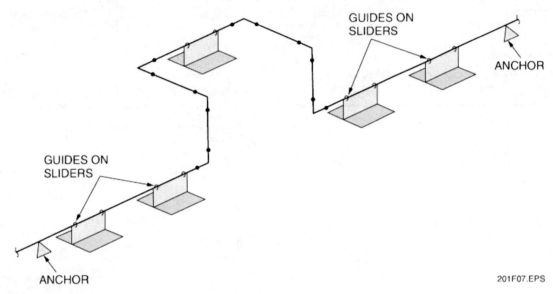

201F07.EPS

Figure 7 ◆ Expansion loop.

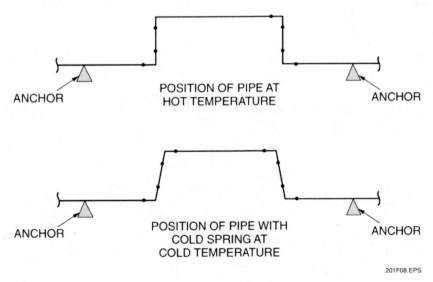

201F08.EPS

Figure 8 ◆ Cold springing.

Heat conservation insulation is used to maintain warm temperatures within pipelines. Cold conservation insulation is used to maintain cold temperatures within pipelines. Materials used for insulation vary and are selected according to the requirements of the application. *Table 3* lists insulating materials for different temperature applications.

Mineral wool or fiberglass insulation is usually wrapped with polyethylene. In underground piping, rigid foam insulation may be used as a primary fill in the haunch area at the bottom of the trench. As in other characteristics of piping systems, insulation is usually specified in the drawings, although it may be only by reference to a local standard specification.

Table 3 Insulating Materials for Different Temperature Applications

Heat Conservation Insulation		Cold Conservation Insulation	
Material	**Temperature Range**	**Material**	**Temperature Range**
Fiberglass	Up to 400°F	Foam rubber	Down to 0°F
85% magnesia	Up to 600°F	Mineral wool	Down to −150°F
Mineral wool	Up to 1,200°F	Rock cork	Down to −250°F
Calcium silicate	Up to 1,200°F	Rigid foam plastic	Down to −400°F

201T03.EPS

1. The choice of piping material is critical for chemical systems because _____.
   a. the temperatures are very high
   b. many materials being transported will attack the piping system
   c. the materials are so expensive
   d. the flow rates are extremely high

2. Acid piping systems are more vulnerable to freezing at lower concentrations.
   a. True
   b. False

3. Utility air systems provide pressure between _____ psig.
   a. 0 and 50
   b. 50 and 125
   c. 125 and 175
   d. 175 and 225

4. Steam used in household heating is usually at less than _____ psi.
   a. 2
   b. 15
   c. 50
   d. 125

5. Steam that is within the pressure range of greater than 200 psi is called _____.
   a. saturated
   b. condensate
   c. superheated
   d. water

6. Water that is safe to drink is called _____ water.
   a. utility
   b. demineralized
   c. potable
   d. cooling

7. Dangerous material piping is identified by the color _____.
   a. red
   b. yellow
   c. green
   d. bright blue

8. Green pipe is in Class _____.
   a. D
   b. S
   c. V
   d. F

9. Which of the following is *not* an accepted response to thermal expansion?
   a. Cold springing
   b. Insulation
   c. Flexibility in layout
   d. Expansion loops

10. Insulation material for piping where the temperature could be as high as 1,200°F or as low as –150°F would be _____.
    a. fiberglass
    b. foam rubber
    c. mineral wool
    d. rock cork

Summary

The purpose of this module is to provide an overview of the types of piping systems pipefitters may encounter. You must always determine what a piping system contains before operating valves, inspecting for leaks, or performing maintenance on the system. Color codes can provide information on hazards and the contents of a system. Different procedures and materials are used for the different types of piping systems, which are chemical, compressed air, fuel oil, steam, and water systems. Engineering specifications will let you know what provisions to make for expansion and insulation, to lengthen the service life of the components, and to make the system operate more efficiently.

Notes

Trade Terms Introduced in This Module

Acid: A chemical compound that reacts with and dissolves certain metals to form salts.

Ambient temperature: The same temperature as the surrounding atmosphere; room temperature.

Caustic: A material that is capable of burning or corroding by chemical action.

Condensate: The liquid product of steam, caused by a loss in temperature or pressure.

Haunch: The area below the middle of the pipe in a bench.

Heat-tracing: The addition of heat-producing wire to the pipe to prevent the material inside from being frozen.

Material safety data sheet (MSDS): A document that describes the composition, characteristics, health hazards, and physical hazards of a specific chemical. It also contains specific information about how to safely handle and store the chemical and lists any special procedures or protective equipment required.

Pounds per square inch gauge (psig): Amount of pressure in excess of the atmospheric pressure level.

Saturated steam: Steam in contact with water.

Slurry: A mixture of liquid and suspended solids, like muddy water or paper pulp.

Superheated steam: Steam used in heavy industrial applications.

Water hammer: An increase in pressure in a pipeline caused by a sudden change in the flow rate. In a steam line, water hammer is caused by condensate blocking the flow of steam at a pipe bend.

Resources & Acknowledgments

Additional Resources

This module is intended to be a thorough resource for task training. The following reference works are suggested for further study. These are optional materials for continued education rather than for task training.

Audel Mechanical Trades Pocket Manual, 1990. Carl Nelson. New York, NY: Macmillan Publishing Company.

The Pipe Fitters Blue Book, 2002. W.V. Graves. Clinton, NC: Construction Trades Press.

Figure Credits

Cianbro Corporation, Module Opener, 201T02

NCCER CURRICULA — USER UPDATE

NCCER makes every effort to keep its textbooks up-to-date and free of technical errors. We appreciate your help in this process. If you find an error, a typographical mistake, or an inaccuracy in NCCER's curricula, please fill out this form (or a photocopy), or complete the online form at **www.nccer.org/olf**. Be sure to include the exact module ID number, page number, a detailed description, and your recommended correction. Your input will be brought to the attention of the Authoring Team. Thank you for your assistance.

Instructors – If you have an idea for improving this textbook, or have found that additional materials were necessary to teach this module effectively, please let us know so that we may present your suggestions to the Authoring Team.

NCCER Product Development and Revision

13614 Progress Blvd., Alachua, FL 32615

Email: curriculum@nccer.org
Online: www.nccer.org/olf

❑ Trainee Guide   ❑ Lesson Plans   ❑ Exam   ❑ PowerPoints   Other _____

Craft / Level: _____   Copyright Date: _____

Module ID Number / Title: _____

Section Number(s): _____

Description: _____

_____

_____

_____

_____

Recommended Correction: _____

_____

_____

_____

_____

Your Name: _____

Address: _____

_____

Email: _____   Phone: _____

08202-06

Drawings and Detail Sheets

08202-06
Drawings and Detail Sheets

Topics to be presented in this module include:

Overview

Drawings are the instructions for the pipefitter. Site plans show where all of the runs are on the job site; line lists and specifications tell the particular material, connections, and fittings for each run of pipe. Notes convey specific information on some aspect that cannot be derived from the drawing itself, and the elevations and sections show how everything goes together. Read all of the drawings, and learn what everything means, and you will be ready to make the system work. Teach yourself how to sketch well, so that you can communicate with everyone involved in a job.

Objectives

When you have completed this module, you will be able to do the following:

1. Identify parts of drawings.
2. Identify types of drawings.
3. Make field sketches.
4. Interpret drawing indexes and line lists.

Trade Terms

Battery limits
Chase
Isometric drawing (ISO)
Line number
Orthographic projection
Piping and instrumentation drawing (P&ID)
Sketch
Spool

Required Trainee Materials

1. Pencil and paper
2. Appropriate personal protective equipment

Prerequisites

Before you begin this module, it is recommended that you successfully complete *Core Curriculum*; *Pipefitting Level One*; and *Pipefitting Level Two*, Module 08201-06.

This course map shows all of the modules in the second level of the *Pipefitting* curriculum. The suggested training order begins at the bottom and proceeds up. Skill levels increase as you advance on the course map. The local Training Program Sponsor may adjust the training order.

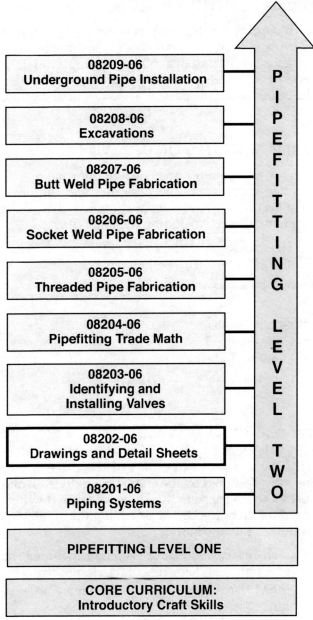

202CMAP.EPS

1.0.0 ◆ INTRODUCTION

A blueprint, or drawing, is a detailed plan of a part or system. A detail sheet is a drawing, at a larger scale, of part of another drawing. Drawings and detail sheets provide the information needed to build or install part of a system or the entire system.

2.0.0 ◆ PARTS OF DRAWINGS

Blueprints contain a great deal of information in a limited space. To do this, designers and engineers use certain conventions and standard ways of drawing objects, including the following:

- Title blocks
- Scales and measurements
- Symbols and abbreviations
- Notes
- Revision blocks
- Coordinates
- Drawing indexes
- Line lists

2.1.0 Title Blocks

The title block contains information used to identify the project or any part of the project. It is usually designed by the engineering division and uses a coding system that designates all drawings used on that project. The following information is found in the title block:

- *Company name* – The title block includes the logo, name, and address of the design firm.
- *Project title* – This is the project name and location.
- *Drawing title* – This explains what is drawn on that sheet. If it is a detail drawing, the part is identified. If the sheet contains an assembly drawing, this block is left blank.
- *Professional stamp* – This is a registered seal of approval used by the engineer or architect.
- *Design supervisor* – This block contains the name or initials of the person who designed the project and the date on which it was signed by the design supervisor.
- *Drawn* – This block contains the name or initials of the person who drafted the drawing and the date on which it was drawn.
- *Checked* – This block contains the name or initials of the person who reviewed the drawing and the date on which it was reviewed.
- *Approved* – This block contains the name or initials of the person who approved the drawing,

usually the project engineer, and the date on which it was approved.

- *Owner's approval* – This block contains the signature of the owner or a representative and the date on which the owner approved the drawing.
- *Scale* – This shows the relationship between the size of the drawing and the size of the actual object. If the drawing is the same size as the object, the scale is full size.
- *Project number* – This is the code number assigned to the project.
- *Sheet number* – This indicates the sequence of this sheet in the series of drawings.
- *Revision number (Rev.)* – This indicates the current drawing. The number in this block can be referenced to the revision block for information about each revision.
- *Key plan* – A key plan is shown if one plan drawing does not include the entire system being designed. The key plan shows where this drawing fits into the overall geographic location. The shaded area shows the location of the drawing. The key plan may be shown on the plan view for reference.

The title block is usually located on the lower right corner of each drawing. *Figure 1* shows a typical title block.

2.2.0 Scales and Measurements

Scales on drawings are usually written in inches or in feet and inches. Measurements of less than 1 foot are written as 0 feet and the number of inches. The scale of a drawing is written in the title block. It shows the size of the drawing as compared to the actual size of the object represented.

The scale of a drawing varies according to the size of the object being drawn and the size of the paper used. The drawing scale is usually indicated by a ratio between the size of the object and the size of the drawing. For example, a scale of 1 inch = 1 foot means that for every inch shown on the drawing, the actual size of the object is 1 foot. The standard scale of piping drawings is ⅜ inch = 1 foot. The letters U.N. stand for "unless noted otherwise on drawing." This indicates that some parts of the drawing may use a different scale.

Graphic scales on drawings (*Figure 2*) show distances that correspond with a unit of length on the ground or with the object represented in the drawing. They are usually placed in or near the title block of the drawing. When a drawing has not been held to a particular scale, and cannot usefully be measured, the scale block will hold the initials N.T.S., for "not to scale."

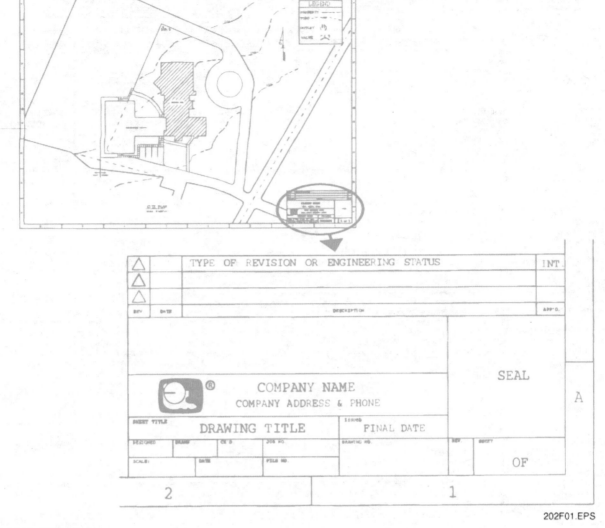

Figure 1 ◆ Title block.

2.3.0 Symbols and Abbreviations

Symbols and abbreviations are used on drawings to show information in a small area. Always check the symbols and abbreviations with the legend.

Symbols are used to show objects, such as valves, fittings, flanges, and connections, that would require too much detail if drawn accurately. These objects are represented on drawings by small, simple line drawings. Symbols vary from job to job. *Figure 3* shows commonly used piping symbols.

Standard abbreviations have been developed for common pipefitting terms. The pipefitter should recognize these abbreviations and the meaning of each. *Table 1* lists commonly used pipefitting abbreviations.

2.4.0 Notes

Notes are provided in addition to drawings to explain items that cannot be easily drawn. These items include general requirements, quality assurance criteria, calibration tolerances, and similar information. Notes may be located in a corner of a blueprint, or they may be listed separately and included with a set of blueprints. *Figure 4* shows typical notes.

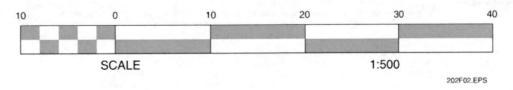

202F02.EPS

Figure 2 ◆ Graphic scale.

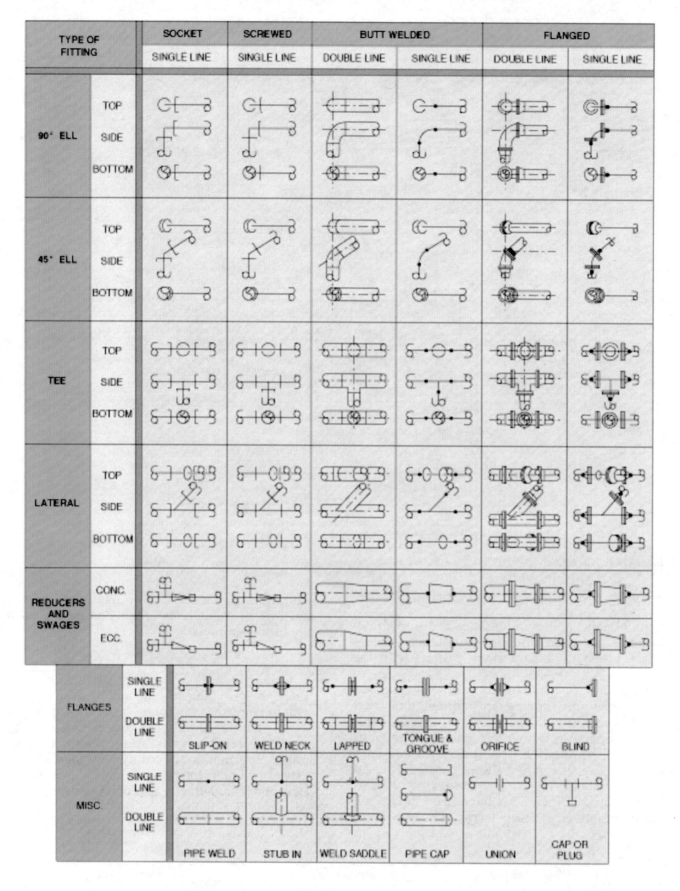

202F03.EPS

Figure 3 ◆ Piping symbols.

NOTES:
1. FOR PIPING GENERAL NOTES AND REFERENCE DRAWINGS SEE DWG 5439-3.
2. ALL PUMPS ARE LOCATED TO CENTER OF DISCHARGE UNLESS OTHERWISE NOTED.

202F04.EPS

Figure 4 ◆ Notes.

Table 1 Pipefitting Abbreviations (1 of 3)

Adapter	ADPT
Air preheating	A
American Iron and Steel Institute	AISI
American National Standards Institute	ANSI
American Petroleum Institute	API
American Society for Testing and Materials International	ASTM
American Society of Mechanical Engineers	ASME
Ash removal water, sluice or jet	AW
Aspirating air	AA
Auxiliary steam	AS
Bench mark	BM
Beveled	B
Beveled end	BE
Beveled, both ends	BBE
Beveled, large end	BLE
Beveled, one end	BOE
Beveled, small end	BSE
Bill of materials	BOM
Blind flange	BF
Blowoff	BO
Bottom of pipe	BOP
Butt weld	BW
Carbon steel or cold spring	CS
Cast iron	CI
Ceiling	CLG
Chain operated	CO
Chemical feed	CF
Circulating water	CW
Cold reheat steam	CR
Compressed air	CA
Concentric or concrete	CONC
Condensate	C
Condenser air removal	AR
Continue, continuation	CONT
Coupling	CPLG

Detail	DET
Diameter	DIA
Dimension	DIM
Discharge	DISCH
Double extra-strong	XX STRG
Drain	DR
Drain funnel	DF
Drawing	DWG
Drip leg or dummy leg	DL
Ductile iron	DI
Dust collector	DC
Each	EA
Eccentric	ECC
Elbolet	EOL
Elbow	ELB
Electric resistance weld	ERW
Elevation	EL
Equipment	EQUIP
Evaporator vapor	EV
Exhaust steam	E
Expansion	EXP
Expansion joint	EXP JT
Extraction steam	ES
Fabrication (dimension)	FAB
Face of flange	FOF
Faced and drilled	F&D
Factory Mutual	FM
Far side	FS
Feed pump balancing line	FB
Feed pump discharge	FD
Feed pump recirculating	FR
Feed pump suction	FS
Female	F
Female	FM
Female pipe thread	FPT
Field support or forged steel	FS

202T01A.EPS

Table 1 Pipefitting Abbreviations (2 of 3)

Field weld	FW
Figure	FIG
Fillet weld	W
Finish floor	F/F
Finish floor	FIN FL
Finish grade	FIN GR
Fitting	FTG
Fitting makeup	FMU
Fitting to fitting	FTF
Flange	FLG
Flat face	FF
Flat on bottom	FOB
Flat on top	FOT
Floor drain	FD
Foundation	FDN
Fuel gas	FG
Fuel oil	FO
Gauge	GA
Galvanized	GALV
Gasket	GSKT
Grating	GRTG
Hanger	HGR
Hanger rod	HR
Hardware	HDW
Header	HDR
Heat traced, heat tracing	HT
Heater drains	HD
Heating system	HS
Heating, ventilating, and air conditioning	HVAC
Hexagon	HEX
Hexagon head	HEX HD
High point	HPT
High pressure	HP
Horizontal	HORIZ
Hot reheat steam	HR
Hydraulic	HYDR
Increaser	INCR
Input/output	I/O
Inside diameter	ID
Insulation	INS
Invert elevation	IE
Iron pipe size	IPS
Isometric	ISO
Issued for construction	IFC

Lap weld	LW
Large male	LM
Length	LG
Long radius	LR
Long tangent	LT
Long weld neck	LWN
Low pressure	LP
Low-pressure drains	DR
Low-pressure steam	LPS
Lubricating oil	LO
Main steam	MS
Main system blowouts	BL
Makeup water	MU
Male	M
Malleable iron	MI
Manufacturer	MFR
Manufacturer's Standard Society	MSS
National Pipe Thread	NPT
Nipolet	NOL
Nipple	NIP
Nominal	NOM
Not to scale	NTS
Nozzle	NOZ
Outside battery limits	OSBL
Outside diameter	OD
Outside screw and yoke	OS&Y
Overflow	OF
Pipe support	PS
Pipe tap	PT
Piping and instrumentation diagram	P&ID
Plain	P
Plain end	PE
Plain, both ends	PBE
Plain, one end	POE
Point of intersection	PI
Point of tangent	PT
Pounds per square inch	PSI
Process flow diagram	PFD
Purchase order	PO
Radius	RAD
Raised face	RF
Raised face slip-on	RFSO

202T01B.EPS

Table 1 Pipefitting Abbreviations (3 of 3)

Raised face smooth finish	RFSF	Temperature or temporary	TEMP
Raised face weld neck	RFWN	That is	I.E.
Raw water	RW	Thick	THK
Reducer, reducing	RED	Thousand	M
Relief valve	PRV-PSV	Thread, threaded	THRD
Ring-type joint	RTJ	Threaded	T
Rod hanger or right hand	RH	Threaded end	TE
		Threaded, both ends	TBE
Safety valve vents	SV	Threaded, large end	TLE
Sanitary	SAN	Threaded, one end	TOE
Saturated steam	SS	Threaded, small end	TSE
Schedule	SCH	Threadolet	TOL
Screwed	SCRD	Top of pipe or top of platform	TOP
Seamless	SMLS	Top of steel or top of support	TOS
Section	SECT	Treated water	TW
Service and cooling water	SW	Turbine	TURB
Sheet	SH	Typical	TYP
Short radius	SR		
Slip-on	SO	Underwriters Laboratories	UL
Socket weld	SW		
Sockolet	SOL	Vacuum	VAC
Stainless steel	SS	Vacuum cleaning	VC
Standard weight	STD WT	Vents	V
Steel	STL	Vitrified tile	VT
Suction	SUCT		
Superheater drains	SD	Wall thickness or weight	WT
Swage	SWG	Weld neck	WN
Swaged nipple	SN	Weldolet	WOL
		Well water	WW
		Wide flange	WF

202T01C.EPS

Notes can be divided into two categories: general notes that apply to many parts of the structure, and local notes that explain a specific part of the construction and are keyed to that point with a line.

2.5.0 Revision Blocks

The revision block is usually located in the lower center of the drawing and is used to record any changes, or revisions, to the drawing (*Figure 5*). All revisions are noted in this block. They are dated and are identified by a letter or number and a description of the revision. A revised drawing is shown by the addition of a letter or number next to the original drawing number in the title block.

2.6.0 Coordinates

Coordinates are a system of numbers used to define the geographic position of a point, line, piece of equipment, or other object on the plant site. A coordinate is the distance a given point is from a base point commonly called a bench mark. Coordinates are written as a letter and number combination. The letter designates the direction of the coordinate, normally north or east, in relation to the bench mark, and the number designates the distance in feet and inches. Coordinates are used to locate major equipment on drawings.

A property plan indicates the position of the originating point or coordinate bench mark. From this point, a grid system is developed and is used

PROJ	NO	REVISION	RVSD	CHKD	APPD	DATE
3483	01	RELEASED FOR CONSTRUCTION		APD	NWS	JULY 06
3483	02	DELETED PART OF LINE 12037		APD	NWS	AUG 06
3483	03	⚠ ADDED WELDING SYMBOL		APD	NWS	AUG 06

202F05.EPS

Figure 5 ◆ Revision block.

by all design groups to coordinate their drawings. The north-south axis and the east-west axis intersect through this originating point, and every point on the drawing can be referenced by the originating point. For example, a point that is 30 feet east of the north-south axis through the originating point has an east coordinate of E. 30'-0". If this same point were 1 foot north of the east-west axis through the originating point, it would have a north coordinate of N. 1'-0". Put together, the two coordinates pinpoint the location of the point as E. 30'-0" and N. 1'-0". *Figure 6* shows a coordinate origin.

With the advent of global positioning system (GPS) equipment, the use of site-specific coordinate systems has become less common. The GPS uses triangulation from satellites to give the latitude and longitude of a point, as well as its height. Latitude is the measurement of displacement north or south of the equator and is measured in degrees, minutes, and seconds. Longitude is the measurement in degrees, minutes, and seconds from an arbitrarily chosen line running through Greenwich Observatory in England. These dimensions are given in the following form: latitude written as, for example, North (N) 30° 15" 30'; followed by longitude written as East (E) 45° 12" 10.3'. Elevation is determined by the GPS coordinate systems as well. A GPS base station is set up during the site survey and layout process, allowing the position of the components of the piping system to be compared to the GPS base with considerable precision.

2.7.0 Drawing Indexes

The drawing index, provided on the first page of the drawings, is a complete list of all drawings included in the set. It is developed from the plot plan. The drawings are usually organized into the following major categories:

- Civil (C)
- Architectural (A)
- Interior/graphics (D)
- Structural (S)
- Mechanical equipment (M)
- Piping and plumbing (P)
- Electrical (E)

The pages within each category are designated by a letter followed by a number. Frequently the civil and architectural drawings are combined, usually being labeled A.

2.8.0 Line Lists

The line list, or line table, provides the following information about piping lines:

- Line numbers in numerical order
- Class or specification of each line and size
- Whether or not each line is insulated
- What each line transports (liquid or gas)
- Starting and termination points
- Design pressure and temperature
- Operating pressure and temperature
- Testing pressure

The piping supervisor usually fills out the line list and selects the piping specifications that can be used with the system.

3.0.0 ◆ TYPES OF DRAWINGS

Construction drawings show as much construction detail as possible. They are presented graphically. Most construction drawings consist of orthographic views and are presented as plans and elevations. The following types of drawings can be used by pipefitters during the construction process:

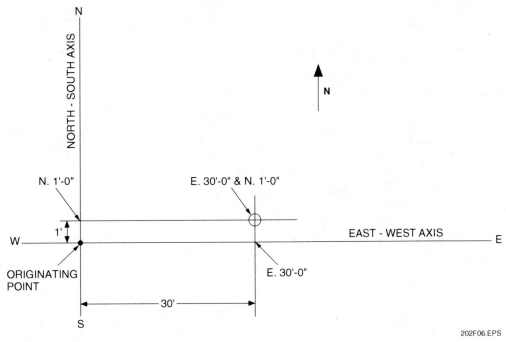

Figure 6 ◆ Coordinate origin.

- Plot plans
- Structural drawings
- Elevation and section drawings
- As-built drawings
- Equipment arrangement drawings
- Piping and instrumentation drawings
- Isometric drawings
- Spool drawings
- Equipment drawings
- Pipe support drawings and detail sheets
- Orthographic drawings

3.1.0 Plot Plans

A plot plan shows the location of a structure on a piece of property and usually indicates where the water and sewer mains are installed. The plot plan includes information on sections of the land where nothing can be built, such as the following:

- *Setback* – This is the minimum distance that must be maintained between the street and the structure.
- *Easements* – These are places where utilities may legally be installed.
- *Side yards* – These areas provide access to rear yards, reduce the possibility of fire transferring from one building to another, and promote ventilation.
- *Battery limits* – The outside perimeter of the object. This means that work beyond these limits, such as utility piping, is not the responsibility of the contractor.

Figure 7 shows a simplified plot plan.

3.2.0 Structural Drawings

Structural drawings indicate how the framework of the building is to be constructed. The framework may include structural steel, cast-in-place concrete, precast concrete, or other material. Structural drawings are coded with the letter S. Depending on the complexity of the building, structural drawings may be combined with architectural drawings, or they may stand alone. *Figure 8* shows a structural drawing.

3.3.0 Elevation and Section Drawings

The elevation, or height, of an object is shown on an elevation view. Elevation views may show the front, rear, or sides of the structure drawn to scale and show an entire unit or area. They are usually drawn looking toward the top or left side of the drawing. Elevation views are sometimes referred to as front, rear, and side elevations. When the direction is known, the direction pictured is listed. *Figure 9* shows elevation drawings.

3.4.0 As-Built Drawings

As-built, or record, drawings are construction drawings that have been revised to show significant changes made during the construction process. These drawings are usually based on changes marked on blueprints and drawings and other data furnished by the contractor or architect.

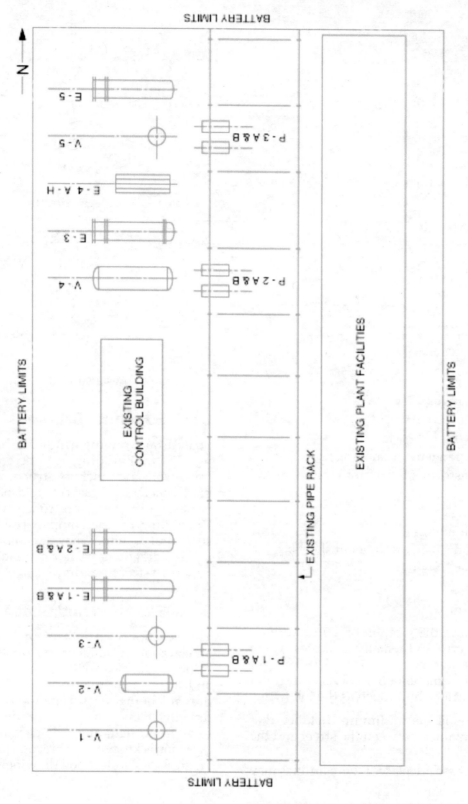

Figure 7 ◆ Plot plan.

202F07.EPS

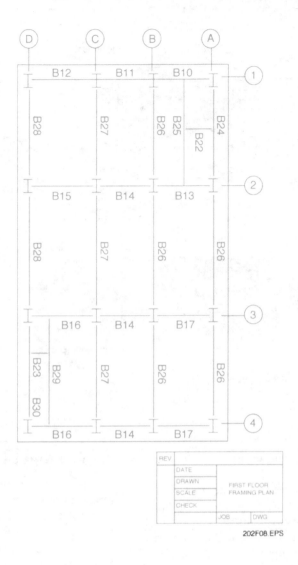

Figure 8 ◆ Structural drawing.

As-built drawings serve as a permanent record of the actual construction details of the project.

3.5.0 Equipment Arrangement Drawings

Equipment arrangement drawings show how the equipment is arranged and the locations of all equipment used with the system. Equipment arrangement drawings can be either plan or section views. Equipment on these drawings is usually drawn to correspond to the basic shape of the item represented. These representations are common to most companies.

3.5.1 Plan Views

A plan view is a view of an object or area from above, looking down on the system (*Figure 10*). The plan view shows the exact locations of the equipment in relation to established reference points, objects, or surfaces. These reference points may be building columns, walls, bench marks, or other equipment.

Plan views are usually laid out with north at the top of the drawing. If the need for space makes it necessary to rotate the drawing, it is rotated so that north is to the left side of the sheet. A key plan is sometimes shown on the plan view for reference. The small rectangle in the upper right corner of the drawing is labeled as a pipe chase. A pipe chase is an enclosed area where pipe or electrical cables are run vertically through the building. This tells you where to put the pipes that run from floor to floor or from floor to ceiling.

3.5.2 Section Views

It is not possible to show all the necessary information in plans and elevations, so a section view must be used. A section view is the result of an imaginary line cut through the building. Most section views are vertical, but horizontal or plan

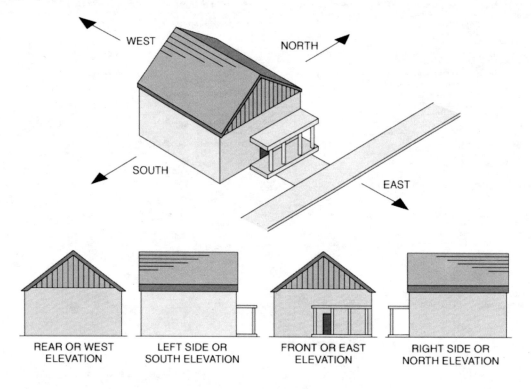

WEST

NORTH

SOUTH

EAST

| REAR OR WEST ELEVATION | LEFT SIDE OR SOUTH ELEVATION | FRONT OR EAST ELEVATION | RIGHT SIDE OR NORTH ELEVATION |

202F09.EPS

Figure 9 ◆ Elevation drawings.

sections may be included. In section views, the parts are usually labeled.

The section view shows how the equipment or parts fit together in the system (*Figure 11*). From this drawing, the height, width, and location of the equipment can be seen. This is accomplished by drawing what the component looks like on the inside. Section drawings are usually drawn on a larger scale than floor plans and elevations.

3.6.0 Piping and Instrumentation Drawings

Piping and instrumentation drawings (P&IDs) are schematic diagrams of a complete system or systems (*Figure 12*). P&IDs show the process flow. They also show all equipment, pipelines, valves, instruments, and controls necessary for the operation of that system. Because their purpose is only to provide a representation of the work to be done, P&IDs are not drawn to scale or dimensioned. Pieces of equipment on P&IDs are usually drawn to correspond with the actual shape of the object. Valves and other parts are usually indicated with symbols. P&IDs often show an area that is enclosed by a dashed line and has the initials VS inside the enclosed area. The VS indicates that all of the equipment within the enclosed area is vendor-supplied; therefore, you are only responsible for the piping up to that area.

3.7.0 Isometric Drawings

An isometric drawing, or ISO, combines the plan and elevation views of a system or part of a system into one drawing (*Figure 13*). Usually, only one pipeline is shown on an isometric drawing, although for simple or duplicate pipelines, more than one line may be shown. An isometric pipe drawing shows the details needed to fabricate and install a section of a piping system. An ISO also usually includes a bill of materials, which lists the types of pipe, flanges, valves, bolts, and gaskets required for the pipe run.

Isometric piping drawings use the same symbols as piping schematics or any other piping drawings. The ISO has arrows that show the direction of flow. To read an ISO, begin at the end of the pipe run nearest the source and work in the direction of flow. Trace the line from the first fitting, and identify all the fittings and components by their symbols.

3.8.0 Spool Drawings

A spool is a prefabricated section of piping. A spool drawing is a representation of a section of piping that is to be fabricated before erection. Spool drawings are usually drawn for each line in the system. They provide the detail dimensions of each line. A spool drawing usually includes the following information:

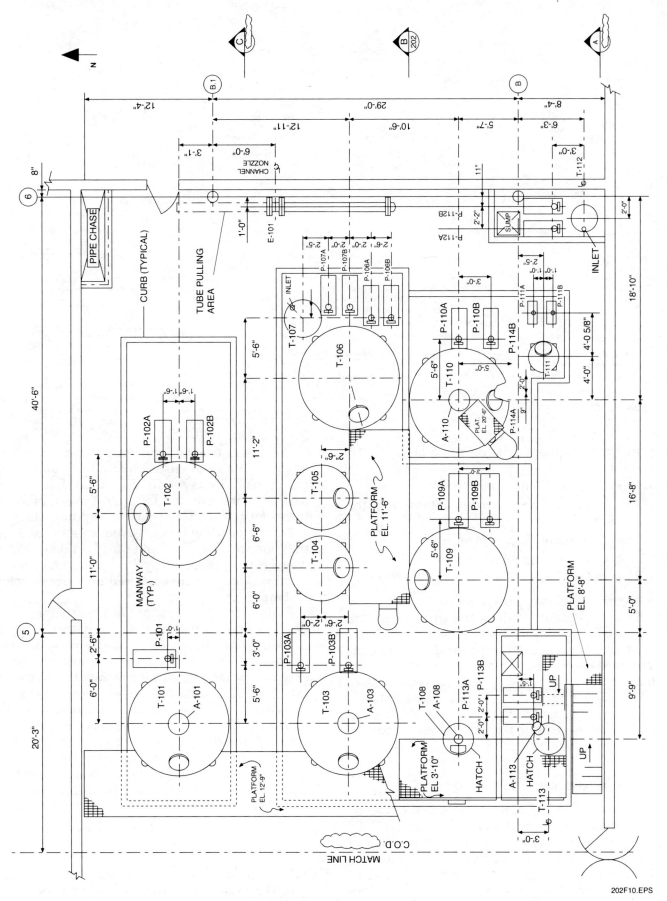

Figure 10 ◆ Plan view.

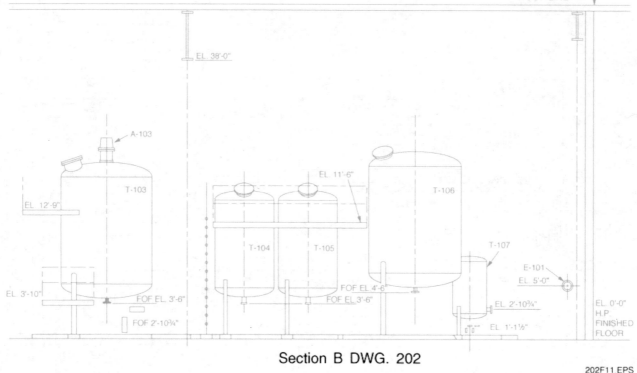

Section B DWG. 202

202F11.EPS

Figure 11 ◆ Section view.

- Instructions the welder needs to fabricate the spool
- Cut lengths of pipe
- Any fittings and flanges required to fabricate the spool
- Materials required, or specifications
- Any special treatment required
- How many spools of the same type are required

Figure 14 shows a spool drawing.

Bolts and gaskets, valves, and instruments are not usually shown on a spool drawing.

Most companies fabricate from spool drawings, and most spools are fabricated in the shop rather than in the field. The size of an individual spool is usually determined by the fabricator's ability to transport the spool to the place of erection. Generally, a spool is limited in dimensions to about 40 feet by 10 feet by 8 feet.

3.9.0 Equipment Drawings

Equipment, or vendor, drawings are supplied by equipment manufacturers. The drawings include the fabrication, erection, and setting drawings that are necessary to install a piece of equipment. They may also include manufacturer's standard drawings or catalog cuts, performance charts brochures, and other data that illustrates items. Samples of some materials are also required.

Pipefitters use equipment drawings to install equipment and to ensure that all the necessary prepiping is in its required location. Equipment drawings are required for almost every product that is fabricated away from the building site.

Equipment drawings are usually prepared after the contract is awarded and construction is under way. They are prepared by the equipment vendors or fabricators and are approved by the contractor and architect.

Equipment drawings are either stamped certified or noncertified by the manufacturer. If the drawings are stamped certified, the manufacturer guarantees that the piece of equipment is made exactly like the dimensions shown in the drawing. If the piece of equipment is different than the drawing shows, the manufacturer is held liable for the equipment. If the drawing is stamped noncertified, the equipment is not guaranteed to the specifications shown in the drawing. Noncertified drawings and equipment are much less expensive than certified drawings and equipment.

3.10.0 Pipe Support Drawings and Detail Sheets

Before a piping system is installed, engineers analyze the system and determine the type, size, and placement of all hangers and supports in the system. Piping drawings are then made, showing

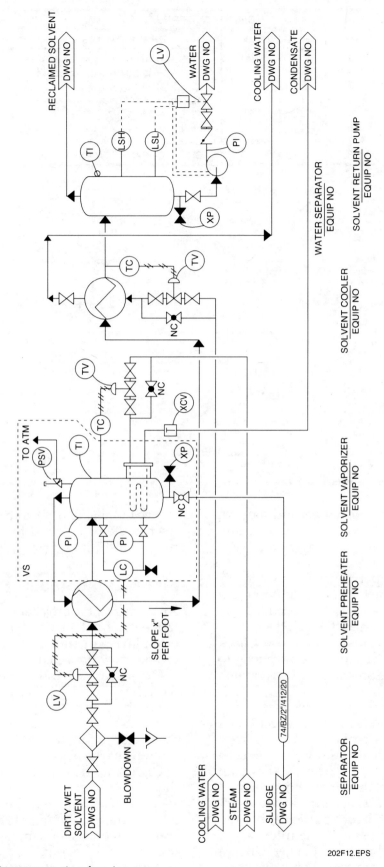

Figure 12 ◆ Piping and instrumentation drawing.

202F12.EPS

BILL OF MATERIALS			
P.M.	REQ'D	SIZE	DESCRIPTIONS
1		1½"	PIPE SCH/40 ASTM-A-120 GR. B
2		¾'	PIPE SCH/40 ASTM-A-120 GR. B
3		1½'	90 ELL ASTM-A-197 BW
4	1	1½"	TEE ASTM-A-197
5	2	¾"	45 ELL ASTM-A-197
6	1	1½' × ¾'	BELL RED. CONC. ASTM-A-197
7	2	1½"	GATE VA. BW ASTM-A105
8	1	1½"	CHECK VA. SWING BW ASTM-A105

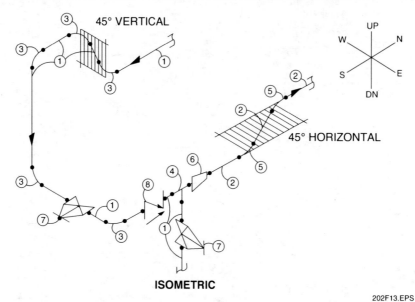

202F13.EPS

Figure 13 ◆ Simplified piping isometric drawing.

the placement of each type of hanger and support. The pipefitter must be able to read and interpret these drawings to install the hangers and supports in the proper place.

Symbols for pipe supports vary from job to job. On piping drawings, the hanger or support has a reference number that refers to the specific detail sheets used on that job. Check the detail sheets to determine the exact type of hanger to use. Check your company's standards and specifications to determine the reference prefixes used on your job site.

Each engineering company also uses pipe hanger and support symbols that are specific to a job site. These symbols appear on plan and isometric drawings and are called out by the reference number as explained above. *Figure 15* shows an example of pipe support drawings and symbols.

Refer to the detail sheets of a pipe hanger symbol to determine the details of a particular hanger. The detail sheets include all of the information needed to fabricate and install a hanger. Often a number of detail sheets are copied onto one page, at the back of a set of drawings, or added to the drawings on which the details are applicable. Notice the symbol in the upper left-hand corner of *Figure 15* that designates a clevis hanger with ref-

erence number 5HR. Before installing this hanger, refer to the detail sheet for 5HR. *Figure 16* shows the detail sheet for 5HR.

3.11.0 Orthographic Drawings

An orthographic projection is a drawing that illustrates the exact shape of an object. It is made by extending perpendicular lines from an object to create a new projected view.

Orthographic drawings can contain one or more separate views. Each view shows the shape of the object as it is seen from a different direction or point of view. By combining two or more of these views, the drafter can provide an accurate drawing of the object.

Orthographic drawings are done at an angle that does not create a distortion caused by perspective. The front view shows the surfaces visible on the front of the object. If the drawing is full scale, each line is the same size as the corresponding edge of the object.

Although the front view may show more information about an object than a pictorial drawing, it does not provide all the information needed to visualize the object. For instance, the width of the object cannot be seen from the front view. The

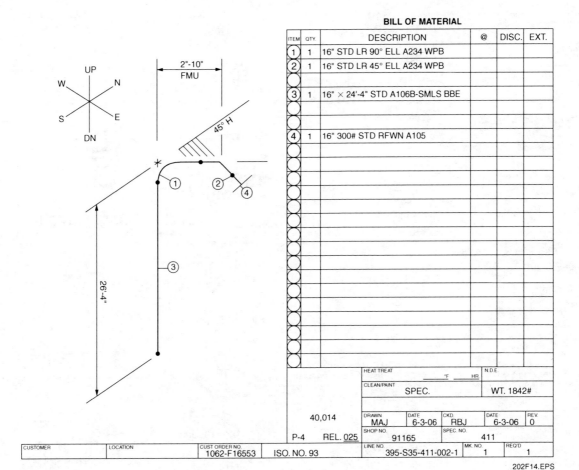

BILL OF MATERIAL

ITEM	QTY	DESCRIPTION	@	DISC.	EXT.
1	1	16" STD LR 90° ELL A234 WPB			
2	1	16" STD LR 45° ELL A234 WPB			
3	1	16" × 24'-4" STD A106B-SMLS BBE			
4	1	16" 300# STD RFWN A105			

HEAT TREAT				N.D.E.	
		°F	HR		
CLEAN/PAINT	SPEC.			WT. 1842#	

	DRAWN MAJ	DATE 6-3-06	CKD. RBJ	DATE 6-3-06	REV. 0
40,014	SHOP NO. 91165		SPEC. NO. 411		
P-4 REL. 025					

CUSTOMER	LOCATION	CUST ORDER NO. 1062-F16553	ISO. NO. 93	LINE NO. 395-S35-411-002-1	MK. NO. 1	REQ'D 1

202F14.EPS

Figure 14 ◆ Spool drawing.

width of the object and any other details needed by the craftworker are drawn on the top of the object. The top view, along with the side view, enables the drafter to convey a more complete representation of the object.

Combining the three views into one drawing produces a group of drawings that resemble a box. Each of the three views is drawn as it would appear to an observer looking directly at the object. The observer is able to see the front, top, and the right side of the object at the same time. By unfolding the box, you can rotate the three views into the same plane. In most cases, these three views correctly relate the exact shape of the object. *Figure 17* shows the relationship of the orthographic views of an object.

4.0.0 ◆ FIELD SKETCHES

The design of a building does not include all the details of the plumbing system, which must be planned by the pipefitter. By consulting the working drawings and sketching the work to be done, pipefitters can check materials and locate difficulties on the project. This saves time and labor.

Even when piping has been designed by others, it is helpful to make sketches of the parts. Gener-

ally, the full set of plans is kept in the field office because it is too large to carry to a specified point of installation. By making a detailed sketch, the pipefitter has a ready reference to take to the work site. Sketches save time and help avoid mistakes.

In addition to reading and interpreting blueprints, pipefitters must make freehand sketches of parts or assembled units to convey information about how parts should be made or assembled. Sketching is a means of conveying an idea rather than a method of making complete, perfect drawings. You can probably think of a situation in which you had to explain to someone what an object looked like or how something worked, and in doing so, you grabbed a pencil and began sketching the object. If you are like most people, your sketch needed more explanation than the object itself. You must develop some basic sketching skills to produce sketches that are clear and easy to read.

To produce clear sketches, you must be able to draw lines, angles, arcs, ellipses, and circles. You must be able to make sketches that are in correct proportion to the object being sketched, including dimensions. As with any skill that you might try to learn, field sketching takes considerable practice. As you learn these skills, practice the procedures

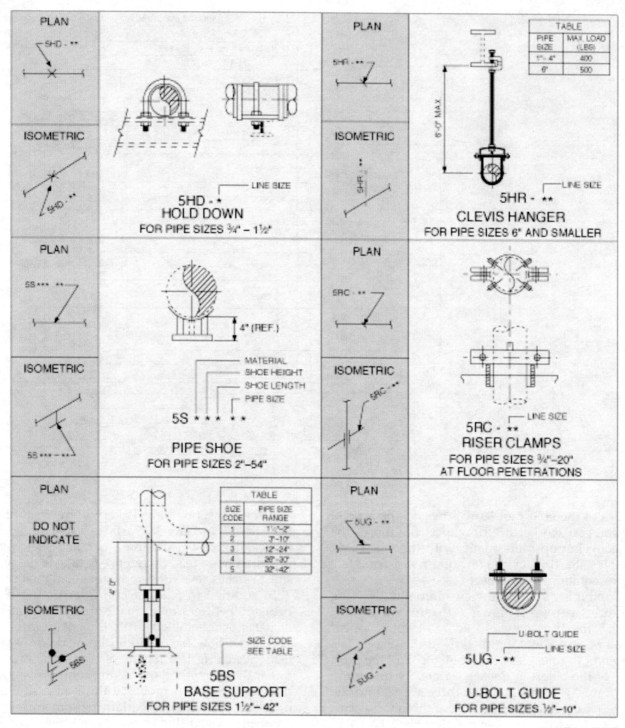

Figure 15 ◆ Symbols for pipe hangers and supports.

202F15.EPS

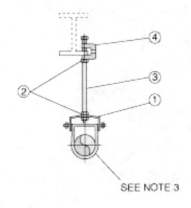

CLEVIS HANGER SELECTION CHART

PIPE SIZE	ROD SIZE	CLEVIS LOAD (LBS)
1"	⅜"	400
1¼"	⅜"	400
1½"	⅜"	400
2"	⅜"	400
2½"	½"	400
3"	½"	400
3½"	½"	400
4"	⅝"	440
5"	⅝"	440
6"	¾"	500

SEE NOTE 3

ASSEMBLY TAG NOS. ARE AS FOLLOWS:

**5HR - **

HANGER ROD DESIGNATION ———— ———— LINE SIZE

FABRICATION NOTES FOR METALLIC APPLICATIONS

1. PARTS EQUAL TO POWER PIPING COMPONENTS MAY BE USED.
2. FIELD TO CUT TO LENGTH REQUIRED AND TO THREAD BOTH ENDS.
3. FOR STAINLESS STEEL PIPE (½"– 6") USE 16-GAUGE STAINLESS STEEL BEARING PLATE.

TABLE 1			
LINE SIZE	CLAMP FIGURE NO.	BEAM BRACKET & BOLT SIZE	ROD DIA. & WELDLESS EYE & HEX NUT SIZE
1	222	½"	⅜"
2	222	½"	⅜"
3	222	½"	½"
4	222	½"	¾"
6	222	¾"	¾"

MARK NO.	DESCRIPTION	TYPE	QUAN. REQ'D
①	ADJUSTABLE CLEVIS HANGER, POWER PIPING FIG. NO. 11 (SEE NOTE 1)	5HR	1
②	HEX NUT SAME SIZE AS ROD DIAMETER, POWER PIPING FIG. NO. 61 (SEE NOTE 1)	5HR	3
③	6'-0" LONG CARBON STEEL ROD, ASTM-A36, NOT THREADED, DIAMETER PER TABLE 1 (SEE NOTE 2)	5HR	1
④	CLAMP WITH RETAINING CLIP	5HR	1

ENGINEERING COMPANY NAME

HANGER RODS

**5HR - **

* = LINE SIZE

SPECIFICATION NUMBER

50201	SHEET	1 OF 1

202F16.EPS

Figure 16 ◆ Detail sheet for 5HR.

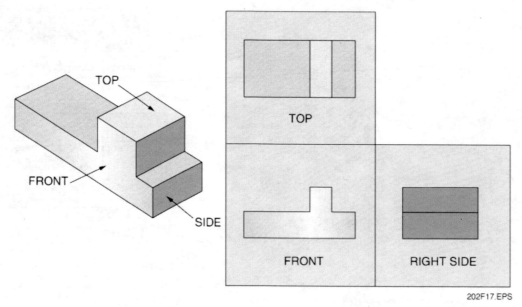

Figure 17 ◆ Orthographic views of object.

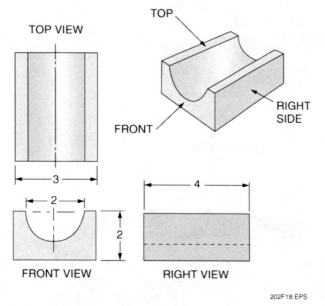

Figure 18 ◆ Three-view orthographic sketch.

so that you will continually do a better job. The following sections explain the basic skills needed to make orthographic and isometric sketches.

4.1.0 Making Orthographic Sketches

Orthographic sketches are the easiest freehand sketches to make. Orthographic sketches show only one side of an object at a time, but they usually show two or three views of the object. The views are developed as in any regular mechanical drawing, and the same types of lines are used. The difference is that orthographic sketches are drawn freehand. Sketches of small parts are made

as near the actual size as possible or to scale. *Figure 18* shows a three-view orthographic sketch.

4.2.0 Making Isometric Sketches

Isometric sketches show two or more sides of the object in a single view. An isometric sketch is developed around three major lines called isometric base lines or axes. The left-side and the right-side isometric base lines each form a 30-degree angle to the horizontal base line. The vertical base line forms a 90-degree angle to the horizontal base line. *Figure 19* shows the isometric base lines.

Follow these steps to make an isometric sketch of a rectangular object:

Step 1 Sketch the three isometric base lines.

Step 2 Mark the length of the object on the right-side isometric base line.

Step 3 Mark the width of the object on the left-side isometric base line.

Step 4 Mark the height of the object on the vertical isometric base line. *Figure 20* shows marking the length, width, and height of the object.

Step 5 Sketch the remaining lines parallel to the three isometric base lines to complete the sketch. *Figure 21* shows completing the sketch.

Step 6 Sketch the dimensions. *Figure 22* shows sketching the dimensions.

Step 7 Erase all unnecessary lines.

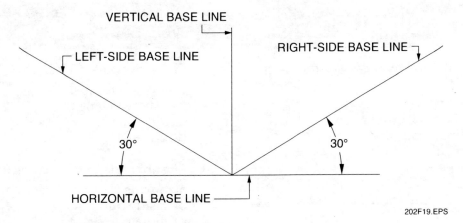

VERTICAL BASE LINE

LEFT-SIDE BASE LINE

RIGHT-SIDE BASE LINE

30°   30°

HORIZONTAL BASE LINE

202F19.EPS

Figure 19 ◆ Isometric base lines.

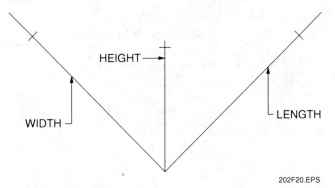

HEIGHT

WIDTH

LENGTH

202F20.EPS

Figure 20 ◆ Marking length, width, and height.

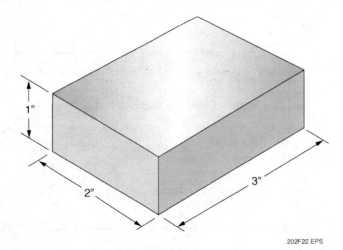

1"

2"

3"

202F22.EPS

Figure 22 ◆ Sketching dimensions.

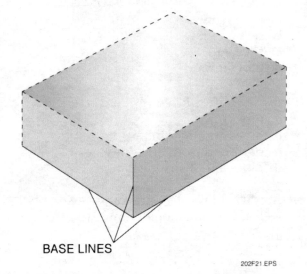

BASE LINES

202F21.EPS

Figure 21 ◆ Completing sketch.

1. Which of the following is *not* found on a title block?
   a. Scale of drawing
   b. Name of design supervisor
   c. Bill of materials
   d. Drawing title

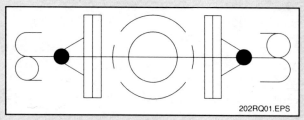

Figure 1

202RQ01.EPS

2. The symbol in *Figure 1* is used to represent a _____.
   a. socket 90-degree ell
   b. flanged 45-degree ell
   c. flanged tee
   d. concentric reducer

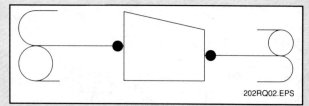

Figure 2

202RQ02.EPS

3. The symbol in *Figure 2* is used to represent a _____.
   a. butt welded 45-degree ell
   b. flanged concentric reducer
   c. butt welded eccentric reducer
   d. flanged tee

4. The abbreviation for a raised-face weld-neck flange is _____.
   a. RFSF
   b. RFWN
   c. RW
   d. RFSO

5. The abbreviation LR means _____.
   a. left and right
   b. leave remainder
   c. long radius
   d. lift ring

6. The abbreviation ID stands for _____.
   a. ductile iron
   b. inside dimension
   c. induction drive
   d. inside diameter

7. The line list will *not* specify _____.
   a. the specifications for a line of pipe
   b. what each line transports
   c. start and termination points
   d. scale of the drawing

8. An elevation view is seen from the top of the structure.
   a. True
   b. False

9. A view of the construction from a line cut through the middle is called a _____.
   a. plan view
   b. section view
   c. slice view
   d. plot view

10. A P&ID drawing shows _____.
    a. pipes only
    b. pipes and their diameters
    c. parts and isometric drawings
    d. pipes and instrumentation

11. An ISO shows the parts of the assembly as well as the direction of flow.
    a. True
    b. False

12. A spool is a _____.
    a. wheel
    b. piping segment
    c. special valve
    d. drawing showing bolts and gaskets

13. Certified equipment drawings do *not* require compliance by equipment to specification.

    a. True
    b. False

14. A single orthographic sketch shows the object from _____.

    a. a 30-degree angle
    b. a 45-degree angle
    c. one side only
    d. the pipefitter's viewpoint

15. An isometric sketch shows the object from _____.

    a. two or more sides at once
    b. the top only
    c. top, end, and side separately
    d. the pipefitter's viewpoint

Summary

A blueprint or drawing is a detailed plan of a part or system. Drawings are the source of information on how a system is to be assembled, what materials are to be used, and how the connections are to be made. Pipefitters must be able to take a set of drawings, read them, and understand the way the system or structure is to function. Notes and specifications are provided to clarify particular details of the system. The better you are at using and interpreting drawings, the more valuable you will be to your company. You will use your skill in reading and interpreting drawings whether you are installing, troubleshooting, or repairing piping.

Notes

Trade Terms Introduced in This Module

Battery limits: The outside perimeter of a project.

Chase: A recess on the inside of a wall, used for piping or electrical lines.

Isometric drawing (ISO): A three-dimensional drawing of a piping system that is not drawn to scale to provide clarity.

Line number: A group of abbreviations that specify size, service, material class/specification, insulation thickness, and tracing requirements of a given piping segment.

Orthographic projection: The projection of a single view of an object.

Piping and instrumentation drawing (P&ID): A schematic flow diagram of a complete system or systems that shows function, instrument, valving, and equipment sequence.

Sketch: A drawing representing the primary features of an object, usually a rough draft or freehand drawing.

Spool: A piping segment of a pipe system or an ISO.

Resources & Acknowledgments

Additional Resources

This module is intended to be a thorough resource for task training. The following reference work is suggested for further study. This is optional material for continued education rather than for task training.

Process Piping Drafting, 1986. Rip Weaver. Houston, TX: Gulf Publishing Company.

Figure Credits

Cianbro Corporation, Module Opener

NCCER CURRICULA — USER UPDATE

NCCER makes every effort to keep its textbooks up-to-date and free of technical errors. We appreciate your help in this process. If you find an error, a typographical mistake, or an inaccuracy in NCCER's curricula, please fill out this form (or a photocopy), or complete the online form at **www.nccer.org/olf**. Be sure to include the exact module ID number, page number, a detailed description, and your recommended correction. Your input will be brought to the attention of the Authoring Team. Thank you for your assistance.

Instructors – If you have an idea for improving this textbook, or have found that additional materials were necessary to teach this module effectively, please let us know so that we may present your suggestions to the Authoring Team.

NCCER Product Development and Revision

13614 Progress Blvd., Alachua, FL 32615

Email: curriculum@nccer.org
Online: www.nccer.org/olf

❏ Trainee Guide   ❏ Lesson Plans   ❏ Exam   ❏ PowerPoints   Other _____

Craft / Level: _____ Copyright Date: _____

Module ID Number / Title: _____

Section Number(s): _____

Description:
_____
_____
_____
_____
_____

Recommended Correction:
_____
_____
_____
_____
_____

Your Name: _____

Address: _____
_____

Email: _____ Phone: _____

08203-06

Identifying and Installing Valves

08203-06
Identifying and Installing Valves

Topics to be presented in this module include:

Overview

Valves are the steering wheels, brakes, and switches of pipe systems. Some valves are on-and-off flow controls, while others regulate the amount of flow. Some valves divert flow from one direction to another. The selection and proper installation of valves is a critical pipefitting skill. To install and use valves properly, you must understand the function of the valve and the characteristics of that particular valve, such as linings and part interactions. Each valve has advantages and limitations with which you must become familiar.

Objectives

When you have completed this module, you will be able to do the following:

1. Identify types of valves that start and stop flow.
2. Identify types of valves that regulate flow.
3. Identify valves that relieve pressure.
4. Identify valves that regulate the direction of flow.
5. Identify types of valve actuators.
6. Explain how to properly store and handle valves.
7. Explain valve locations and positions.
8. Explain the factors that influence valve selection.
9. Interpret valve markings and nameplate information.

Trade Terms

Actuator	Head loss
Angle valve	Kinetic energy
American Society for Testing Materials International (ASTM International)	Packing
	Phonographic
	Plug
Ball valve	Plug valve
Body	Positioner
Bonnet	Relief valve
Butterfly valve	Seat
Check valve	Thermal transients
Control valve	Throttling
Corrosive	Torque
Deformation	Trim
Disc	Valve body
Elastomeric	Valve stem
Galling	Valve trim
Gate valve	Wedge
Globe valve	Wire drawing
	Yoke bushing

Required Trainee Materials

1. Pencil and paper
2. Appropriate personal protective equipment

Prerequisites

Before you begin this module, it is recommended that you successfully complete *Core Curriculum*; *Pipefitting Level One*; and *Pipefitting Level Two*, Modules 08201-06 and 08202-06.

This course map shows all of the modules in the second level of the *Pipefitting* curriculum. The suggested training order begins at the bottom and proceeds up. Skill levels increase as you advance on the course map. The local Training Program Sponsor may adjust the training order.

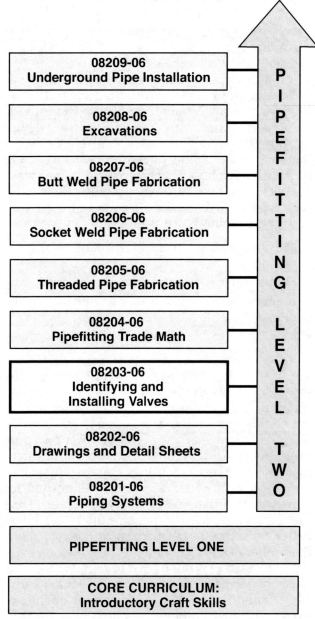

203CMAP.EPS

1.0.0 ◆ INTRODUCTION

Valves are devices that control the flow of fluids or gases through a piping system. While the designs of valves vary, all valves have two common features: a passageway through which fluid or gas flows and a moveable part that opens and closes the passageway. A valve can be used to provide on-off service only, can act as a throttling device to allow different flow rates, can relieve excess pressure in a pipeline, or can prevent reversal of flow through a line. There are basically four ways to control the flow through a piping system, and each type of valve uses one or more of these methods. These methods are as follows:

- Moving a disc or plug into or against a passageway
- Sliding a flat, round surface across a passageway
- Rotating an opening inside a shaft across the passageway
- Moving a flexible material into the passageway

Valves are made of a variety of materials, including bronze, iron, carbon steel, aluminum, alloy steels, and polyvinyl chloride (PVC). The application of the valve usually dictates what type of material is used, and there is often more than one material that meets the needs of an application. Bronze or iron with bronze trim is used for valves in air or water services. Iron is also used for valves in low-pressure steam services. Steel is used for valves that regulate noncorrosive products. Ductile iron, which is less expensive than steel, offers better resistance to mechanical or thermal shock than other metals.

Stainless steel is often used to regulate the flow of corrosive materials and in services that require sterile conditions. Most systems that operate at extremely low temperatures also use stainless steel valves because they are less brittle at low temperatures than iron or carbon steel. Highly corrosive chemicals require specialized valve bodies that are made of or lined with plastic, rubber, ceramics, or special alloys.

This module introduces the types of valves used to start and stop flow, regulate flow, relieve pressure, and regulate the direction of flow. Various types of valve actuators are also explained.

2.0.0 ◆ VALVES THAT START AND STOP FLOW

Many valves are designed to operate either completely open or completely closed. These valves are not practical for throttling or controlling the flow of fluid through a piping system. The valves used to start and stop flow through a system include the following types:

- Gate
- Knife
- Ball
- Plug
- Three-way

2.1.0 Gate Valves

Gate valves are used to start or stop fluid flow, but not to regulate or throttle flow. The term gate is derived from the appearance of the disc in the flow stream, which is similar to a gate. *Figure 1* shows a gate valve. The disc is completely removed from the flow stream when a gate valve is fully open. The disc offers virtually no resistance to flow when the valve is open, so there is little pressure drop across an open gate valve. When the valve is fully closed, a disc-to- seal ring contact surface exists for 360 degrees and good sealing is provided. With proper mating of the disc-to-seal ring, very little or no leakage occurs across the disc when the gate valve is closed.

Gate valves are not used to regulate flow because the relationship between valve stem movement and flow rate is nonlinear; that is, stem movement does not change the flow rate much when the gate is open. Operating with the valve in a partially open position can cause disc and seat wear, which will eventually lead to valve leakage.

The primary consideration in the application of a gate valve in comparison to a globe valve is that the gate valve represents much less flow restriction than the globe valve. This reduced flow restriction is the result of straight-through body construction and the design of the disc. A gate valve can be used for a wide variety of fluids and provides a tight seal when closed. The major disadvantages of using a gate valve as opposed to a globe valve are as follows:

- It is not good for throttling applications.
- It is prone to vibration in the partially open state.
- It is more subject to seat and disc wear than a globe valve.
- Repairs, such as lapping and grinding, are generally more difficult to accomplish.

Gate valves are available with a variety of fluid-control elements. Classification of gate valves is usually made by the type of fluid-control element used.

The fluid-control elements are available as the following:

- Solid wedge
- Flexible wedge
- Split wedge
- Double disc (parallel disc)

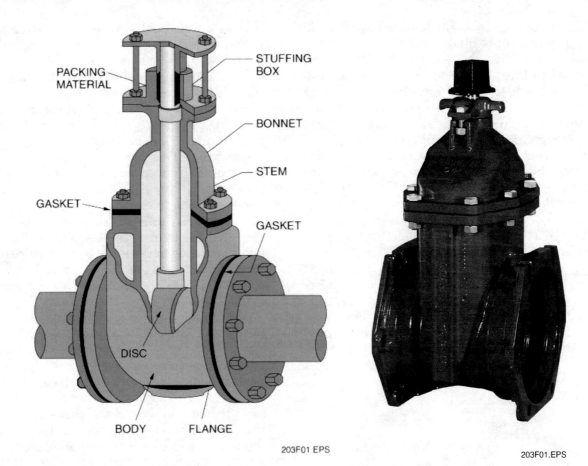

Figure 1 ◆ Gate valve.

Solid, flexible, and split wedges (*Figure 2*) are used in valves with inclined seats, while the double discs are used in valves with parallel seats. Regardless of the style of wedge or disc used, they are all replaceable. In services where solids or high velocity may cause rapid erosion of the seat or disc, these components should have a high surface hardness as well as replaceable seats and discs. If the seats are not replaceable, any damage would require refacing of the seat. Valves being used in corrosive service should always be specified with renewable seats.

The solid or single wedge shown in *Figure 3* is the most commonly used disc because of its simplicity and strength. A valve with this type of wedge may be installed in any position. It is suitable for almost all fluids and is most practical for turbulent flow. The majority of solid wedge gate valves have resilient seats, which provide a better sealing action. Resilient seats are made out of rubber, ethylene propylene diene monomer (EPDM), or other compressible materials.

The flexible wedge (*Figure 4*) is a one-piece disc with a cut around the perimeter to improve the ability to match error or change in angle between the seats. The cut varies in size, shape, and depth. A shallow, narrow cut gives little flexibility but retains strength. A deeper and wider cut, or a

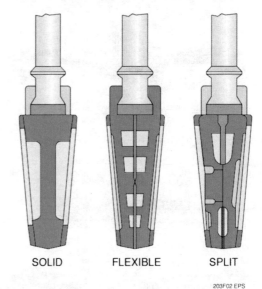

Figure 2 ◆ Wedge types.

cast-in recess, leaves little material at the center, allowing more flexibility but potentially compromising strength and risking permanent set. A correct profile of the disc half in the flexible wedge design can provide uniform deflection properties at the disc edge, so that the wedging force applied in seating will force the disc seating surface uniformly and tightly against the seat.

Gate valves used in steam systems have flexible gates. This prevents binding of the gate within the valve when the valve is in the closed position. When steam lines are heated, they will expand, causing some distortion of valve bodies. If a solid gate fits snugly between the seats of a valve in a cold steam system, when the system is heated and pipes elongate, the seats will compress against the gate, wedging the gate between them and clamping the valve shut. A flexible gate, by contrast, flexes as the valve seat compresses it. This prevents clamping.

The major problem associated with flexible gates is that water tends to collect in the body neck. Under certain conditions, the admission of steam may cause the valve body neck to rupture, the bonnet to lift off, or the seat ring to collapse. It is essential that correct warming-up procedures be followed to prevent this. Some very large gate valves also have a three-position vent and bypass valve installed. This valve allows venting of the bonnet either upstream or downstream of the valve and has a position for bypassing the valve.

Split wedges, as shown in *Figure 5*, have a ball-and-socket design, which are self-adjusting and self-aligning to both seating surfaces. The disc is free to adjust itself to the seating surface if one half of the disc is slightly out of alignment because of foreign matter lodged between the disc half and the seat ring. This type of wedge (disc) is suitable for handling non-condensing gases and liquids, particularly corrosive liquids, at normal temperatures. The freedom of movement of the discs inside the valve prevents them from binding, even though the valve may be closed when the seats are hot. Expansion and contraction of the parts of the valve will not prevent proper operation. This type of valve should be installed with the stem in the vertical position.

The parallel disc (*Figure 6*) was also designed to prevent valve binding due to thermal transients (changes of temperature). Both low-pressure iron valves and high-pressure steel types have this disc. The wedge surfaces between the parallel-faced disc halves press together under stem thrust and spread the discs to seal against the seats. The tapered wedges may be part of the disc halves or may be separate elements. The lower wedge must bottom out on a rib at the valve bottom so that the stem can develop seating force. In one version (*Figure 7*), the wedge contact surfaces are curved to keep the point of contact as close to perfect as possible.

In other parallel-disc gates (*Figures 8* and *9*), the two halves do not move apart under wedge action.

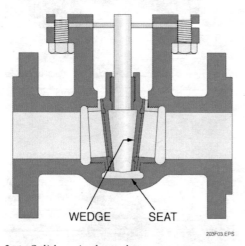

Figure 3 ◆ Solid or single wedge.

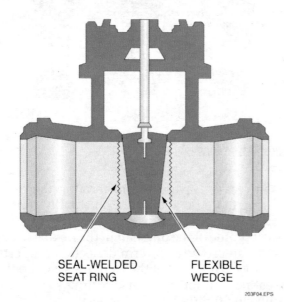

Figure 4 ◆ Flexible wedge.

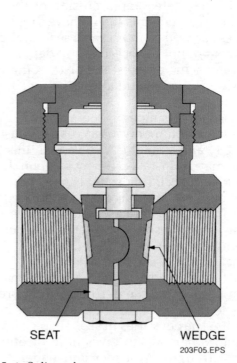

Figure 5 ◆ Split wedge.

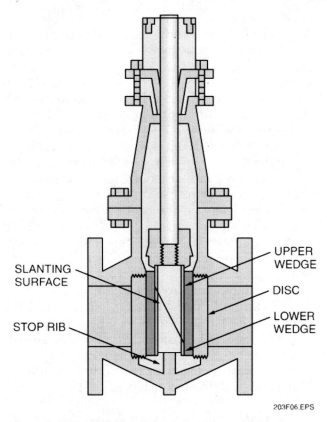

Figure 6 ◆ Parallel-disc gate valve.

SLANTING SURFACE

STOP RIB

UPPER WEDGE

DISC

LOWER WEDGE

203F06.EPS

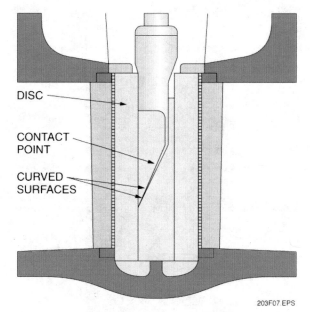

DISC

CONTACT POINT

CURVED SURFACES

203F07.EPS

Figure 7 ◆ Parallel disc.

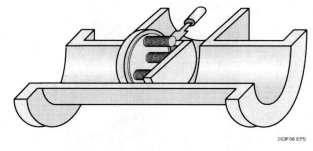

203F08.EPS

Figure 8 ◆ Spring-loaded parallel-disc gate valve (cutaway).

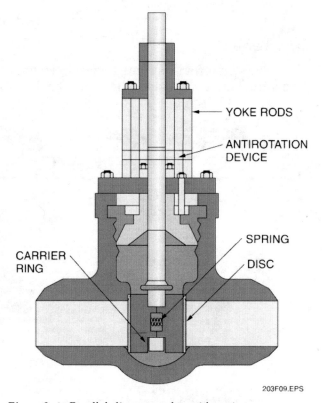

YOKE RODS

ANTIROTATION DEVICE

SPRING

DISC

CARRIER RING

203F09.EPS

Figure 9 ◆ Parallel-disc gate valve with spring.

Instead, the upstream pressure holds the downstream disc against the seat. A carrier ring lifts the discs. A spring or springs hold the discs apart and seated when there is no upstream pressure.

Another design found on parallel gate discs provides for sealing only one port. In these designs, the high-pressure side pushes the disc open (relieving disc) on the high-pressure side, but forces the disc closed on the low-pressure side (*Figure 10*). With such designs, the amount of seat leakage tends to decrease as differential pressure across the seat increases.

These valves usually have a flow direction marking to show which side is the high-pressure (relieving) side. Make sure these valves are not installed backwards in the system.

Some parallel-disc gate valves used in high-pressure systems are equipped with an integral bonnet vent/bypass line. A three-way valve is used to position the line to bypass in order to equalize pressure across the discs prior to opening. When the gate valve is closed, the three-way valve is positioned to vent the bonnet to one side or the other. This prevents moisture from accumulating in the bonnet. The three-way valve is positioned to the high-pressure side of the gate valve when the gate valve is closed to ensure that flow does not bypass the isolation valve. The high pressure acts against spring compression and forces one gate off its seat. The three-way valve vents this flow back to the pressure source.

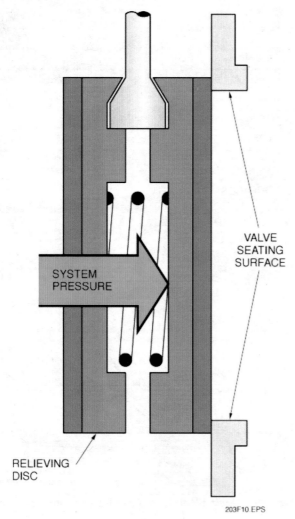

Figure 10 ◆ Relieving disc gate valve.

2.1.1 Valve Stem

The valve stem provides a link between the handwheel and the gate. Often, the pipefitter has to attach the handwheel to the stem after the valve is installed. Some handwheels bolt onto the stem; others slip over a keyway cut into the stem and are then secured to the stem by a nut or machine screw. Handwheels vary according to the type of stem they fit. Three general types of stems include the rising stem, the nonrising stem, and the outside screw-and-yoke (OS&Y) stem (*Figure 11*).

When a rising stem valve is opened, both the handwheel and the stem rise. The height of the stem gives an approximation of how far the valve is open. These stems can only be installed where there is sufficient clearance for the handwheel to rise. Rising stems come in contact with the fluid in the valve and are only used with fluids that will not harm the threads, such as hydrocarbons, water, and steam.

Nonrising stems are suitable in spaces where space is limited. As the name implies, neither the stem nor the handwheel rises when the valve is opened. A spindle inside the valve body turns when the handwheel is turned and raises the stem; therefore, stem wear is held to a minimum. This type of stem also comes in contact with the fluid in the valve.

The OS&Y stem is suitable for corrosive fluids because the stem does not come in contact with the fluid in the line. As the handwheel is turned, the stem moves up through the handwheel. The height of the stem indicates how far the valve is open.

2.2.0 Knife Gate Valves

A special type of gate valve is the knife gate valve (*Figure 12*). This valve serves in slurry and waste lines and in other low-pressure applications. It is also used extensively in paper mills. The sharp edge of the disc bottom is easily forced closed in

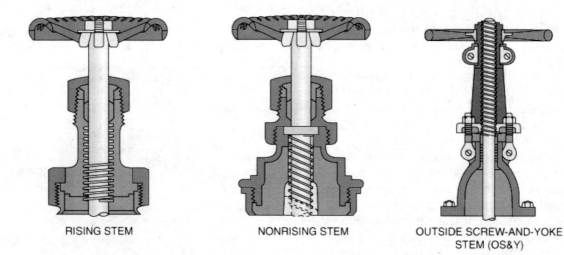

Figure 11 ◆ Valve stems.

RISING STEM

NONRISING STEM

OUTSIDE SCREW-AND-YOKE STEM (OS&Y)

contact with a metal or *elastomeric* seat. When open, the disc is in the air after passing through a full-width packing box.

Knife valves have a disc that can cut through deposits and flow-stream solids such as resin slurry. These valves, like most gate valves, can be positioned by manual, electrical, pneumatic, and hydraulic actuation.

The positioning of a knife gate valve is critical. If the valve is used to isolate equipment, the seat side of the valve must face in the direction of flow. If the valve is used to isolate flow, the seat side must be opposite the direction of flow.

2.3.0 Ball Valves

Ball valves, as the name implies, are stop valves that use a ball to stop or start the flow of fluid. They are rotary action valves. The ball performs the same function as the disc in the globe valve. However, it turns instead of traveling up and down. When the actuator is operated to open the valve, the ball rotates to a point where the hole through the ball is in line with the valve body inlet and outlet. When the valve is shut, which requires only a 90-degree rotation for most valves, the ball is rotated so that the hole is perpendicular to the flow openings of the valve body, and flow is stopped.

Most ball valves are quick-acting, requiring only a 90-degree turn to operate the valve either completely open or closed. *Figure 13* shows the major components of a ball valve. The ball valve, in general, is the least expensive of any valve configuration. In early designs having metal-to-metal seating, the valves could not provide bubble-tight sealing and were not fire-safe. With the development of elastomeric materials and with advances in plastics, the original metallic seats have been replaced with more effective sealing materials such as fluorinated polymers, nylon, neoprene, and Buna-N. Ball valves have the advantage of low maintenance costs.

In addition to quick on-off operation, ball valves are compact, require no lubrication, and provide tight sealing with low torque. With a soft seat on both sides of the ball, most ball valves provide equally effective sealing of flow in either direction. Many designs permit adjustment for wear.

Because conventional ball valves have relatively poor throttling characteristics, they are not generally satisfactory for this service. In a throttling

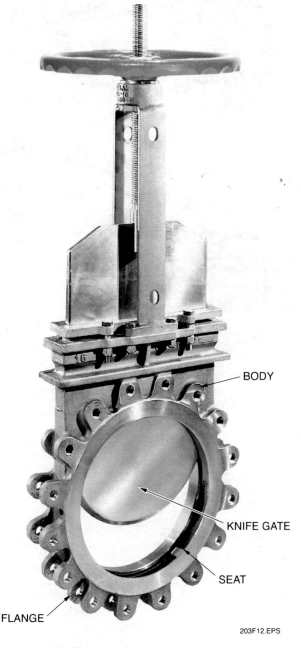

Figure 12 ◆ Knife gate valve.

203F12.EPS

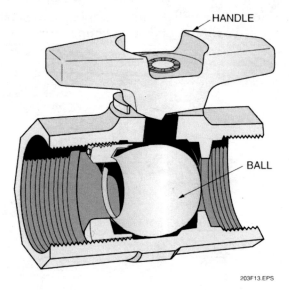

Figure 13 ◆ Ball valve.

203F13.EPS

position, the partially exposed seat rapidly erodes because of the high-velocity flow. However, a ball valve has been developed with a spherical surface-coated plug off to one side in the open position. This plug rotates into the flow passage until it blocks the flow path completely. Seating is accomplished by the eccentric movement of the plug. The valve requires no lubrication and can be used for throttling service.

Ball valves are designed on the simple principle of floating a polished ball between two plastic seating surfaces, permitting free turning of the ball. Because the plastic is subject to deformation under load, some means must be provided to hold the ball against at least one seat. This is normally accomplished through spring pressure, differential line pressure, or a combination of both. *Figure 14* shows a combination of line pressure and a spring on the ball.

Ball valves are available in the venturi (*Figure 15*) and full-port patterns. The latter has a ball with a bore equal to the inside diameter of the pipe. Balls are usually metal in metallic bodies with **trim** (seats) produced from elastomeric materials. All-plastic designs are also available. Seats and balls are replaceable.

Ball valves are available in top-entry (*Figure 16*) and split-body (end-entry) types (*Figure 17*). In the former, the ball and seats are inserted through the top, while in the latter, the ball and seats are inserted from the ends.

The resilient seats for ball valves are made from various elastomeric materials. The most common seat materials are virgin tetrafluoroethylene (TFE), filled TFE, nylon, Buna-N, neoprene, and combinations of these materials. Because of the presence of the elastomeric materials, these valves cannot be used at elevated temperatures. To overcome this disadvantage, a graphite seat has been developed that will permit operation up to 1,000°F. Typical maximum operating temperatures of the

valves at their full pressure ratings are shown in *Table 1* for various seat materials. Another valve operating on the same principle as the ball is the cone valve, in which a conical plug, with a port in the middle, can be turned to obstruct or permit flow. This is a very fast-operating valve, requiring only a half-turn to work.

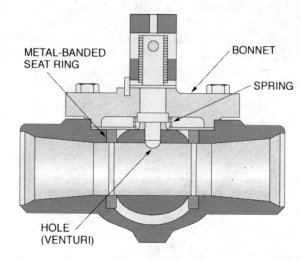

203F15.EPS

Figure 15 ◆ Venturi-type ball valve.

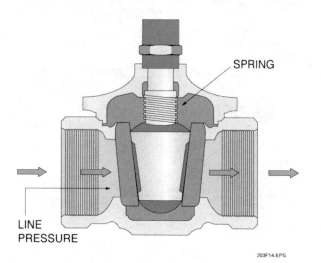

203F14.EPS

Figure 14 ◆ Ball valve with slanted seals.

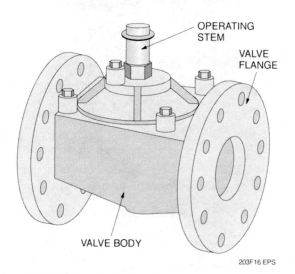

203F16.EPS

Figure 16 ◆ Top-entry ball valve.

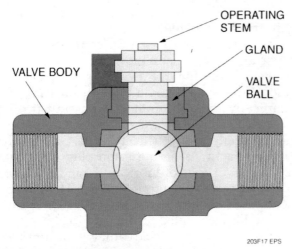

Figure 17 ◆ Split-body ball valve.

Table 1 Maximum Operating Temperature for Seat Materials

Seat Material	Operating Temperature
TFE (Virgin)	450°F
Filled TFE	400°F
Buna-N	180°F
Neoprene	180°F

203T01.EPS

CAUTION

Use care when choosing the seat material to make sure it is compatible with the materials being handled by the valve. Be sure that the seat can handle the temperature range of the process.

2.4.0 Plug Valves

Plug valves are used to stop or start fluid flow. They are rotary-action valves. The name is derived from the shape of the disc, which resembles a plug. A plug valve is shown in *Figure 18*.

The body of a plug valve is machined to receive the tapered or cylindrical plug. The disc is a solid plug with a bored passage at a right angle to the longitudinal axis of the plug. In the open position, the passage in the plug lines up with the inlet and outlet ports of the valve body. When the plug is turned 90 degrees from the open position, the solid part of the plug blocks the ports and stops fluid flow.

Plug valves are available in either a lubricated or non-lubricated design and with a variety of styles of port openings through the plug. There are numerous plug designs.

An important characteristic of the plug valve is its easy adaptation to multiport construction.

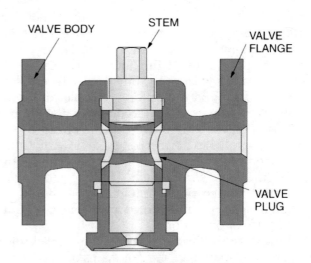

Figure 18 ◆ Plug valve.

Multiport valves are widely used. Their installation simplifies piping. They provide a much more convenient operation than multiple gate valves do. They also eliminate pipe fittings. The use of a multiport valve eliminates the need for two, three, or even four conventional shutoff valves, depending on the number of ports in the plug valve.

Plug valves are normally used only for on-off operations, particularly where frequent operation of the valve is necessary. These valves are not normally recommended for throttling service because, as with the gate valve, a great percentage of flow change occurs near shutoff at high velocity. However, a diamond-shaped port has been developed for throttling service.

Multiport valves are particularly good on transfer lines and for diverting services. A single multiport valve may be installed in lieu of three or four gate valves or other types of shutoff valves.

Many of the multiport configurations do not permit complete shutoff of flow, however. In most cases, one flow path is always open. These valves are intended to divert the flow to one line while shutting off flow from the other lines. If complete shutoff of flow is required, a suitable multiport valve should be used, or a secondary valve should be installed on the main feed line ahead of the multiport valve to permit complete flow shutoff.

It should also be noted that in some multiport configurations, simultaneous flow to more than one port is also possible. Great care should be taken in specifying the port arrangement to guarantee proper operation.

Plugs are either round or cylindrical with a taper. They may have various types of port openings, each with a varying degree of free area relative to the corresponding pipe size (*Figure 19*).

- Rectangular port is the standard-shaped port with a minimum of 70 percent of the area of the corresponding size of standard pipe.
- Round port means that the valve has a full round opening through the plug and body, of the same shape as standard pipe.
  - Full port means that the area through the valve is equal to or greater than the area of standard pipe.
  - Standard opening means that the area through the valve is less than the area of standard pipe. These valves should be used only where restriction of flow is unimportant.
- Diamond port means that the opening through the plug is diamond-shaped. This has been designed for throttling service. All diamond port valves are venturi, restricted-flow type.

Clearances and leakage prevention are the chief considerations in plug valves. Many plug valves are of all-metal construction. In these versions, the narrow gap around the plug can permit leakage. If the gap is reduced by sinking the taper plug deeper into the body, actuation torque will climb rapidly, and galling can occur.

Lubrication remedies this. A series of grooves around the port openings in the plug or body is supplied with grease prior to actuation, not only to lubricate the plug motion but also to seal the gap (*Figure 20*). Grease injected into a fitting at the stem top travels down through a check valve in the passageway and then past the plug top to the grooves on the plug and down to a well below the plug.

The lubricant must be compatible with the temperature and nature of the fluid. The most common substances controlled by plug valves are gases and liquid hydrocarbons. Some water lines have these valves too, if lubricant contamination is not a serious danger. This type can go to 24-inch size, with pressure capability of 6,000 psig. Steel and iron bodies are available. The plug can be cylindrical or tapered.

The correct choice of lubricant is extremely important for successful lubricated plug valve performance. In addition to providing adequate lubrication to the valve, the lubricant must not react chemically with the material passing through the valve. The lubricant must not contaminate the material passing through the valve, either. All manufacturers of lubricated plug valves have developed a series of lubricants that are compatible with a wide range of media. Their recommendations should be followed regarding which lubricant is best suited for the service.

To overcome the disadvantages of lubricated plug valves, two basic types of non-lubricated plug valves were developed. A non-lubricated valve may be a lift-type, or it may have an elastomer sleeve or plug coating that eliminates the need

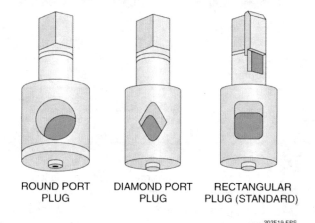

ROUND PORT PLUG   DIAMOND PORT PLUG   RECTANGULAR PLUG (STANDARD)

203F19.EPS

Figure 19 ◆ Plug valve – port types.

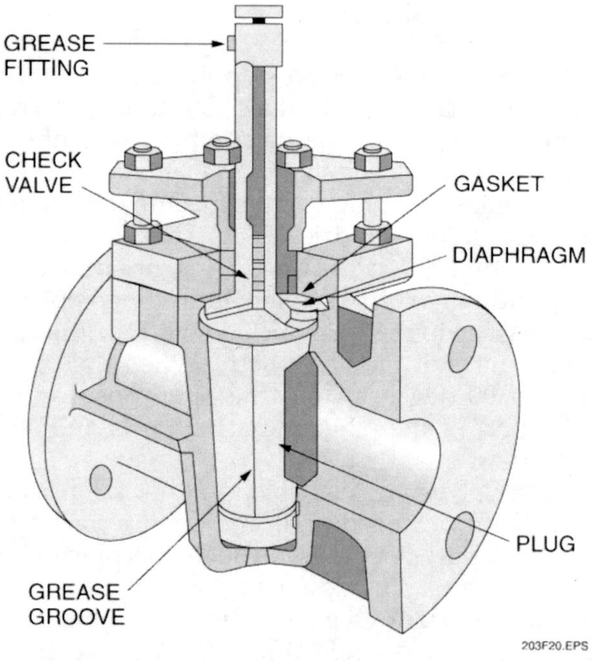

GREASE FITTING

CHECK VALVE

GASKET

DIAPHRAGM

PLUG

GREASE GROOVE

203F20.EPS

Figure 20 ◆ Lubricated taper plug valve.

to lubricate the space between the plug and seat. Non-lubricated plug valves are very commonly used in sewage plants.

Lift-type valves provide a means of mechanically lifting the tapered plug slightly from its seating surface to permit easy rotation. The mechanical lifting can be accomplished through either a cam (*Figure 21*) or an external lever.

A typical non-lubricated plug valve with an elastomer sleeve is shown in *Figure 22*. In this particular valve, a sleeve of TFE surrounds the plug. It is retained and locked in place by the metal body. This results in a continuous primary seal between the sleeve and the plug, both while the plug is rotated and when the valve is in either the open or closed position. The TFE sleeve is durable and essentially inert to all but a few rarely encountered chemicals. It also has a low coefficient of friction and therefore is self-lubricating.

Lubricants are available in stick form and in bulk. Stick lubrication is usually employed when a small number of valves are in service or when they are widely scattered throughout the plant. However, for a large number of valves, gun lubrication is the most convenient and economical solution.

Valves are usually shipped with an assembly lubricant. This assembly lubricant should be removed and the valve completely relubricated with the proper lubricant before the valve is put into service.

Regular lubrication is critical for best results. Extreme care should be taken to prevent any foreign matter from entering the plug when inserting new lubricant.

2.5.0 Three-Way Valves

Three-way valves are multiport plug valves that are installed at the intersection of three lines. They are used to direct the flow between two of the lines only and to block off the third line. Situations requiring three-way valves include alternating the connections of two supply lines to a common delivery line or diverting a line into either of two possible directions. Three-way valves are designed so that when the plug is turned from one position to another, the channels previously connected are completely closed off before the new channels begin to open. This design prevents mixture of fluids or loss of pressure. Three-way valves come in several different arrangements. They can be two- or three-port valves with stops that limit the turning of the plug to two, three, or four positions (*Figure 23*).

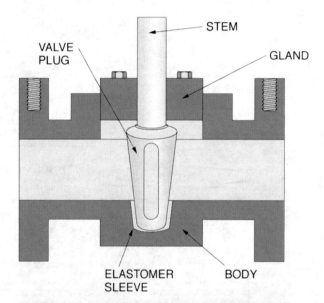

203F22.EPS

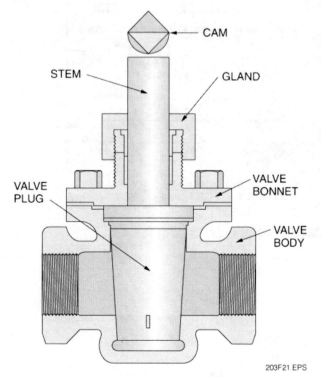

203F21.EPS

Figure 21 ◆ Cam-operated and non-lubricated plug valve.

Figure 22 ◆ Typical non-lubricated plug valve with elastomer sleeve.

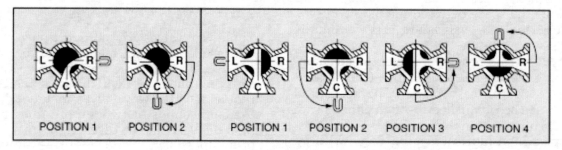

TWO-PORT, TWO POSITIONS

THREE-PORT, FOUR POSITIONS

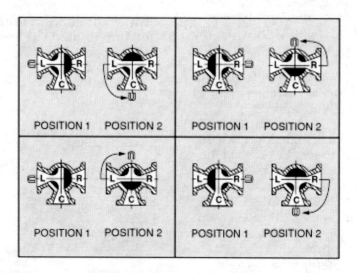

THREE-PORT, TWO POSITIONS

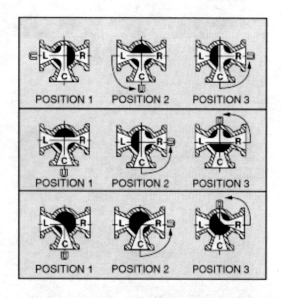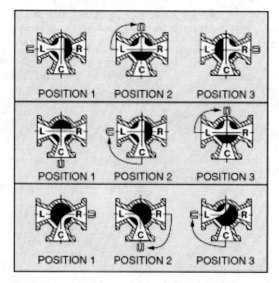

THREE-PORT, THREE POSITIONS

203F23.EPS

Figure 23 ◆ Three-way valve applications.

3.0.0 ◆ VALVES THAT REGULATE FLOW

Many valves are designed to provide accurate flow control through a system. The valves used to regulate flow through a system include the following:

- Globe
- Angle
- Y-type
- Needle
- Butterfly
- Diaphragm
- Control

3.1.0 Globe Valves

Globe valves are used to stop, start, and regulate fluid flow. They are important elements in power plant systems and are commonly used as the standard against which other valve types are judged.

As shown in *Figure 24*, the globe valve disc can be totally removed from the flow path or it can completely close the flow path. The essential principle of globe valve operation is the perpendicular movement of the disc away from the seat. This causes the space between disc and seat ring to gradually close as the valve is closed. This characteristic gives the globe valve good throttling ability, which permits its use in regulating flow. Therefore, the globe valve is used not only for start-stop functions, but also for flow-regulating functions.

It is generally easier to obtain very low seat leakage with a globe valve as compared to a gate valve. This is because the disc-to-seat ring contact is more at right angles, which permits the force of closing to tightly seat the disc.

Globe valves can be arranged so that the disc closes against the direction of fluid flow (flow-to-open) or so that the disc closes in the same direction as fluid flow (flow-to-close). When the disc closes against the direction of flow, the kinetic energy of the fluid impedes closing but aids opening of the valve. When the disc closes in the same direction as flow, the kinetic energy of the fluid aids closing but impedes opening of the valve. This characteristic makes the globe valve preferable to the gate valve when quick-acting stop valves are necessary.

Along with its advantages, the globe valve has a few drawbacks. Although valve designers can eliminate any or all of the drawbacks for specific services, the corrective measures are expensive and often narrow the valve's scope of service. The most evident shortcoming of the simple globe valve is the high head loss on the downstream side, a loss of pressure from the two or more right-angle turns of flowing fluid.

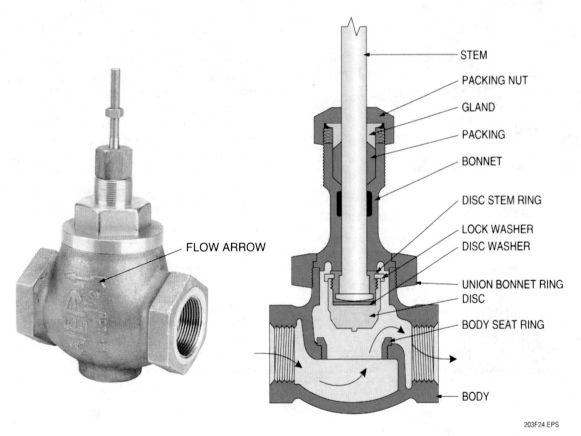

FLOW ARROW

STEM
PACKING NUT
GLAND
PACKING
BONNET
DISC STEM RING
LOCK WASHER
DISC WASHER
UNION BONNET RING
DISC
BODY SEAT RING
BODY

203F24.EPS

Figure 24 ◆ Globe valve.

High pressure losses in the globe valve can cost thousands of dollars a year for large high-pressure lines. The fluid-dynamic effects from the pulsation, impacts, and pressure drops in traditional globe valves can damage valve trim, stem packing, and actuators. Troublesome noise can also result. In addition, large sizes require considerable power to operate, which may make gearing or levers necessary.

Another drawback is the large opening needed for assembly of the disc. Globe valves are often heavier than other valves of the same flow rating. The cantilever mounting of the disc on its stem is also a potential trouble source. Each of these shortcomings can be overcome, but only at costs in dollars, space, and weight.

The angle valve of *Figure 25* is a simpler modification of the basic globe form. With the ports at right angles, the disc can be a simple flat plate. Fluid can flow through with only a single 90-degree turn, discharging more symmetrically than the discharge from an ordinary globe. Installation advantages also may suggest the angle valve. It can replace an elbow, for example.

For moderate conditions of pressure, temperature, and flow, the angle valve closely resembles the ordinary globe. Many manufacturers have interchangeable trim and bonnets for the two body styles, with the body differing only in the outlet end. The angle valve's discharge conditions are so favorable that many high-technology control valves use this configuration. Like straight flow-through globe valves, they are self-draining and tend to prevent solid buildup inside the valve body.

Like valve bodies, there are also many variations of disc and seat arrangements for globe valves. The three basic types are shown in *Figure 26*.

The ball-shaped disc shown in *Figure 26(A)* fits on a tapered, flat-surfaced seat and is generally used on relatively low-pressure, low-temperature systems. It is generally used in a fully open or shut position, but it may be employed for moderate throttling of a flow.

Figure 26(B) shows one of the proven modifications of seat/disc design, a hard nonmetallic insert ring on the disc to make closure tighter on steam and hot water. The composition disc resists erosion and is sufficiently resilient and cut resistant to close on solid particles without serious permanent damage.

The composition disc is renewable. It is available in a variety of materials that are designed for different types of service, such as high- and low-temperature water, air, or steam. The seating surface is often formed by a rubber O-ring or washer.

The plug-type disc, *Figure 26(C)*, provides the best throttling service because of its configuration. It also offers maximum resistance to galling, wire drawing, and erosion. Plug-type discs are available in a variety of specific configurations, but in general they all have a relatively long, tapered configuration. Each of the variations has specific types of applications and certain fundamental characteristics. *Figure 27* shows the various types.

The equal percentage plug, as its name indicates, is used to allow precise control of flow through the valve. For example, if you turn the actuator 30

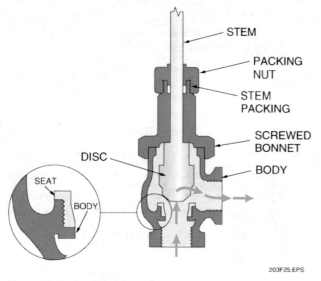

Figure 25 ◆ Angle globe valve.

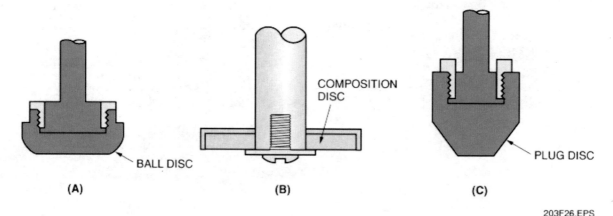

Figure 26 ◆ Seat/disc arrangements.

percent of its 90-degree swing, flow will increase by 30 percent.

V-port plugs have a V-shaped port cut in the plug. The flow does not change direction, but is throttled because the port is smaller than the flow chamber. Unlike other ports, this valve allows the flow to be controlled with only moderate changes of pressure.

Needle plugs are used primarily for instrumentation applications and are seldom available in valves over 1 inch in size. These plugs lower pressure and flow. The threads on the stem are usually very fine. For that reason, the opening between the disc and seat does not change rapidly with stem rise. This permits closer regulation of flow.

All of the plug configurations are available in either a conventional globe valve design or the angle valve design. When the needle plug is used, the valve name changes to needle valve. In all other cases the valves are still referred to as globe valves with a specific type of disc.

Globe and angle valves should be installed so that the pressure is under the disc (flow-to-open). This promotes easy operation. It also helps to protect the packing and eliminates a certain amount of erosive action on the seat and disc faces. However, when high temperature steam is the medium being controlled, and the valve is closed with the pressure under the disc, the valve stem, which is now out of the fluid, contracts on cooling. This action tends to lift the disc off the seat, causing leaks that eventually result in wire drawing, small erosion channels on seat and disc faces. Therefore, in high-temperature steam service, globe valves

may be installed so that the pressure is above the disc (flow-to-close).

3.2.0 Y-Type Valves

Y-type valves (*Figure 28*) are a cross between a gate valve and a globe valve. They provide the straight-through flow with minimum resistance of the gate valve and the throttling ability and flow control of a globe valve. The Y-type valve produces lower pressure drop and turbulence than a standard globe and is preferred over a globe valve for use in corrosive services. Y-type valves are also used in many high-pressure applications, such as boiler systems. All of the disc and seat designs available in globe valves are also available in Y-type valves.

3.3.0 Butterfly Valves

A butterfly valve has a round disc that fits tightly in its mating seat and rotates 90 degrees in one direction to open and allow fluid to pass through the valve (*Figure 29*). The butterfly valve can be operated quickly by turning the handwheel or hand lever one-quarter of a turn, or 90 degrees, to open or close the valve. These valves can be used completely open, completely closed, or partially open for noncritical throttling applications.

Butterfly valves are typically used in low to medium pressure and low to medium temperature applications. They generally weigh less than other types of valves because of their narrow body design. When the butterfly valve is equipped with a hand lever, the position of the hand lever indicates whether the valve is open, closed, or partially open. When the lever is parallel with the flow line through the valve, the valve is open. When the lever is perpendicular to the flow line through the valve, the valve is closed. Butterfly valves that are 12 inches in diameter and larger are usually equipped with a handwheel or gear-operated actuator because of the large amount of fluid flowing through the valve and the great amount of pressure pushing against the seat when the valve is being closed.

Butterfly valves have an arrow stamped on the side, indicating the direction of flow through the valve. They must be installed in the proper flow direction, or the seat will not seal and the valve will leak. Three types of butterfly valves include the wafer, wafer lug, and two-flange valve.

3.3.1 Wafer Valves

The wafer-type butterfly valve (*Figure 30*) is designed for quick installation between two flanges. No gasket is needed between the valve and the flanges because the valve seat is lapped

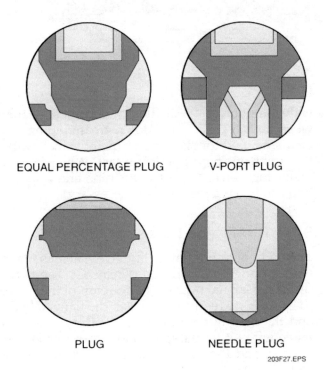

EQUAL PERCENTAGE PLUG   V-PORT PLUG

PLUG   NEEDLE PLUG

203F27.EPS

Figure 27 ◆ Globe valve plugs.

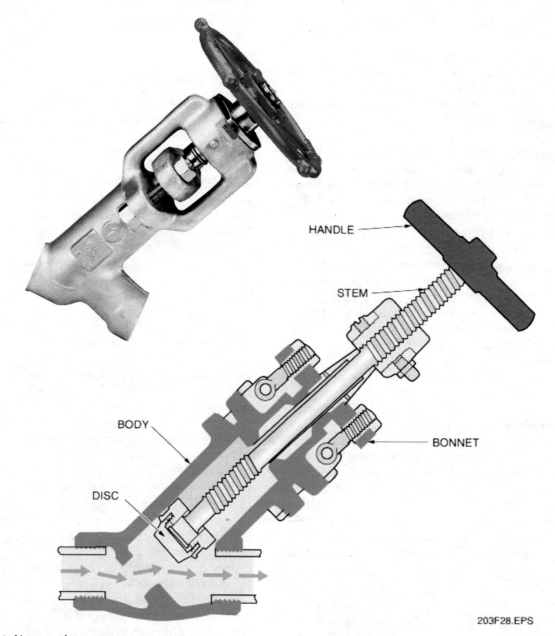

HANDLE

STEM

BODY

DISC

BONNET

203F28.EPS

Figure 28 ◆ Y-type valve.

over the edges of the body to make contact with the valve faces. Bolt holes are provided in larger wafer valves to help line up the valve with the flanges. During installation of the valves, the valve must be in the open position prior to torquing to prevent damage to the valve seat.

3.3.2 Wafer Lug Valves

Wafer lug valves (*Figure 31*) are the same as wafer valves except that they have bolt lugs completely around the valve body. Like the wafer valve, no gasket is needed between the valve and the flanges because the valve seat is lapped over the edges of the body to make contact with the valve faces. The lugs are normally drilled to match ANSI 150-pound steel drilling templates. The lugs on some

wafer lug valves are drilled and tapped so that when the valve is closed, downstream piping can be dismantled for cleaning or maintenance while the upstream piping is left intact. When tapped wafer lug valves are used for pipe end applications, only one pipe flange is necessary.

3.3.3 Two-Flange Valves

The body of a two-flange butterfly valve is made with a flange on each end (*Figure 32*). The valve seat is not lapped over the flange ends of the valve, so gaskets are required between the flanged body and the mating flanges. The two-flange valve is made with either flat-faced flanges or raised-face flanges, and the mating flanges must match the valve flanges.

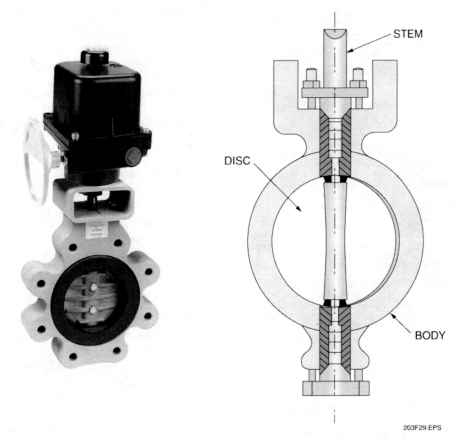

STEM

DISC

BODY

203F29.EPS

Figure 29 ◆ Butterfly valve.

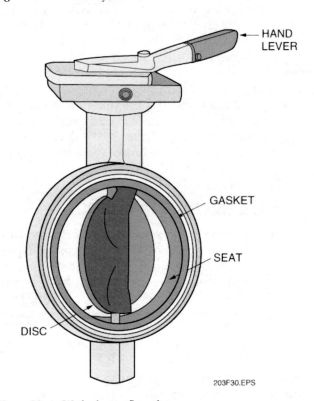

HAND LEVER

GASKET

SEAT

DISC

203F30.EPS

Figure 30 ◆ Wafer butterfly valve.

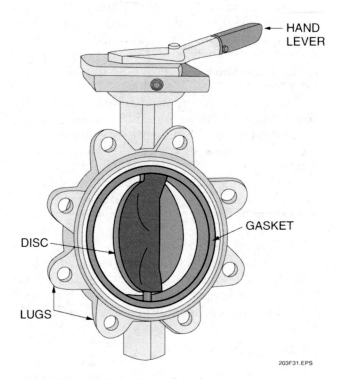

HAND LEVER

GASKET

DISC

LUGS

203F31.EPS

Figure 31 ◆ Wafer lug butterfly valve.

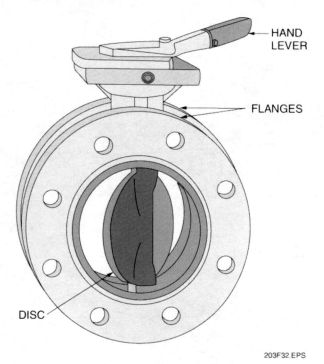

Figure 32 ◆ Two-flange butterfly valve.

203F32.EPS

3.4.0 Diaphragm Valves

A diaphragm valve is one in which flow through the valve is controlled by a flexible disc that is connected to a compressor by a stud molded into the disc. The valve stem moves the compressor up and down, regulating the flow. Diaphragm valves have no seats because the body of the valve acts as the seat for the flexible disc. The operating mechanism of the valve does not come in contact with the material within the pipeline.

These valves can be used fully opened, fully closed, or for throttling service. They are good for handling slurries, highly corrosive materials, or materials that must be protected from contamination. Many types of fluids that would clog other types of valves will flow through a diaphragm valve. *Figure 33* shows a weir-type diaphragm valve. In addition to the types shown, there is a straight-through type. A special case of the diaphragm valve is the pinch valve, in which the flexible lining of the opening is pinched closed by moving bars at the top and bottom of the valve. The linings are replaceable, and prevent exposure of the metal parts of the valve to the substance passing through.

3.5.0 Needle Valves

Needle valves, as shown in *Figure 34*, are used to make relatively fine adjustments in the amount of fluid allowed to pass through an opening. The needle valve has a long, tapering, needle-like point on the end of the valve stem. This needle acts as a disc. The longer part of the needle is

smaller than the orifice in the valve seat. It therefore passes through it before the needle seats. This arrangement permits a very gradual increase or decrease in the size of the opening and thus allows a more precise control of flow than could be obtained with an ordinary globe valve. Needle valves are often used as component parts of other, more complicated valves. For example, they are used in some types of reducing valves. Most constant-pressure pump governors have needle valves to minimize the effects of fluctuations in pump discharge pressure. Needle valves are also used in some components of automatic combustion control systems where very precise flow regulation is necessary.

4.0.0 ◆ CONTROL VALVES

Control valves (*Figure 35*) are variations of angle or globe valves that are controlled by pneumatic, electronic, or hydraulic actuators. They operate in a partially open position and are most commonly used for pressure limiting or flow control. Control valves are precision-built for increased accuracy in flow control. The control valve is usually smaller than the pipeline to avoid throttling and constant wear of the valve seat. The actuators for control valves are explained in more detail later in the module.

Control valves are often used with pressure-sensing elements that measure the pressure drop at a given point in the pipeline. The flow rate is directly related to the pressure drop. The sensing element sends a pressure signal to a controller that compares the pressure drop with the pressure drop for the desired flow rate. If the actual flow is different, the controller adjusts the valve through the valve actuator to increase or decrease the flow. *Figure 36* shows a schematic for a control valve in a pipeline.

5.0.0 ◆ VALVES THAT RELIEVE PRESSURE

Valves used to relieve pressure in a pipeline, tank, or vessel are known as safety valves and pressure-relief valves (*Figure 37*). These are installed in pipelines to prevent excess pressure from rupturing the line and causing an accident. Both types are adjustable and operate automatically.

5.1.0 Safety Valves

Safety valves are normally used in steam, air, or other gas service pipelines. They operate in the closed position until the pressure in the line rises above the preset pressure limit of the valve. At this point, the valve opens fully to relieve the pressure and remains open until the pressure drops, at which point the valve snaps shut. Because of this

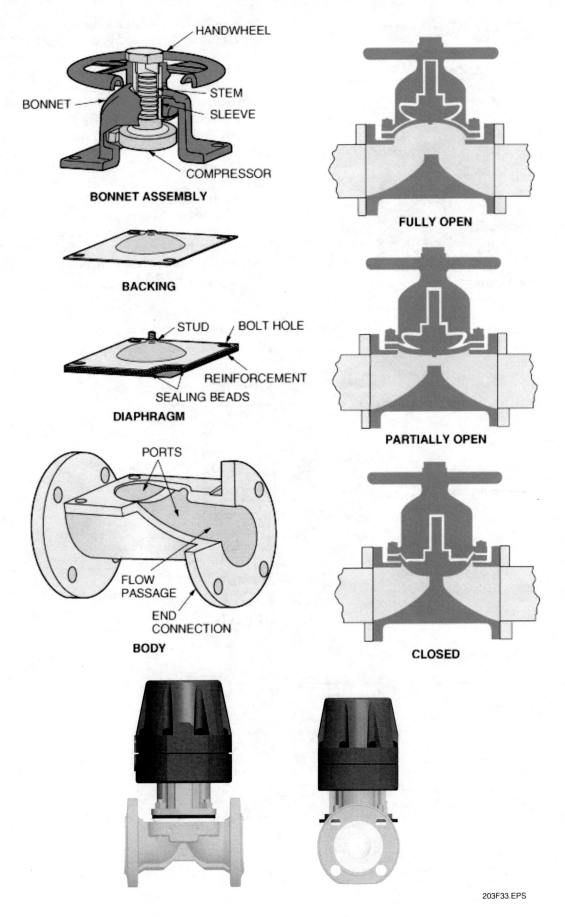

BONNET ASSEMBLY

HANDWHEEL

BONNET

STEM

SLEEVE

COMPRESSOR

BACKING

DIAPHRAGM

STUD

BOLT HOLE

REINFORCEMENT

SEALING BEADS

BODY

PORTS

FLOW PASSAGE

END CONNECTION

FULLY OPEN

PARTIALLY OPEN

CLOSED

203F33.EPS

Figure 33 ◆ Diaphragm valve.

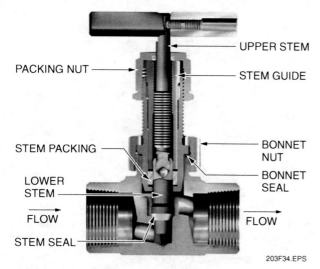

Figure 34 ◆ Needle valve.

PACKING NUT

UPPER STEM

STEM GUIDE

STEM PACKING

BONNET NUT

LOWER STEM

BONNET SEAL

FLOW

FLOW

STEM SEAL

203F34.EPS

fully open and tight-closing feature, safety valves are commonly referred to as pop-off valves. *Figure 38* shows a cutaway view of a safety valve.

Safety valves are most commonly installed in the vertical position. For air or gas service, these valves can be installed upside down to allow moisture to collect, which seals the surfaces. Safety valves should be mounted directly to the tank or vessel being protected using connecting piping and block valves. Before the safety valve can close, its discharge pressure must drop several pounds below the opening pressure. A safety valve has an adjustment to control the closing of the valve, depending on the amount of pressure drop after the valve has opened. This adjustment is calibrated by the manufacturer and should not be adjusted in the field.

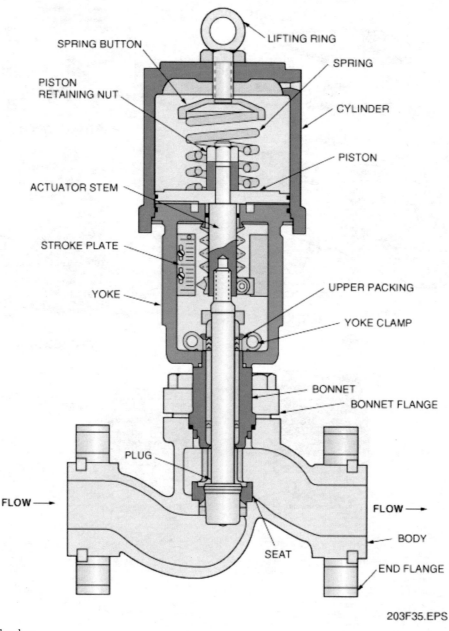

SPRING BUTTON

LIFTING RING

PISTON RETAINING NUT

SPRING

CYLINDER

ACTUATOR STEM

PISTON

STROKE PLATE

YOKE

UPPER PACKING

YOKE CLAMP

PLUG

BONNET

BONNET FLANGE

FLOW

FLOW

BODY

SEAT

END FLANGE

203F35.EPS

Figure 35 ◆ Control valve.

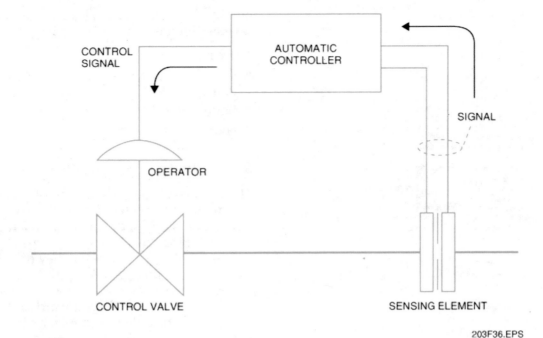

Figure 36 ◆ Control valve schematic.

203F37.EPS

Figure 37 ◆ Pressure-relief valves.

5.2.0 Pressure-Relief Valves

Relief valves are normally used to relieve pressure in liquid services. They are operated by pressure acting directly on the bottom of a disc that is held on its seat by a spring. The amount of spring pressure on the disc is determined at the factory when the valve is manufactured. When liquid pressure becomes great enough to overcome the force of the spring on the disc, the disc rises, allowing the liquid to escape. After enough liquid has escaped to allow the pressure to drop, the system pressure becomes too low to overcome the force of the spring on the disc. The spring then pushes the disc back onto its seat, stopping the flow of liquid through the valve. The force of the spring is called the compression of the valve. *Figure 39* shows a pressure-relief valve.

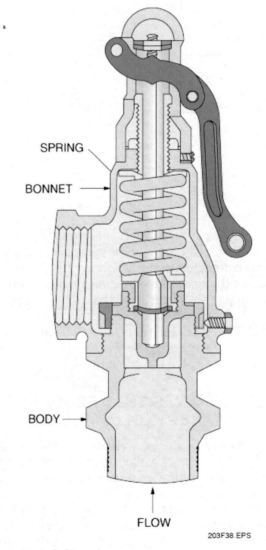

203F38.EPS

Figure 38 ◆ Safety valve.

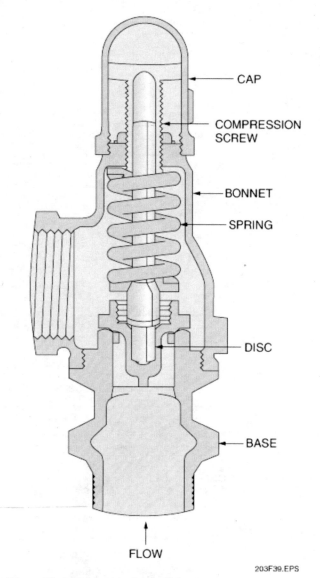

Figure 39 ◆ Pressure-relief valve.

CAP

COMPRESSION SCREW

BONNET

SPRING

DISC

BASE

FLOW

203F39.EPS

Spring-loaded relief valves are most commonly installed with the stem in the vertical position. The piping connected to these valves must be as large or larger than the valve inlet and outlet openings. You can use a reducer to reduce down to the inlet of a relief valve, but you cannot increase the line size to the inlet of the relief valve. The piping must also be well-supported so that the line strains will not cause the valve to leak at the seat.

6.0.0 ◆ VALVES THAT REGULATE DIRECTION OF FLOW

Valves that regulate the direction of flow are called check valves. Check valves regulate flow by preventing the backflow of liquids or gases in a pipeline. Check valves are intended to permit the flow in only one direction, which is indicated by an arrow stamped on the side of the valve. The force of flow and the action of gravity cause the valve to

open and close. Check valves are manufactured in several different designs, including the following:

- Swing
- Lift
- Ball
- Butterfly
- Foot

6.1.0 Swing Check Valves

The most commonly used type of check valve is the swing check valve (*Figure 40*). This valve has a disc that swings, or pivots, to open and close the valve. Since check valves provide straight line flow through the valve body, they offer less resistance to flow than other types of check valves. Swing check valves can be mounted horizontally or vertically with the flow upward. If this valve is used in a vertical flow, a counterweight is needed to make sure it does not stick open.

The disc, which is hinged at the top, seats against a machined seat in the tilted wall bridge opening. The disc swings freely from the fully closed position to the fully open position, which provides unobstructed flow through the valve. The flow through the valve causes the disc to stay open, and the amount that the disc is open depends on the volume of liquid moving through the valve. Gravity or reversal of flow causes the disc to close, preventing backflow through the valve.

Swing check valves can also be equipped with an outside lever and counterweight to assist the valve in closing. These valves are often referred to as weighted check valves. A special case of this valve is the flap valve, used in some slurry and waste pipe-to-tank applications to prevent backflow. The flap valve does not have an enclosing housing, only a flap and a seat. It is essentially the self-closing end of the pipe line.

6.2.0 Lift Check Valves

The lift check valve (*Figure 41*) is designed like a globe valve, with an indirect line of flow through the valve. The lift check valve also works automatically by line pressure and prevents the reversal of flow through the valve. A lift check valve contains a disc that is held against the disc seat by gravity or by the pressure of the fluid flowing in the opposite direction through the valve. When fluid flows through the valve in the correct direction, the force of the fluid lifts the disc away from the seat to allow the fluid to pass through the valve. Lift check valves can be designed for horizontal or vertical applications and must only be used for the application they are designed for.

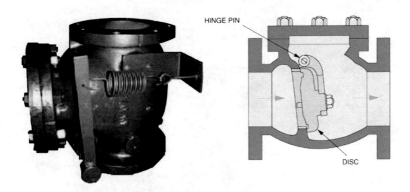

ANGLE CHECK VALVE

STRAIGHT-THROUGH CHECK VALVE
203F40.EPS

Figure 40 ◆ Swing check valves.

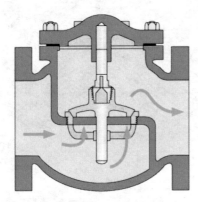

HORIZONTAL LIFT CHECK

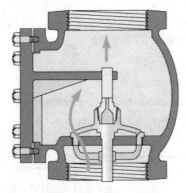

VERTICAL LIFT CHECK
203F41.EPS

Figure 41 ◆ Lift check valve.

6.3.0 Ball Check Valves

Ball check valves (*Figure 42*) operate in the same manner as lift check valves, except they use a ball instead of a disc to allow the fluid to pass through the valve. Ball check valves that are not spring-loaded should only be installed in the vertical position. Gravity and the reversal of flow cause the ball to rest against the valve seat, restricting flow through the valve.

Ball check valves are recommended for use in lines in which the fluid pressure changes rapidly since the action of the ball is practically noiseless.

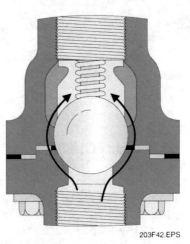

203F42.EPS

Figure 42 ◆ Ball check valve.

The ball rotates during operation and therefore tends to wear at a uniform rate over its entire surface. The ball valve also stops flow reversal more rapidly than other types of check valves.

6.4.0 Butterfly Check Valves

A butterfly check valve has two vanes that resemble the wings of a butterfly. These vanes fold back from a central hinge to open and allow flow but close to prevent backflow. *Figure 43* shows a butterfly check valve.

NOTE

If the butterfly check valve is used in the horizontal position, it must have a spring to assist in closing.

6.5.0 Foot Valves

Foot valves (*Figure 44*) are used at the bottom of the suction line of a pump to maintain the prime of the pump. This type of valve operates similarly to the lift check valve. On the inlet side of the foot valve is a strainer that keeps foreign material out

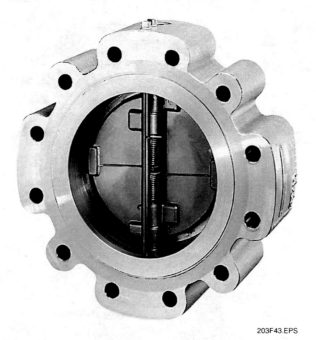

Figure 43 ◆ Butterfly check valve.

203F43.EPS

203F44.EPS

Figure 44 ◆ Foot valve.

of the line. The weight of the liquid in the pipeline between the pump and the foot valve keeps the suction pipeline full when the pump is shut down. The weight pushes the seats closed in the valve, preventing the liquid from flowing out through the valve. When the pump starts, the force of the suction causes the liquid coming into the valve to push against the outside of the seats to open the valve.

> **NOTE**
>
> If the foot valve is used in the horizontal position, it must have a spring to assist in closing.

7.0.0 ◆ VALVE ACTUATORS

The primary purpose of a valve actuator, also known as an operator, is to provide automatic control of a valve or to reduce the effort required to manually operate a valve. A standard handwheel or hand lever attached to a valve stem is one type of actuator commonly used on smaller valves that do not require great effort to operate. Larger valves must be equipped with some other type of actuator.

In today's automated industries and central control stations, many valves are mechanically operated and powered by electric, pneumatic, or hydraulic actuators or operators. These actuators are used when a valve is remote from the main working area, when the frequency of operation of the valve would require unreasonable human effort, or when rapid opening or closing of the

valve is required. There are many types of valve actuators in use today, including the following:

- Gear operators
- Chain operators
- Pneumatic and hydraulic actuators
- Electric or air motor-driven actuators

7.1.0 Gear Operators

Gear operators minimize the effort required to operate large valves that work at unusually high pressures. Gear operators are also used to connect to valves located in inaccessible areas. Three basic types of gear operators are spur-gear operators, bevel-gear operators, and worm-gear operators.

7.1.1 Spur-Gear Operators

Spur-gears are used to connect parallel shafts. They have straight teeth cut parallel to the shaft axis and use gears to transfer motion and power from one shaft to another parallel shaft. Spur-gear operators can be as simple as one gear transmitting power to another gear attached to the valve stem, or they can consist of many gears, depending on how much power needs to be transmitted. *Figure 45* shows a spur-gear operator.

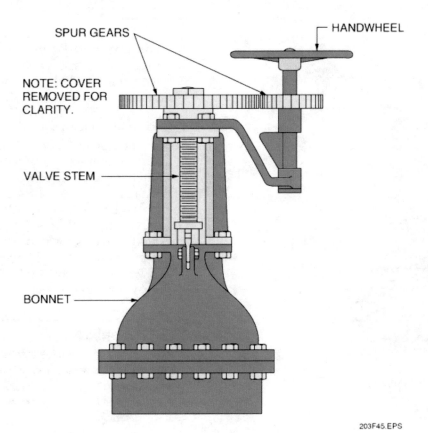

Figure 45 ◆ Spur-gear operator.

203F45.EPS

Although you may see valves with open gear trains in older installations, this design is no longer permitted by OSHA. Gear-driven valves in current use have enclosed gear boxes like the one shown in *Figure 46*.

7.1.2 Bevel-Gear Operators

Bevel-gears are used to transmit power between two shafts that intersect at a 90-degree angle. The bevel gear resembles a cone, in that the teeth on each gear are angled at 45 degrees to mesh with the gears on the mating gear. Bevel-gear operators transmit power from a handwheel and a shaft that is perpendicular to the valve stem (*Figure 47*).

7.1.3 Worm-Gear Operators

Worm gears are used in butterfly valves to transmit power at right angles (*Figure 48*). This means that the shafts of the connecting gears are at 90-degree angles to each other. The worm is a cylindrical-shaped gear similar to a screw, and is the driver. The worm gear is the larger, circular gear in the assembly. Worm-gear operators transmit power from the handwheel, which is attached to the worm, to the worm gear, which is attached to the valve stem. Worm gears are set up like bevel gears, with a side-mounted handwheel. A large

203F46.EPS

Figure 46 ◆ Enclosed gear box valve.

speed ratio is obtainable with a worm-gear operator because the operator allows rapid opening and closing of large valves with minimum effort. It takes only one-quarter turn of the handwheel to open or close the valve.

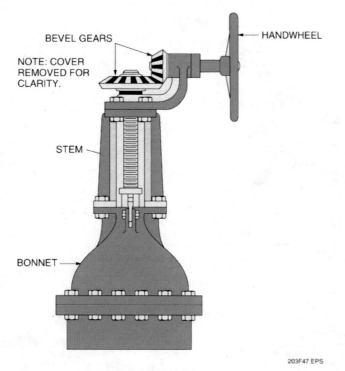

Figure 47 ◆ Bevel-gear operator.

7.2.0 Chain Operators

A chain operator (*Figure 49*) is installed in some situations when the valve is mounted too high to reach the handwheel. The stem of the valve is mounted with a chain wheel, and the chain is brought to within 3 feet of the working floor level. The handwheel is frequently attached to a lanyard to prevent accidental operation.

Chain operators are used only when they are absolutely necessary. Universal-type chain wheels that attach to the regular valve handwheel have been blamed for many industrial accidents. In corrosive atmospheres where an infrequently operated valve is located, the attaching bolts of this type of chain wheel have been known to fail. Chain wheels attach to the stem and replace the regular valve handwheel.

> **WARNING!**
> OSHA now requires that the wheels for chain operators be secured with a safety cable (lanyard). This is done to prevent them from falling if their fasteners have been weakened by rust or corrosion.

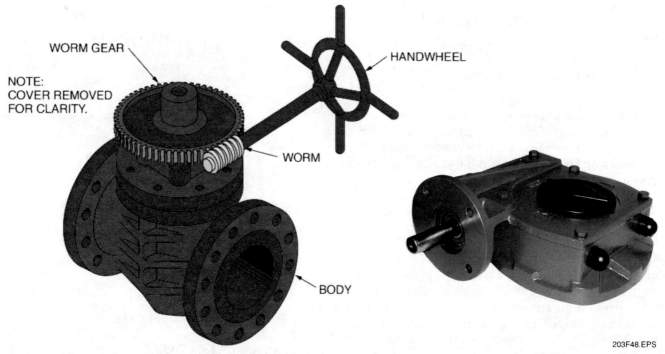

Figure 48 ◆ Worm-gear operator.

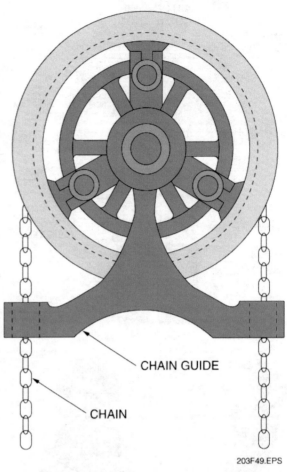

CHAIN GUIDE

CHAIN

203F49.EPS

Figure 49 ◆ Chain operator.

7.3.0 Pneumatic and Hydraulic Actuators

Pneumatic and hydraulic actuators operate in basically the same manner, except the pneumatic actuator operates off air pressure and the hydraulic actuator operates off fluid pressure. A cylinder assembly that contains a piston is attached to the valve stem. Air or fluid pressure above and below the piston moves the piston and the valve stem up and down, opening and closing the valve. Many pneumatic and hydraulic actuators also have spring-loaded pistons in either the naturally open or naturally closed position, depending on which position is considered the fail-safe position. These springs allow the valve to fail open or fail closed. If the control medium is lost, the spring will force the valve to the selected fail-safe position, which is determined as part of the system design process. This safety feature is required so that the valve will return to its safe position in case of pressure failure. *Figure 50* shows a spring-loaded pneumatic valve actuator.

WARNING!

The springs in pneumatic valve actuators are under strong tension and can cause serious injury if let loose. Do not disassemble a valve operator unless you have been trained for that work.

7.4.0 Electric or Air Motor-Driven Actuators

Electric or air motor-driven valve actuators contain a motor that is linked through reduction gears to the valve stem. The electric motor is equipped with electrical limit switches that shut off the motor when the motor has turned the valve stem as far as it can go in either direction. This prevents unnecessary wear and tear on the motor. Electric valve actuators are usually equipped with a handwheel for controlling the valve manually if the power fails. Actuators should only be removed, turned, and adjusted by qualified technicians.

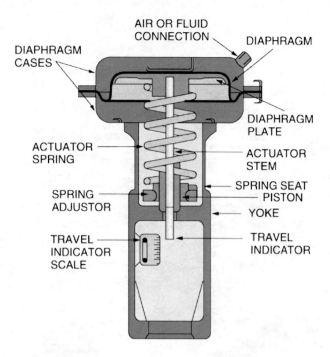

DIAPHRAGM CASES
AIR OR FLUID CONNECTION
DIAPHRAGM
DIAPHRAGM PLATE
ACTUATOR SPRING
ACTUATOR STEM
SPRING ADJUSTOR
SPRING SEAT PISTON
YOKE
TRAVEL INDICATOR SCALE
TRAVEL INDICATOR

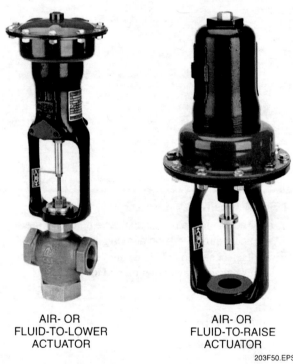

AIR- OR FLUID-TO-LOWER ACTUATOR

AIR- OR FLUID-TO-RAISE ACTUATOR

203F50.EPS

Figure 50 ◆ Spring-loaded pneumatic valve actuators.

8.0.0 ◆ STORING AND HANDLING VALVES

Once a valve has been received at the job site, it must be properly stored until it is installed in the pipeline. If the valve has been abused in storage, it may fail after installation. Caution must be taken to ensure that the valves are handled properly and safely.

8.1.0 Safety Considerations

When working with and around valves, you must be alert and work cautiously to ensure your own safety and the safety of your coworkers. Follow these guidelines when handling or working around valves:

- Be aware of all pinch points.
- Do not stand under a load.
- Watch for overhead power lines and other equipment.
- Do not overload temporary work platforms.
- Ensure that final supports and hangers are in place before installing a large valve.
- Never operate any valve in a live system without proper authorization.
- Always use a spud wrench or drift pin when aligning bolt holes in a flanged valve. Never align bolt holes with your fingers.
- Never stand in front of a safety valve relief discharge.

8.2.0 Storing Valves

Most valve manufacturers wrap the valves in protective wrapping for shipment. This wrapping should remain on the valve until installation. Follow these guidelines when storing and handling valves:

- Clearly label all valves with identification tags or stenciled placards.
- Store all valves on appropriate hardwood dunnage to keep the valves off the ground. Never store valves on the ground.
- Store all valves on the basis of their compatibility with other valves. For example, do not store stainless steel valves with carbon steel valves because the rust from carbon steel can cause the stainless steel to corrode.
- Store valves in an area where corrosive fumes, freezing weather, and excessive water can be kept to a minimum.
- Store the valves in an area where no objects can fall and damage them.
- Do not remove any tags from valves.
- Ensure that all open ends have end protectors.
- Store valve handles with their mating valves.
- When storing valves outside, always lay them on their side so that water cannot get trapped inside the valve. If water inside the valve freezes, it can shatter the body material.

8.3.0 Rigging Valves

Special precautions must also be taken when rigging valves. Follow these guidelines when rigging valves:

- Clean and protect all threads and weld ends before lifting.
- Select your rigging equipment based on the weight of the valve.
- Do not rig a valve by the stem, handle, or actuator, or through the body opening. Rig only to the valve body.
- Rig stainless steel valves using nylon straps only.
- Install a tag line to control the lift.
- Rig for the proper position of the valve in the final installation.
- Rig to allow for installation of bolts and nuts during installation.
- Remove all shipping materials before installation.

9.0.0 ◆ INSTALLING VALVES

Proper installation of valves is very important to the efficient operation of the piping system. Valves can be installed in piping systems with welded, threaded, or flanged joints. The procedure for installing a valve is the same as for installing a fitting. However, there are added factors to consider when installing valves.

The location of a valve in a piping system is very important. Most piping drawings indicate exactly where to install the valve. The direction in which the stem and handwheel are to be located when working with small-bore piping is often the pipefitter's responsibility. When working with large-bore piping, the stem direction and orientation are normally shown on the drawings. Because of regular maintenance procedures, the valve must be easily accessible when possible. The height of the valve must be accessible without causing hazards. The best installation height for valve handles is between 2 feet and 4 feet 6 inches off the floor or working level. Valves installed below and above this area (up to 6 feet 6 inches) create either a tripping hazard or a face hazard; therefore, the valve handwheel must have some type of guard around it. *Figure 51* shows the order of preference for valve location.

Valves that are installed with the stem in the upright position tend to work best. The stem can be rotated down to the horizontal position, but should not point down. If the valve is installed with the stem in the downward position, the bon-

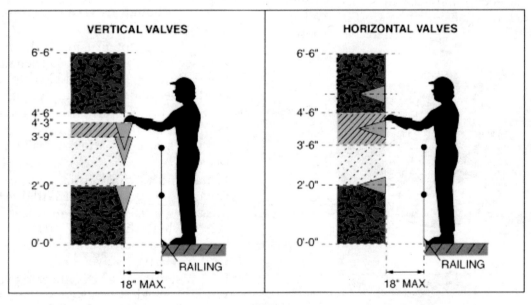

Figure 51 ◆ Order of preference for valve location.

net acts like a trap for sediment, which may cut and damage the stem. Also, if water is trapped in the bonnet in cold weather, it may freeze and crack the body of the valve.

Another factor that must be considered when installing valves is the direction of flow through the valve. Butterfly, safety, pressure-relief, and some other valves either have arrows indicating direction of flow stamped on the side, or the ports are labeled as the inlet or the outlet. When the valve is not marked, you must determine which side of the disc you want the pressure against. Gate valves can have the pressure on either side. Globe valves should be installed so that the pressure is below the disc unless pressure above the disc is required in the job specifications (*Figure 52*).

10.0.0 ◆ VALVE SELECTION, TYPES, AND APPLICATIONS

Because of the diverse nature of valves, with valve types overlapping each other in both design and application, the valve selection process must be examined. This section discusses valve selection, valve types, and valve applications.

10.1.0 Valve Selection

With valve selection, there are many factors that must be taken into consideration. Cost is often an overriding factor, although experience has shown

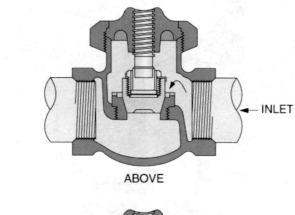

ABOVE

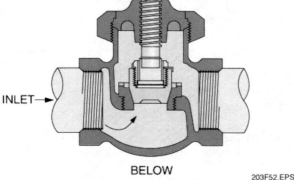

BELOW

203F52.EPS

Figure 52 ◆ Pressure above and below disc.

that sparing expense now may result in additional expense later. When selecting a valve during system design, overall system performance must be taken into consideration. Questions that must be asked include these:

- At what temperature will the system be operating? Are there any internal parts that would be adversely affected by the temperature? Valves designed for high temperature steam systems are not necessarily suited for the extreme low temperatures that may be found in a liquid nitrogen system.

- At what pressure (or vacuum) will this valve be operating? How does the temperature affect the valve's pressure rating? System integrity is a major concern on any system. The valve must be rated at or above the maximum system pressure anticipated. Due to factors such as valve design, packing construction, and end attachments, the valve is often considered a weak point in the system.

- Are there any sizing constraints? It seems obvious that a 2-inch valve would not be installed in a 10-inch pipe, but what may not be obvious is how the yoke size, actuator, or positioner figures in the scenario. Valve manufacturers provide dimensional tables to aid in valve selections.

- Will this valve be used for on-off or throttle application? Throttle valves are generally globe valves, although in some applications a ball valve or butterfly valve may be used.

- To what type of erosion will the valve be exposed? Will it require hardened seats and discs? Will it be throttled close to its seat and need a special pressure drop valve?

- What kind of pressure drop is allowed? Globe valves exhibit the largest pressure drop or head loss characteristics, whereas ball valves exhibit the least.

- What kind of differential pressure will this valve be operated against? Will this differential pressure be used to seat or unseat the valve? Will the high differential pressure deform the body or disc and bind the valve? Will this also require a bypass valve?

- How will this valve be connected to the system? Will it be welded, screwed, or flanged? Should it be butt welded or socket welded? Will it be union threaded or pipe threaded? Should the flanges be raised, flat, phonographic, male/female, or tongue-and-groove?

- In what type of environment will the valve be installed? Is it a dirty environment where an exposed stem would score the yoke bushing and cause premature failure? Is it a clean environment where a different stem lubricant should be used?

- What kind of fluid is being handled? Is it hazardous in such a way that packing leakage may be detrimental? Is it corrosive to the packing or to the valve itself?
- What is the life expectancy required? Will it require frequent maintenance? If so, is it easily repairable, or does the cost of labor justify replacement instead of repair?

If an installed valve is to be replaced, a valve identical to the one removed should be installed. If that valve is no longer manufactured, valve selection should be made in the same fashion as for a new application. The valve dimensions are the limiting factor unless piping alterations can be made. Several questions should be asked:

- Are the system parameters the same as when the system was designed, or has the system intent changed?
- Have any problems been noted since system fabrication that could be remedied by installing a different valve design at this time?
- With what type of operator should the new valve be fitted? Is the new valve compatible with the installed operator?

10.2.0 Valve Types and Applications

As noted in the valve selection discussion, there are many factors that determine the application and/or type of valve to be used:

- The temperature at which the system will be operating
- The sizing constraints or the pressure at which the system will be operating
- The fluid or material that is in the system
- The environment in which the valve will have to operate
- The actuator the valve will use

11.0.0 ◆ VALVE MARKINGS AND NAMEPLATE INFORMATION

Before the present system of valve and flange coding, manufacturers had their own systems. With the development of components rated at higher temperatures and pressures, in conjunction with more stringent regulations, a standard was needed. The Manufacturers Standardization Society (MSS) first developed SP-25 in 1934. In 1978, SP-25 was revised to incorporate all the changes that had developed since 1934. To preclude errors in cross-referencing, the American National Standards Institute (ANSI) and the American Society for Testing Materials International (ASTM International) have adopted the MSS marking system.

Two markings that are frequently used on valves are the flow direction arrow, indicating which way the flow is going, and the bridgewall marking, shown in *Figure 53*. The bridgewall marking is usually found on globe valves and is an indication of how the seat walls are angled in relation to the inlet and outlet ports of the valve. Specifically, it shows whether the wall of the seat on the inlet side angles up or down. The wall of the seat on the outlet side of the globe valve will always be angled opposite to the angle on the inlet side, as indicated in *Figure 54*.

Not all globe valves are designed with angled bridgewalls. However, some applications may specifically require the process to enter either on the top side or the bottom side of the disc in a globe valve.

If the process enters on the top side (bridgewall angled up on the inlet), the force of the process will assist in the closing of the valve. However, if the process enters on the bottom side (bridgewall angled down on the inlet), the force of the process will assist in the opening of the valve.

Markings for flow are not normally used on gate, plug, butterfly, or ball valves. If a gate valve has a flow arrow, it is because the gate valve has a double gate. Double-gate valves are capable of relieving fluid pressure in the event that a high pressure difference exists across the shut gate. Standard practice is for the outlet-side gate to

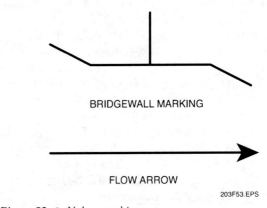

BRIDGEWALL MARKING

FLOW ARROW

203F53.EPS

Figure 53 ◆ Valve markings.

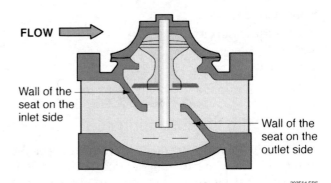

FLOW

Wall of the seat on the inlet side

Wall of the seat on the outlet side

203F54.EPS

Figure 54 ◆ The meaning of bridgewall markings on valves.

relieve to the inlet side. This type of valve is used for specific applications. Therefore, system plans should be consulted for correct valve orientation.

There are normally two identification sets: one permanently embossed, welded, or cast into the valve body, and the other a valve identification plate (*Figure 55*). Typically, as a minimum, the following information will be included within the two sets:

- Rating designation markings
- Material designation markings
- Melt identification markings
- Trim identification markings (if applicable)
- Size markings
- Thread identification markings (if applicable)

11.1.0 Rating Designation

The rating designation of a valve gives the pressure and temperature rating as well as the type of service. *Table 2* shows commonly used service designations.

The product rating may be designated by the class numbers alone, as with a steam pressure rating or pressure class designation. The ratings for products that conform to recognized standards, but are not suitable for the full range of pressures or temperatures of those standards, may be marked as appropriate. The numbers and letters representing the pressure rating at the limiting conditions may also be shown.

The rating designation for products that do not conform to recognized national product standards may be shown by numbers and letters representing the pressure ratings at maximum and minimum temperatures. If desired, the rating designation may be shown as the maximum pressure followed by cold working pressure (CWP) and the allowed pressure at the maximum temperature (for example, 2,000 CWP 725/925°F). Other typical designations are given as the first letter of the system for which they are designated:

- A – Air service
- G – Gas service
- L – Liquid service
- O – Oil service
- W – Water service
- DWV – Drainage waste and vent service

11.2.0 Trim Identification

Trim identification marking is required on the identification plate for all flanged-end and butt welding end steel or flanged-end ductile iron

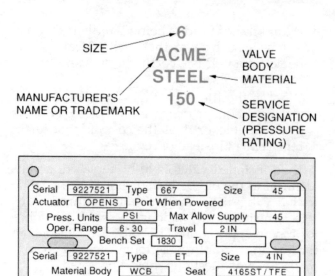

Figure 55 ◆ Valve nameplate.

203F55.EPS

Table 2 Valve Service Designations

Correspond to Steam Working Pressure (SWP)	Correspond to Cold Working Pressure (CWP)
Steam pressure (SP)	Water, oil pressure (WO)
Working steam pressure (WSP)	Oil, water, gas pressure (OWG)
	Water, oil, gas pressure (WOG)
Steam (S)	Gas, liquid pressure (GLP)
	Working water pressure (WWP)
	Water pressure (WP)

203T02.EPS

body valves with trim material that is different than the body material. Symbols for materials are the same. The identification plate may be marked with the word trim followed by the appropriate material symbol.

Trim identification marking for gate, globe, angle, and cross valves, or valves with similar design characteristics, consists of three symbols. The first indicates the material of the stem. The second indicates the material of the disc or wedge face. The third indicates the material of the seat face. The symbol may be preceded by the words stem, disc, or seat, or it may be used alone. If used alone, the symbols must appear in the order given.

Plug valves, ball valves, butterfly valves, and other quarter-turn valves require no trim identification marking unless the plug, disc, closure member, or stem is of different material than the body. In such cases, trim identification symbols on the nameplate first indicate the material of the stem and then the material of the plug, ball, disc, or closure member.

Those valves with seating or sealing material different than the body material must have a third symbol to indicate the material of the seat. In these cases, symbol identification must be preceded by the words stem, disc (or plug, ball, or gate, as appropriate) and the word seat. If used alone, the symbols must appear in the order given.

11.3.0 Size Designation

Size markings are in accordance with the product-referenced marking requirements. For size designation for products with a single nominal pipe size of the connecting ends, the word nominal indicates the numerical identification associated with the pipe size. It does not necessarily correspond to the inside diameter of the valve, pipe, or fitting.

Products with internal elements that are the equivalent of one pipe size, or are different than the end size, may have dual markings unless otherwise specified in a product standard. Unless these exceptions exist, the first number indicates the connecting end pipe size. The second indicates the minimum bore diameter, or the pipe size corresponding to the closure size (for example, 6 × 4, 4 × 2½, 30 × 24).

At the manufacturer's option, triple marking size designation may be used for valves. If triple size designation is used, the first number must indicate the connecting end size at the other end. For example, 24 × 20 × 30 on a valve designates a size 24 connection, a size 20 nominal center section, and a size 30 connection.

Fittings with multiple outlets may be designated at the manufacturer's option in a run × run × outlet size method. For example, 30 × 30 × 24 on a fitting designates a product with size 30 end connections and a nominal size 24 connection between.

11.4.0 Thread Markings

Fittings, flanges, and valve bodies with threaded connecting ends other than American National Standard Pipe Thread or American National Standard Hose Thread will be marked to indicate the type. The style or marking may be the manufacturer's own symbol provided confusion with standard symbols is avoided. Fittings with left-hand threads must be marked with the letters LH on the outside wall of the appropriate opening.

Marking of products with ends threaded for API casing, tubing, or drill pipe must include the nominal size, the letters API, and the thread type symbol as listed in *Table 3*.

Marking of products using other pipe threads must include the following:

- Nominal pipe, tubing, drill pipe, or casing size
- Outside diameter of pipe, tubing, drill pipe, or casing
- Name of thread
- Number of threads per inch

11.5.0 Valve Schematic Symbols

The last and most important aspect of valve identification is the ability to identify different types of valves from blueprints and schematics. In general, the symbols that denote various control valves, actuators, and positioners are standard symbols as shown in *Figure 56*. However, in certain cases these symbols vary, depending on site-specific prints. The legend of a typical system print or schematic will show the symbols that represent all components on the drawing.

Table 3 Examples of Threaded-Type Symbols

Namr/Description	Symbol
Casing – Short round thread	CSG
Casing – Long round thread	LCSG
Casing – Buttress thread	BCSG
Casing – Extreme-line	XCSG
Line pipe	LP
Tubing – Non-upset	TBG
Tubing – External-upset CSG	UP TBG

203T03.EPS

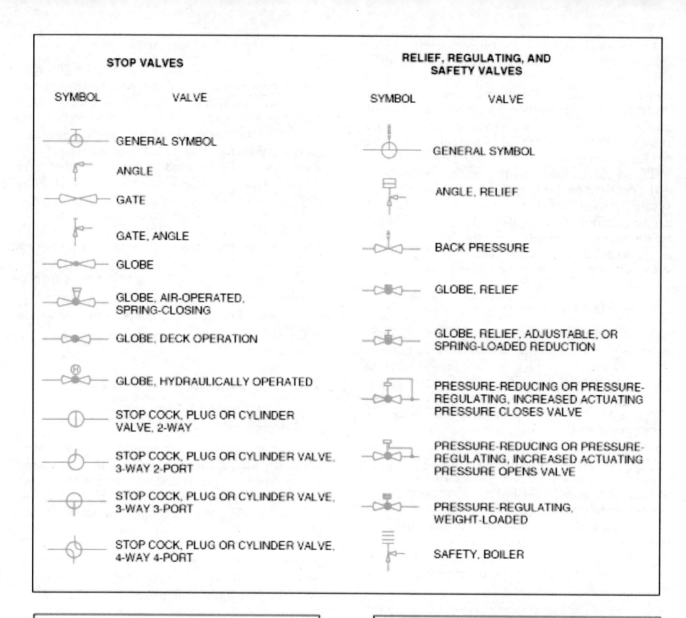

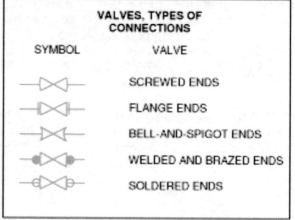

Figure 56 ◆ Typical piping system schematic symbols.

203F56.EPS

1. Gate valves should *not* be used to _____.
   a. start flow
   b. stop flow
   c. throttle flow
   d. control an on/off process

2. If you wanted to start and stop low-pressure two-way flow in a paper mill, you would use a _____ valve.
   a. check
   b. butterfly
   c. knife gate
   d. needle

3. Which of the following valves is a rotary-action valve?
   a. Gate
   b. Needle
   c. Globe
   d. Ball

4. Which of the following valves is easily adapted to multiport construction?
   a. Gate
   b. Plug
   c. Globe
   d. Knife

5. A globe valve that closes against the direction of fluid flow is said to be _____.
   a. flow-to-close
   b. flow-to-open
   c. fail-to-open
   d. fail-to-close

6. One disadvantage of the globe valve is _____.
   a. high head loss
   b. inability to regulate flow
   c. high seat leakage
   d. inability to serve in quick-acting start-stop applications

7. When looking at an illustration of valve plug and seat installation, if the taper of the plug goes from wide to narrow from top to bottom, the plug must travel _____.
   a. up to close, down to open
   b. down to close, up to open
   c. down to close, down to open
   d. up to close, up to open

8. Which valve's discharge conditions are so favorable that many control valves are based on it?
   a. Gate
   b. Plug
   c. Globe
   d. Butterfly

9. A needle valve may be used when an application requires _____.
   a. fine adjustment of flow
   b. coarse adjustment of flow
   c. on-off control
   d. slurry flow

10. Relief valves are normally used in liquid services.
    a. True
    b. False

11. A swing check valve requires a counterweight if used in a _____ application.
    a. horizontal
    b. vertical
    c. steam
    d. liquid

12. The fail-safe position of any automatically actuated valve is the closed position.
    a. True
    b. False

13. When you choose among valves, you would find the largest pressure drop in the _____ valve.
    a. globe
    b. gate
    c. ball
    d. butterfly

14. If a gate valve has a flow arrow indicated on it, it is because the gate valve has a _____.
    a. single gate
    b. double gate
    c. bridgewall marking
    d. vent port

15. Fittings that are marked LH on the outside wall indicate _____.
    a. low heat
    b. liquid hydrogen only
    c. left-hand threads
    d. low hardening material

Summary

Piping systems use valves to start and stop flow, regulate flow, relieve pressure, and prevent reversal of flow. The type of valve used depends on the type of piping system, the nature of the fluid in the system, the temperature and pressure of the fluid in the system, and the desired operation of the system. Valves are manufactured from the same materials as the piping systems. Proper storage, handling, and installation of valves ensure safe and efficient operation of the system.

Valves are an integral part of the operation and control of fluid flow systems. Each device has characteristics that make it specially suited for certain applications. The applications for which valves are used vary widely and include fine control of flow, temperature regulation, pressure regulation, and flow isolation.

Proper selection of the control valve for a specific system is determined by several factors such as system design pressure and temperature, piping size, and system flow conditions.

Actuators and positioners are energy transmission devices that move the valve stem. They may use any of several different energy sources to perform their functions. These devices provide a means through which valves may be operated remotely or against extremely high differential pressures.

The information presented in this module provides detailed descriptions of the construction and operation of various control valves, actuators, and positioners. This information will prove invaluable in the proper selection and correct installation of any of these valves, actuators, and positioners in a fluid system.

Notes

Actuator: The part of a regulating valve that converts electrical or fluid energy to mechanical energy to position the valve.

Angle valve: A type of globe valve in which the piping connections are at right angles.

American Society for Testing Materials International (ASTM International): Founded in 1898, a scientific and technical organization formed for the development of standards on the characteristics and performance of materials, products, systems, and services.

Ball valve: A type of plug valve with a spherical disc.

Body: The main part of the valve. It contains the disc, seat, and valve ports. The body of the valve is directly connected to the piping by threaded, welded, or flanged ends.

Bonnet: The part of a valve containing the valve stem and packing.

Butterfly valve: A quarter-turn valve with a plate-like disc that stops flow when the outside area of the disc seals against the inside of the valve body.

Check valve: A valve that allows flow in one direction only.

Control valve: A globe valve automatically controlled to regulate flow through the valve.

Corrosive: Causing the gradual destruction of a substance by chemical action.

Deformation: A change in the shape of a material or component due to an applied force or temperature.

Disc: Part of a valve used to control the flow of system fluid.

Elastomeric: Elastic or rubberlike. Flexible, pliable.

Galling: An uneven wear pattern between trim and seat that causes friction between the moving parts.

Gate valve: A valve with a straight-through flow design that exhibits very little resistance to flow. It is normally used for open/shut applications.

Globe valve: A valve in which flow is always parallel to the stem as it goes past the seat.

Head loss: The loss of pressure due to friction and flow disturbances within a system.

Kinetic energy: Energy of motion.

Packing: Material used to make a dynamic seal, preventing system fluid leakage around a valve stem.

Phonographic: When referring to the facing of a pipe flange, serrated grooves cut into the facing, resembling those on a phonograph record.

Plug: The moving part of a valve trim (plug and seat) that either opens or restricts the flow through a valve in accordance with its position relative to the valve seat, which is the stationary part of a valve trim.

Plug valve: A quarter-turn valve with a ported disc.

Positioner: A field-based device that takes a signal from a control system and ensures that the control device is at the setting required by the control system.

Relief valve: A valve that automatically opens when a preset amount of pressure is exerted on the valve disc.

Seat: The part of a valve against which the disc presses to stop flow through the valve.

Thermal transients: Short-lived temperature spikes.

Throttling: The regulation of flow through a valve.

Torque: A twisting force used to apply a clamping force to a mechanical joint.

Trim: Functional parts of a pump or valve, such as seats, stem, and seals, that are inside the flow area.

Valve body: The part of a valve containing the passages for fluid flow, valve seat, and inlet and outlet connections.

Valve stem: The part of a valve that raises, lowers, or turns the valve disc.

Valve trim: The combination of the valve plug and the valve seat.

Wedge: The disc in a gate valve.

Wire drawing: The erosion of a valve seat under high velocity flow through which thin, wire-like gullies are eroded away.

Yoke bushing: The bearing between the valve stem and the valve yoke.

Resources & Acknowledgments

Additional Resources

This module is intended to be a thorough resource for task training. The following reference works are suggested for further study. These are optional materials for continued education rather than for task training.

Choosing the Right Valve. New York, NY: Crane Company.

Piping Pointers; Application and Maintenance of Valves and Piping Equipment. New York, NY: Crane Company.

The Piping Guide, 1980. San Francisco, CA: Syentek Books Company, Ltd.

Figure Credits

American Flow Control, 203F01 (photo), 203F40 (straight-through check valve)

Velan Valve Corp., 203F12, 203F28 (photo), 203F46

Marwin Valve, A Division of Richards Industries, 203F15 (photo)

Val-Matic Valve and Manufacturing Corp., 203F18 (photo)

Lumaco Sanitary Valves, 203F22 (photo)

Dwyer Instruments, Inc., 203F24, 203F29, 203F50

ITT Engineered Valves, 203F33 (photo)

Parker Hannifin Corp., 203F34

©2002 Swagelok Company, 203F37

Crispin Valve Co., 203F40 (angle check valve), 203F44

Crane Valves North America, 203F43

Flowserve Corporation, 203F48 (photo)
Provided by Flowserve Corporation, a global leader in the fluid motion and control business. More information on Flowserve and its products can be found at www.flowserve.com.

Babbitt Steam Specialty Company, 203F49 (photo)

Cianbro Corporation, Module opener

NCCER CURRICULA — USER UPDATE

NCCER makes every effort to keep its textbooks up-to-date and free of technical errors. We appreciate your help in this process. If you find an error, a typographical mistake, or an inaccuracy in NCCER's curricula, please fill out this form (or a photocopy), or complete the online form at **www.nccer.org/olf**. Be sure to include the exact module ID number, page number, a detailed description, and your recommended correction. Your input will be brought to the attention of the Authoring Team. Thank you for your assistance.

Instructors – If you have an idea for improving this textbook, or have found that additional materials were necessary to teach this module effectively, please let us know so that we may present your suggestions to the Authoring Team.

NCCER Product Development and Revision

13614 Progress Blvd., Alachua, FL 32615

Email: curriculum@nccer.org

Online: www.nccer.org/olf

❏ Trainee Guide ❏ Lesson Plans ❏ Exam ❏ PowerPoints Other _____

Craft / Level: _____ Copyright Date: _____

Module ID Number / Title: _____

Section Number(s): _____

Description: _____

_____

_____

_____

_____

Recommended Correction: _____

_____

_____

_____

_____

Your Name: _____

Address: _____

_____

Email: _____ Phone: _____

08204-06

Pipefitting Trade Math

08204-06
Pipefitting Trade Math

Topics to be presented in this module include:

Overview

Pipefitters need to use mathematics every day to make decisions about connections. The basic geometric equations show us the relationships between the figures that we see, such as areas and volumes. The mathematical relationships between the sides of triangles allow us to determine the unknown length of a pipe from the other two sides of the connections. Circles and cylinders are ways of understanding the pipes we work with. Rectangles and rectangular solids are tools for understanding machine pads and tanks. Tables present useful information in condensed form for easy use. As you advance in pipefitting, you will find yourself using your mathematical skills more and more.

Objectives

When you have completed this module, you will be able to do the following:

1. Identify and explain the use of special measuring devices.
2. Use tables of weights and measurements.
3. Use formulas to solve basic problems.
4. Solve area problems.
5. Solve volume problems.
6. Solve circumference problems.
7. Solve right triangles using the Pythagorean theorem.

Trade Terms

Adjacent side
Apex
Arithmetic numbers
Circle
Circumference
Cubic
Cylinder
Exponent
Factors
Formula
Hypotenuse
Literal numbers

Opposite side
Perpendicular
Pi
Pyramid
Radius
Rectangular
Run
Set
Solid
Sphere
Travel
Volume

Required Trainee Materials

1. Pencil and paper
2. Scientific calculator

Prerequisites

Before you begin this module, it is recommended that you successfully complete *Core Curriculum*; *Pipefitting Level One*; and *Pipefitting Level Two*, Modules 08201-06 through 08203-06.

This course map shows all of the modules in the second level of the *Pipefitting* curriculum. The suggested training order begins at the bottom and proceeds up. Skill levels increase as you advance on the course map. The local Training Program Sponsor may adjust the training order.

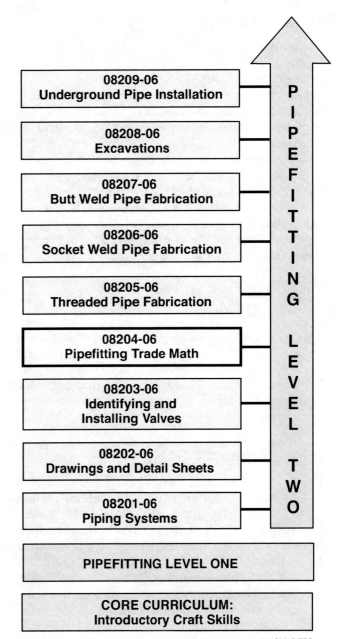

204CMAP.EPS

1.0.0 ◆ INTRODUCTION

In addition to the basic math skills you may have learned in school, a pipefitter needs advanced mathematical skills. You will use these skills to determine the volume of pipes and tanks, and to determine simple and rolling offsets. This module explains how to use an architect's scale and an engineer's scale and how to use tables of weights and measurements; solve basic problems using formulas; use basic geometry to solve area, volume, perimeter, and circumference problems; and use the Pythagorean theorem. All of these skills are applied to your everyday duties as a pipefitter, and the use of mathematics can become a valuable tool that will make your job easier.

2.0.0 ◆ SPECIAL MEASURING DEVICES

Special measuring devices that a pipefitter must be familiar with include the architect's scale and the engineer's scale (*Figure 1*). These tools are useful when reading blueprints and orthographic drawings and when converting scale drawings to actual size.

2.1.0 Architect's Scale

The architect's scale is an all-purpose scale that has many uses. It has a full-size scale of inches, divided into sixteenths, and a number of reduced-size scales in which inches or fractions of an inch represent feet. An architect's scale has 12 separate scales divided into different increments (*Table 1*).

Each scale has a number located at the end of the scale. The numbers at each end of the scale designate the size of the increments on the scale. If the numbers on the left end designate the scale you are using, the scale is read from left to right. If the numbers on the right end of the scale designate the scale you are using, the scale is read from right to left. For example, on the quarter-inch scale, each

quarter of an inch designates one foot. *Figure 2* shows an enlarged view of the quarter-inch scale.

Since the ¼-inch mark is at the right end of the scale, the ¼-inch scale is read from right to left. The scale has long vertical marks that represent feet. The even-numbered feet are labeled, and the odd-numbered feet are not. The fully divided scale to the right of the zero represents inches.

2.2.0 Engineer's Scale

The engineer's scale has several scales, each of which is divided into 10, 20, 30, 40, 50, or 60 parts. These scales can be used to check drawings. If a blueprint is drawn to the 50 scale, meaning that 1 inch equals 50 feet or 50 meters, you would use the scale marked 50. These scales make it easier to determine quick measurements without having to use math to calculate distances. *Figure 3* shows the scales on the engineer's scale.

It is okay to check a given measurement with a scale rule, but you must not rely on scaled dimensions for actually building a structure. Scaled dimensions are only approximations and are not detailed enough for construction. The printing of

Table 1 Scales on Typical Architect's Scale

Scale	Relation of Scale to Object
16	Full Scale
3	3" = 1'
1½	1½" = 1'
1	1" = 1'
¾	¾" = 1'
½	½" = 1'
⅜	⅜" = 1'
¼	¼" = 1'
³⁄₁₆	³⁄₁₆" = 1'
⅛	⅛" = 1'
³⁄₃₂	³⁄₃₂" = 1'
¹⁄₁₆	¹⁄₁₆" = 1'

204T01.EPS

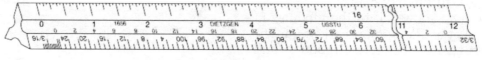

ARCHITECT'S SCALE

ENGINEER'S SCALE

204F01.EPS

Figure 1 ◆ Architect's and engineer's scales.

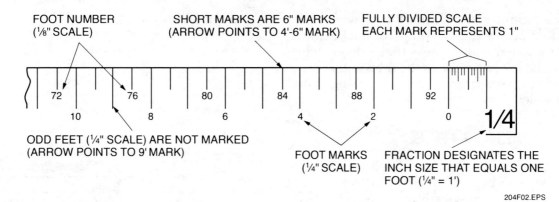

Figure 2 ◆ Enlarged view of quarter-inch scale.

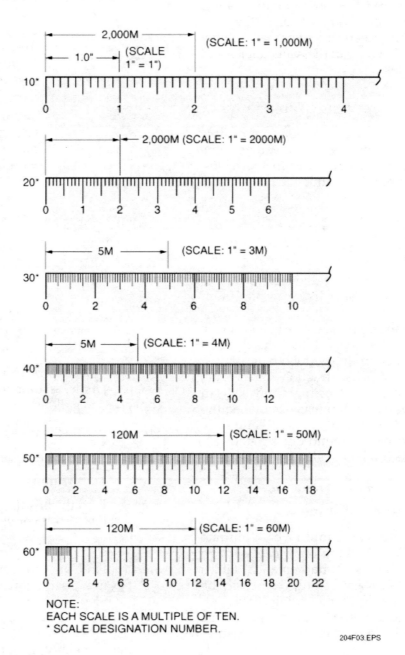

NOTE:
EACH SCALE IS A MULTIPLE OF TEN.
* SCALE DESIGNATION NUMBER.

204F03.EPS

Figure 3 ◆ Scales on engineer's scale.

blueprints can slightly shrink the drawing, causing inaccuracy, or the drawing may not be drawn to the indicated scale. This is highly likely when changes have been made to the drawing. Because of this, always use a written dimension rather than a scaled dimension.

3.0.0 ◆ USING TABLES

Tables consist of two or more parallel columns of data. They can be read quickly and can present large amounts of data clearly and concisely. Handbooks of tables are frequently useful as information references and for solving mathematical problems. While tables vary in form, they are read following the same basic steps. The following sections explain comparative value tables and mathematical tables.

3.1.0 Comparative Value Tables

The simplest types of tables provide comparative values of related quantities. These values come from the definitions of quantities in the tables. Comparative value tables include the following:

- Tables of measure
- Tables of weight
- Multiplication tables
- Tables of money

One type of comparative value table is a table of linear measures. *Table 2* lists sample linear measures.

When mathematical problems are being solved, it is sometimes necessary to know the decimal equivalent of a fraction. In the following sample table, the first column lists fractions of an inch; the second column lists the decimal equivalents in inches, and the third column lists decimal equivalents in millimeters. If any of these values are known, the others can be found quickly and easily. *Table 3* lists decimal equivalents of some common fractions.

3.2.0 Mathematical Tables

Mathematical tables can simplify or eliminate the long calculations that are often necessary in mathematical problems. If mathematical tables are used, solutions for larger units can easily be calculated. For example, the following conversion table for English and metric cubic measurements converts basic units into other units of measurement (*Table 4*).

Table 2 Sample Linear Measures

12 inches = 1 foot
3 feet = 1 yard
1,760 yards = 1 mile

204T02.EPS

Table 3 Decimal Equivalents of Common Fractions

Fraction (inches)	Decimal Equivalent	
	English (inches)	Metric (millimeters)
1/64	0.015625	0.3969
1/32	0.03125	0.7938
3/64	0.046875	1.1906
1/16	0.0625	1.5875
5/64	0.078125	1.9844
3/32	0.09375	2.3813
7/64	0.109375	2.7781
1/8	0.1250	3.1750
9/64	0.140625	3.5719
5/32	0.15625	3.9688
11/64	0.171875	4.3656
3/16	0.875	4.7625
13/64	0.203125	5.1594
7/32	0.21875	5.5563
15/64	0.234375	5.9531
1/4	0.250	6.3500

204T03.EPS

Example:

1 cubic centimeter = 0.06102 cubic inches
1 gallon = 0.1337 cubic feet

Follow these steps to use mathematical tables. As an example, assume that 7 cubic feet are being converted to gallons.

Step 1 Locate the necessary table.

Step 2 Find the unit quantity in the first column.

7 units

Step 3 Find the heading over the column where the correct conversion is listed.

Cubic Feet to Gallons

Step 4 Find the number in this column that is in the same row with the units located in Step 2.

Answer: 52.36

If the number of units to be converted is not on this type of table, it can be calculated with simple addition or multiplication. For example, if 40 cubic feet must be converted to cubic meters, the

Table 4 Conversion of English and Metric Cubic Measurements

Unit	Cubic Inches to Cubic Centimeters	Cubic Centimeters to Cubic Inches	Cubic Feet to Cubic Meters	Cubic Meters to Cubic Feet	Cubic Yards to Cubic Meters	Cubic Meters to Cubic Yards	Gallons to Cubic Feet	Cubic Feet to Gallons
1	16.39	0.06102	0.0283	35.31	0.7646	1.308	0.1337	7.481
2	32.77	0.1220	0.0566	70.63	1.529	2.616	0.2674	14.960
3	49.16	0.1831	0.0849	105.90	2.294	3.924	0.4010	22.440
4	65.55	0.2441	0.1133	141.30	3.058	5.232	0.5347	29.920
5	81.94	0.3051	0.1416	176.60	3.823	6.540	0.6684	37.400
6	98.32	0.3661	0.1699	211.90	4.587	7.848	0.8021	44.880
7	114.70	0.4272	0.1982	247.20	5.352	9.156	0.9358	52.360
8	131.10	0.4882	0.2265	282.50	6.116	10.460	1.069	59.840
9	147.50	0.5492	0.2549	371.80	6.881	11.770	1.203	67.320

204T04.EPS

solution for 4 cubic feet can be found. If the steps given in this section are followed, 4 cubic feet can be converted to 0.1133 cubic meters. Since 4 × 10 = 40, 0.1133 can be multiplied by 10 to get 1.133, which is the solution to converting 40 cubic feet to meters.

These comparative values are handy in several ways. If a tool is digital, such as a programmable lathe or cut-off saw, it may not be possible to instruct it to cut a length of pipe to 36¼ inches. However, the comparative value table will tell you that ¼ inch is the same as 0.250 inches, which the tool will be able to recognize.

Suppose you need to look up ⅛ inch in the comparative value table:

Step 1 Look down through the fraction column until you come to ⅛.

Step 2 Look across that row to the English (inches) value, and you will see that the decimal equivalent of ⅛ is 0.125.

Step 3 If you needed the metric equivalent, you would look to the Metric (millimeters) column, and you would find the value to be 3.1750 millimeters.

4.0.0 ◆ USING FORMULAS

Mathematics in pipefitting is frequently a matter of applying rules about the relationships between measurements. The rules are usually stated as formulas, statements that use letters to represent quantities. By using letters, we can state the relationships between numbers, such as the area of a circle and its width, or how much a pipe can hold as it relates to its diameter and length. This means that, by learning a few formulas, and by learning how to apply them, we can easily figure out the information we need to know to solve the

relationships between the piping systems we are to build and repair. This section explains the following principles of using mathematical formulas in pipefitting:

- Symbolism
- Expressing rules as formulas
- Factors
- Powers
- Square roots
- Evaluating formulas

4.1.0 Symbolism

Symbols are the language of mathematics. In blueprints, the symbols represent parts of the pipe assembly, such as valves and fittings. In mathematical formulas, symbols represent numbers. In blueprints, we use pictures; in mathematics, we use letters and numbers. Letters used to represent numbers in formulas are called literal numbers, as opposed to the arithmetic numbers we are used to. A literal number can represent a single arithmetic number, a wide range of numerical values, or all numerical values, depending on its function in a particular formula.

The multiplication sign × is not used in formulas since it can be mistaken for the letter X. No sign of operation is required when a literal number is multiplied by an arithmetic number or when two or more literal numbers are multiplied. For instance, $5 \times h$ is written $5h$; $b \times h$ is written bh; $4 \times h \times w$ is written $4hw$.

Parentheses, raised dots, or asterisks are often used instead of the multiplication sign when numerical numbers are multiplied. For instance, 3×6 may be written $3(6)$, $3 \cdot 6$, or $3*6$; and $\frac{1}{2} \times 6.3 \times 8$ may be written $\frac{1}{2}(6.3)(8)$, $\frac{1}{2} \cdot 6.3 \cdot 8$, or $\frac{1}{2}*6.3*8$.

4.2.0 Expressing Rules as Formulas

A formula is a short method of writing a mathematical rule. In formulas, the values from rules are represented by letters or symbols. A letter used to replace a value is often the first letter of the word representing that value. If that letter has already been used to represent a value in the formula, it cannot be used again to represent a different value. When a rule is changed into a formula, the values and mathematical operations are replaced with letters or symbols and mathematical signs. *Figure 4* shows changing a rule into a formula.

Since the letters or symbols in a formula represent the numbers in specific problems, these numbers can be written into the formula to replace the letters or symbols. This is called substitution. Usually, the values of all but one of the letters are known. These values can be used in the formula to find the unknown value. Follow these steps to solve a formula by substitution:

Step 1 Determine the formula needed to solve the problem. For this example, assume that the radius of a circle is 6 inches and the diameter is unknown. The formula shown in *Figure 4* can be used to solve the problem.

$$D = 2r$$

Step 2 Substitute the known quantities for the letters or symbols in the formula.

$$D = 2 \times 6$$

Step 3 Perform the mathematical operation needed to solve the problem.

$$12 = 2 \times 6$$

Step 4 Check the answer.

$$12 \div 2 = 6$$

Step 5 Label the answer with the correct unit of measure.

$$D = 12 \text{ inches}$$

If the diameter in the example was known and the radius was the unknown value, the formula would read 12 = 2r. It could be solved by dividing the diameter by 2, so that $12 \div 2 = 6$.

> **NOTE**
>
> All measurements used in a formula must be the same type of unit.

4.3.0 Factors

When multiplying two or more numbers to find a given number, the numbers being multiplied are factors of the given number. For instance, the factors of the number 12 are 1 and 12, 2 and 6, and 3 and 4, since $1 \times 12 = 12$, $2 \times 6 = 12$, and $3 \times 4 = 12$.

4.4.0 Powers

A power is the product of two or more equal factors. For example, 4×4 is the second power of 4; $6 \times 6 \times 6$ is the third power of 6; and $h \times h \times h \times h$ is the fourth power of h. For example, 5^4 is the same as $5 \times 5 \times 5 \times 5$. The exponent 4 tells you that 5 is taken as a factor four times and is read five to the fourth power; therefore, $5^4 = 625$.

4.5.0 Roots

The square root of a number, if multiplied by itself, equals the number of which it is the root. The radical sign ($\sqrt{\ }$) indicates a root of a number. The index number, which is written above and to the left of the radical sign, indicates the number of times that a root is to be taken as an equal factor to produce the given number. The index number for a square root is 2. The 2 is usually omitted in the square root sign. Therefore, the square root of 16 is written $\sqrt{16}$. It asks the question, "What number multiplied by itself equals 16?" Since $4 \times 4 = 16$, 4 is the square root of 16. So $\sqrt{16} = 4$.

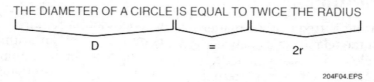

THE DIAMETER OF A CIRCLE IS EQUAL TO TWICE THE RADIUS

D   =   2r

204F04.EPS

Figure 4 ◆ Changing a rule into a formula.

Sections 1.0.0–4.2.0

1. The formula for gas mileage is mpg = miles ÷ gallons. If I drove 63 miles on 3 gallons of gas, my mpg is _____.

   a. 11
   b. 21
   c. 31
   d. 189

2. A pipefitter is earning $22.50 per hour on the job. The formula for his paychecks is P = 22.5 × hours. If he worked 40 hours, he would earn _____.

   a. $180
   b. $900
   c. $1,800
   d. $2,250

3. The outside diameter of a 4-inch Schedule 40 pipe is 4½ inches. What is the radius of the pipe?

   a. 2 inches
   b. 2¼ inches
   c. 4½ inches
   d. 9 inches

4. A coil being made in a shop is bent to a 4-foot radius. What is the diameter of the coil?

   a. 2 feet
   b. 8 feet
   c. 12 feet
   d. 16 feet

5. A class 150 flange on 6-inch steel pipe has a 9½-inch diameter bolt circle. What is the radius of the bolt circle?

   a. 3 inches
   b. 3½ inches
   c. 4½ inches
   d. 4¾ inches

When the square root of a fraction is to be computed, both the numerator and the denominator must be enclosed within the radical sign. For instance, $\sqrt{\frac{5}{6}}$ indicates that the square root of the complete fraction is to be taken. To compute the square root of a fraction, you must find the square root of both the numerator and the denominator.

Example: Compute the square root of $\frac{16}{36}$:

$\sqrt{16} = 4$

$\sqrt{36} = 6$

$\sqrt{\frac{16}{36}} = \frac{4}{6} = \frac{2}{3}$ reduced

Root expressions are sometimes written with two or more operations within the radical sign. For instance, the expression $\sqrt{l + h}$ is read as "the square root of l + h." It is computed by first adding l + h and then computing the square root of the sum. Apply the following procedure to solve problems which involve operations within the radical sign:

Step 1 Perform the operations within the radical sign first.

Step 2 Compute the root.

$\sqrt{8 + 17}$

Add the numbers within the radical sign.

$8 + 17 = \sqrt{25}$

$\sqrt{25} : 5 \times 5 = 25,$

so $\sqrt{8 + 17} = 5$

Any power of a number can be determined by multiplying the number by itself the number of times shown in the exponent. That is, the fourth power of a number, stated as n to the fourth power, is n × n × n × n. By the same reasoning, the fourth root of a number is that number that, if multiplied by itself four times, equals the number you want the root of. The cube of a number is that number to the third power (n × n × n). The cube root of a number is the number that is multiplied by itself three times to obtain that number. For example, 2^3, read as two cubed, equals 8. The cube root of 27 is 3 because 3 × 3 × 3 equals 27.

Section 4.5.0

1. What is the value of 6 to the third power?
   a. 2
   b. 18
   c. 36
   d. 216

2. What is the square root of 9?
   a. 2
   b. 3
   c. 4½
   d. 81

3. The third power of 8 is 8 × 8 × 8.
   a. True
   b. False

4. 2 to the fourth power is _____.
   a. 4
   b. 6
   c. 16
   d. 23

5. The exponent 3 means that the number is to be multiplied by itself 3 times.
   a. True
   b. False

4.6.0 Evaluating Formulas

Normally, formulas contain two or more arithmetic operations. It is essential that you perform the operations in the proper order to solve the formula. You should perform the operations in the following order:

Step 1 Perform the operations in parentheses first. If there are parentheses within parentheses, perform the operation within the innermost parentheses first.

Step 2 Perform the operations for powers and roots as they occur.

Step 3 Perform multiplication and division operations from left to right in the order in which they occur.

Step 4 Perform addition and subtraction operations from left to right in the order in which they occur.

One way to remember this order of operations is to remember PEMDAS: parentheses; exponent (powers and roots); multiply; divide; add; subtract.

To determine the numerical value of one letter in an expression when the numerical value of the other letter is known, use the following procedure:

Step 1 Write the expression.

Step 2 Replace each letter in the expression with its numerical value, and add a multiplication sign where multiplication is needed.

Step 3 Perform the operations in the proper order.

Example:

Find the perimeter of a rectangle when h equals 6 inches and w equals 4 inches.

Step 1 Write the expression.

$$p = 2h + 2w$$

Step 2 Replace the letters in the expression with their numerical value, and add a multiplication sign where multiplication is needed.

$$p = 2(6) + 2(4)$$

Step 3 Perform the operations in the proper order.

$$p = 12 + 8$$
$$p = 20 \text{ inches}$$

Another use for the formula for the perimeter of a rectangle would be to determine one of the sides if you know the perimeter of another side. Then the formula would state: $p = 2l + 2w$. If the perimeter (p) is 18 and one of the sides is 4, then you could state the formula as $18 = 2l + 2(4)$ or $18 = 2l + 8$. You can do the same operation to both sides of the equation, so you subtract the 8 from both sides: $18 - 8 = 2l + 8 - 8$. Now you have $10 = 2l$. Divide both sides by 2, and you have the length of the unknown side as $l = 5$. Remember, you can do the same operation on both sides of an equation and the two sides will still be equal.

5.0.0 ◆ SOLVING AREA PROBLEMS

Area is the amount of plane (flat) surface in a closed space. The area of any given surface can be calculated for practical applications, such as the following:

- Room layout
- Materials estimates
- Cost estimates
- Sizes of stock or parts

Area is measured by using standard areas of smaller units, such as the square inch and square foot. A square inch is a surface enclosed by a square that is 1 inch on each side. A square foot is enclosed by a square that is 1 foot on each side. There are 144 square inches in 1 square foot. These units can be used to solve problems in finding the area of rectangles, triangles, and circles.

5.1.0 Finding Area of Rectangles

A rectangle is a four-sided figure. Its opposite sides are equal, and all sides are joined at right angles. To find the area of a rectangle, it is necessary to find the number of surface units it contains. For example, if a rectangle contains 3 rows of square inches and there are 4 square inches in each row, the rectangle contains 12 square inches. *Figure 5* shows this rectangle.

The area of any rectangle can be found by multiplying the length by the width, or A = lw. Area is always expressed in square units. Remember that you can obtain the length or width if you know the area and either the length or width. In the case of a rectangle that is 2 feet long with an area of 10 square feet, the equation A = lw becomes 10 = 2w. Divide both sides by 2, and you get 5 = w.

In an area formula, all measurements must be expressed in or converted to the same type of unit.

For example, if a room is 20 feet, 6 inches long and 15 feet, 9 inches wide, the area of floor is found as follows:

A = lw

A = 20 feet, 6 inches × 15 feet, 9 inches

A = 20.5 feet × 15.75 feet

A = 322.875 square feet

A square is a rectangle with four equal sides. A square that is 6 inches on each side is called a 6-inch square. The formula for finding the area of a rectangle can be used to find the area of a square. However, since each side is the same length, the length of any side can also be multiplied by itself, or $A = S^2$. This is read as the area equals the side squared. For example, the area of the 6-inch square is found as follows:

$A = S^2$

A = 6 × 6

A = 36 square inches

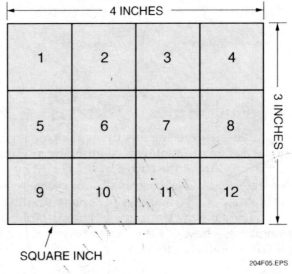

Figure 5 ◆ Rectangle.

Sections 5.0.0–5.1.0

1. What is the area of a rectangle that is 2½ feet by 4½ feet?
   a. 2 sq ft
   b. 8 sq ft
   c. 11¼ sq ft
   d. 12 sq ft

2. What is the area of a room that is 20 feet long by 15 feet wide?
   a. 250 sq ft
   b. 300 sq ft
   c. 600 sq ft
   d. 750 sq ft

3. What is the length of a rectangular platform that is 12 feet wide and has an area of 180 square feet?
   a. 15 ft
   b. 90 ft
   c. 190 ft
   d. 900 ft

4. One side of a square is 6 inches, so its area is _____.
   a. 6 sq in
   b. 12 sq in
   c. 36 sq in
   d. 216 sq in

5. The pad under a pump is 4½ feet wide and 5 feet long. What is its area?
   a. 22½ sq ft
   b. 25 sq ft
   b. 29 sq ft
   d. 40 sq ft

5.2.0 Finding Area of Triangles

A triangle is a three-sided figure with three angles. The area of any triangle can be found by multiplying the product of the base and the height by one half, or $A = \frac{1}{2}bh$. The base of a triangle can be any one of its sides. The height is perpendicular to the base. *Figure 6* shows a triangle. For example, if a triangle has a base of 15 feet and a height of 12 feet, the area is found as follows:

$A = \frac{1}{2}bh$

$A = \frac{1}{2} \times 15 \times 12$

$A = 90$ square feet

If the base and the area are given, then you can plug them into the formula, and you would have 90 = 15h/2. First multiply both sides by 2, which gives you 180 = 15h. Now divide both sides by 15, and you will find 12 = h. The height is 12 feet.

5.3.0 Finding Area of Circles

The area of a circle can be found by multiplying pi (pronounced pie) by the product of the radius multiplied by itself, or $A = \pi r^2$. For example, if a pipe has an inside radius of 12 inches, the cross-sectional area of the pipe is found as follows:

$A = \pi r^2$

$A = 3.1416 \times 12 \times 12$

$A = 452.39$ or 452.4 square inches

This formula can also be used to find the pressure exerted on a piston. If the piston has a diameter of 6 inches and the pressure exerted on the piston head is 160 pounds per square inch, the total pressure is found as follows:

$A = \pi r^2$

$r = D \div 2 = 6 \div 2 = 3$

$A = 3.1416 \times 3 \times 3$

$A = 28.2744$ square inches

Pressure = 160 pounds per square inch

Pressure per square inch × area = total pressure

$160 \times 28.2744 = 4{,}523.904$

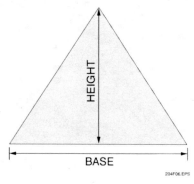

204F06.EPS

Figure 6 ◆ Triangle.

Sections 5.2.0–5.3.0

1. What is the area of a triangle that is formed when two of the legs, joined by a right angle, are each 4 feet long?

   a. 4 sq ft
   b. 8 sq ft
   c. 16 sq ft
   d. 32 sq ft

2. How much canvas will you need to cover the end of a triangular tent that is 6 feet high and 6 feet wide at the bottom?

   a. 6 sq ft
   b. 12 sq ft
   c. 18 sq ft
   d. 36 sq ft

3. You have a 50-square-foot piece of tarp to put over the end of a 10-foot wide pipe rack. How high can the pipe be stacked?

   a. 5 ft
   b. 10 ft
   c. 20 ft
   d. 40 ft

4. How much area will be covered by a triangular pad, if all three sides are 10 feet, and the distance from the middle of one side to the opposite angle is 8.8 feet (round to the closest foot)?

   a. 44 sq ft
   b. 50 sq ft
   c. 88 sq ft
   d. 100 sq ft

5. How many feet of CDX sheathing will you need to cover the gable of your house? The triangle is 50 feet across and 10 feet high.

   a. 60 sq ft
   b. 100 sq ft
   c. 250 sq ft
   d. 500 sq ft

6. What is the area of the raised face of a 6-inch blind flange?

   a. 16.33 sq in
   b. 28.27 sq in
   c. 36 sq in
   d. 113 sq in

7. What is the area of a manhole cover that is 24 inches in diameter?

   a. 6.1 sq in
   b. 73.9 sq ft
   c. 144 sq in
   d. 452.39 sq in

8. How many square inches of plywood will be needed to fabricate a circular cover for one end of a coil of pipe 24 inches outside diameter? Assume that you need to have a 1-inch overhang all around.

   a. 6 sq ft
   b. 73.93 sq in
   c. 169 sq in
   d. 530.93 sq in

9. A circular tank is 11,310 sq ft. What is the radius?

   a. 6 ft
   b. 60 ft
   c. 120 ft
   d. 3,600 ft

10. If each person standing in a circus ring covers 2 square feet of ring, how many people can get into a 50-foot diameter ring?

   a. 25
   b. 100
   c. 205
   d. 981

Note: Use the value 3.1416 for pi.

6.0.0 ◆ SOLVING VOLUME PROBLEMS

Volume is the amount of space a solid figure occupies. Every solid figure has three dimensions: length, width, and height. The volume of any solid figure can be calculated for practical applications involving such objects as the following:

- Tanks
- Boxes
- Pipes
- Ducts
- Buildings

Volume, like area, is measured in standard units. A basic unit of measure for volume is the cubic inch. A cubic inch is the space occupied by a solid figure that is 1 inch long, 1 inch wide, and 1 inch high. The following sections explain using cubic measures to find the volume of the following:

- Rectangular solids
- Cylinders
- Spheres
- Pyramids
- Cones

6.1.0 Finding Volume of Rectangular Solids

A block with rectangular sides is a rectangular solid (*Figure 7*). To find the volume of a rectangular solid, it is necessary to find the number of cubic units it contains. For example, if a rectangular solid contains 4 cubic inches on each of three layers and the layers are 2 cubic inches deep, the rectangular solid contains 24 cubic inches.

The volume of any rectangular solid can be found by multiplying the product of the length

and width by the height, or V = lwh. Volume is always expressed in cubic units.

For example, an excavation is 60¾ feet long, 28½ feet wide, and 15 feet deep; the problem is to find the amount of earth removed. The volume is found as follows:

V = lwh

V = 60¾ × 28½ × 15

V = 60.75 × 28.5 × 15

V = 25,970.625 or 25,971 cubic feet

Volume problems can also be manipulated to find more than one of the dimensions. If you know the length, width, and volume, the height can be solved for. Another way of setting this equation up is to isolate the unknown on one side, with the known quantities on the other side. The equation for the volume of a rectangle (V = lwh) can be set up to determine the height by dividing both sides by lw, giving V/lw = h. You could also find the width by solving w = V/lh.

This formula can be used to find any of its variables if the other three variables are known. For example, a scrap box is to be built from steel plate to fit into a 10-foot by 12-foot space, and it must have a 480-cubic-foot volume. The height needed is found as follows:

V = lwh

480 = 10 × 12 × h

480 = 120 × h

480 ÷ 120 = 4

Height = 4 feet

6.2.0 Finding Volume of Cylinders

A cylinder (*Figure 8*) is a solid figure with two identical circular bases. The height of a cylinder is the perpendicular distance between the two bases.

To find the volume of a cylinder, the area of one of the circular bases must first be calculated using the formula $A = \pi r^2$. The area is then multiplied by the height of the cylinder, or $V = \pi r^2 h$. For example, the volume of a cylinder with a radius of 4 inches and a height of 8 inches is calculated as follows:

$V = \pi r^2 h$

V = 3.1416 × 4 × 4 × 8

V = 402.1248 or 402 cubic inches

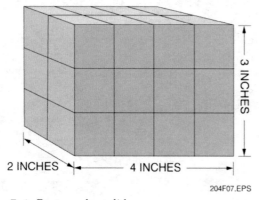

3 INCHES

2 INCHES   4 INCHES

204F07.EPS

Figure 7 ◆ Rectangular solid.

Sections 6.0.0–6.1.0

1. You are pouring a rectangular pad of concrete 6 inches deep by 4 feet wide by 6 feet long. How many 1-cubic-foot bags of concrete mix will you need?

    a. 10
    b. 12
    c. 24
    d. 144

2. What is the volume of a solid concrete block that is 16 inches by 8 inches by 8 inches?

    a. 48 cu in
    b. 128 cu in
    c. 144
    d. 1,024 cu in

3. How many cubic feet are there in a cubic yard?

    a. 3 cu ft
    b. 9 cu ft
    c. 18 cu ft
    d. 27 cu ft

4. How many cubic feet of concrete will be required to pour a pump base pad 18 inches high by 10 feet by 20 feet?

    a. 180 cu ft
    b. 200 cu ft
    c. 300 cu ft
    d. 3,600 cu ft

5. You have to figure the capacity of the pump for a lift station. The rectangular sump is 10 feet wide by 12 feet long, and is to be 25 feet deep. What is the volume?

    a. 2,400 cu ft
    b. 3,000 cu ft
    c. 12,000 cu ft
    d. 30,000 cu ft

Since π is a constant (a number that does not change), you can find the radius of a cylinder from the volume and the height. If you measured and found out that it took 402 cubic inches of liquid to fill a tank and the height of the tank is 8 inches, the equation would become:

$402 = 3.1416r^2(8)$

$402 \div (3.1416 \times 8) = r^2$

$402 \div 25.1328 = r^2$

$r = \sqrt{16} = 4$

If a fuel oil tank with a diameter of 16 feet and a height of 18 feet is filled with oil, the volume of oil is found as follows:

$D \div 2 = r$

$16 \div 2 = 8$

$V = \pi r^2 h$

$V = 3.1416 \times 8 \times 8 \times 18$

$V = 3,619.1232$ or $3,619$ cubic feet

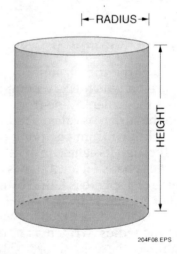

|← RADIUS →|

HEIGHT

204F08.EPS

Figure 8 ♦ Cylinder.

If a cubic foot contains 7½ gallons and oil costs $0.98 per gallon, the cost to fill the tank is found as follows:

$3,619 \times 7.5 = 27,142.5$ gallons

$27,142.5 \times 0.98 = \$26,599.65$

Section 6.2.0

Use the value 3.1416 for pi.

1. You need to choose a container to drain a section of pipe into. The pipe's inside diameter is 6 inches, and the section is 100 feet long. How much liquid, in cubic feet, do you need to be prepared for?

   a. 19.635 cu ft
   b. 34 cu ft
   c. 135.717 cu ft
   d. 200 cu ft

2. How much oil will be lost if the 4-foot-diameter pipe running the 700 miles from Mosul, in Northern Iraq, to the ports in Kuwait is drained by terrorists all at once? Hint: 1 mile is equal to 5,280 feet.

   a. 4,645 cu ft
   b. 46,445 cu ft
   c. 46,445,414 cu ft
   d. 93,000,000 cu ft

3. A natural gas pipeline runs from Russia across the Ukraine. If the pipe is 3 feet in diameter, and 500 miles long, how many cubic feet of gas can it hold at one time?

   a. 1,866 cu ft
   b. 186,611 cu ft
   c. 18,661,104 cu ft
   d. 37,000,000 cu ft

4. What is the diameter, in feet, of a barrel that is 3½ feet tall and holds approximately 11 cubic feet of water?

   a. ½ ft
   b. 1 ft
   c. 2 ft
   d. 4 ft

5. In a chemical plant, you have to fill a cylindrical tank with liquid. If the tank is 20 feet in diameter and 10 feet high, what is the volume of fluid it will hold?

   a. 314.16 cu ft
   b. 1,256.6 cu ft
   c. 1,257 cu ft
   d. 3,141.6 cu ft

6.3.0 Finding Volume of Spheres

A sphere is a three-dimensional figure with a curved surface on which every point is an equal distance from the center. A line from the center to any point on the surface is called a radius. A diameter is a straight line from one point on the edge of a sphere, through its center, to another point on the edge of the sphere. *Figure 9* shows a sphere.

The volume of a sphere is found by multiplying the product of 4 times pi by the radius cubed and then dividing that product by 3, or $V = 4\pi r^3 \div 3$. The cube of the radius, written as r^3, means $r \times r \times r$. For example, the volume of a sphere with a radius of 3 inches is found as follows:

$$V = \frac{4 \times 3.1416 \times 3 \times 3 \times 3}{3}$$

$$V = 339.2928 \div 3$$

$$V = 113.0976 \text{ or } 113 \text{ square inches}$$

If a spherical tank used to store gas is 50 feet in diameter, the volume of the tank is found as follows:

Diameter ÷ 2 = radius

$50 \div 2 = 25$

$V = 4\pi r^3 \div 3$

$$V = \frac{4 \times 3.1416 \times 25 \times 25 \times 25}{3}$$

$V = 196,350 \div 3$

$V = 65,450$ cubic feet

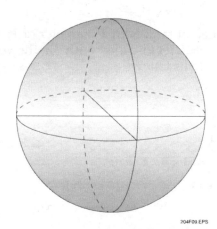

204F09.EPS

Figure 9 ◆ Sphere.

Section 6.3.0

Use the value 3.1416 for pi.

1. A ball 12 inches in diameter contains _____ of air.
   a. 37.6992 cu in
   b. 150 cu in
   c. 904.752 cu in
   d. 1,809.54 cu in

2. A spherical gasoline tank has been drained. If the result was 65,450 cubic feet of gasoline, what was the diameter?
   a. 25 ft
   b. 45 ft
   c. 50 ft
   d. 100 ft

3. A spherical reactor containment vessel, 200 feet in diameter, must be filled with coolant very quickly. How many cubic feet of coolant will be required?
   a. 3,141,600 cu ft
   b. 4,188,800 cu ft
   c. 12,566,400 cu ft
   d. 418,880,000 cu ft

4. A spherical bladder for water for firefighting is 20 feet in diameter. What is its volume in cubic feet?
   a. 314.16 cu ft
   b. 418.88 cu ft
   c. 3,141.6 cu ft
   d. 4,188.8 cu ft

5. A spherical chamber in a valve, 8 inches in diameter, must be included in your drainage calculations. What is its volume?
   a. 67 cu in
   b. 268.0832 cu in
   c. 536.32 cu in
   d. 804 cu in

6.4.0 Finding Volume of Pyramids

A pyramid is a solid figure with a base and three or more triangular faces that taper to one point opposite the base. This point is called the **apex**. The height is a straight line from the apex to the base. *Figure 10* shows a pyramid.

The volume of a pyramid can be found by multiplying the area of its base by the height and then dividing the product by 3, or $V = Ah \div 3$. For example, if the area of the base is 24 square inches and the height is 15 inches, the volume is found as follows:

$V = Ah \div 3$

$V = 24 \times 15 \div 3$

$V = 360 \div 3$

$V = 120$ cubic inches

If the rectangular base of a pyramid has a length of 22 millimeters and a width of 17 millimeters and the height of the pyramid is 42 millimeters, the volume is found as follows:

Area = lw

Area = 22 × 17

Area = 374

$V = Ah \div 3$

$V = \dfrac{374 \times 42}{3}$

$V = \dfrac{15,708}{3}$

V = 5,236 cubic millimeters

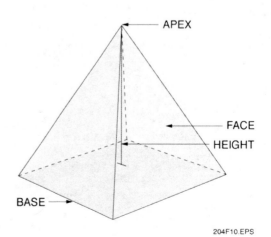

Figure 10 ◆ Pyramid.

204F10.EPS

Section 6.4.0

1. While doing a takeoff for a water main, you find you are to figure the concrete for a pyramidal thrust block with a 2-foot by 2-foot base and a height of 18 inches. How many cubic feet of concrete do you need for each thrust block?

   a. 1 cu ft
   b. 2 cu ft
   c. 6 cu ft
   d. 18 cu ft

2. You need to figure the concrete to be pumped into a pyramidal form to support a pipe coil. The form is 6 feet by 6 feet at the base, and is 6 feet high. How much concrete will you need?

   a. 36 cu ft
   b. 72 cu ft
   c. 144 cu ft
   d. 216 cu ft

3. You have to calculate the piping system to pump concrete into a dam. The dam is basically an upside down pyramid 300 feet long, 100 feet wide, and 200 feet deep. What is its volume?

   a. 600,000 cu ft
   b. 1,200,000 cu ft
   c. 2,000,000 cu ft
   d. 6,000,000 cu ft

4. You need a thrust block for the riser for a fire hydrant. The pyramidal form is 1 foot high and 18 inches by 16 inches at the base. How many cubic inches of concrete will you need?

   a. 96 cu in
   b. 288 cu in
   c. 1,152 cu in
   d. 3,456 cu in

5. The Egyptian Pyramid of Khufu has a 756-foot by 756-foot base. Its volume is 91,636,272 cubic feet. What is its height?

   a. 200 ft
   b. 340 ft
   c. 481 ft
   d. 565 ft

6.5.0 Finding Volume of Cones

A cone (*Figure 11*) is a solid figure with a curved surface. One end of the surface is the apex, and the other is a circular base. Like the pyramid, the height of a cone is a straight line from the apex to the base.

The volume of a cone is found by multiplying pi times the radius of the base squared, times the height, and then dividing the product by 3, or $V = \pi r^2 h \div 3$

For example, the volume of a conical tank with a base radius of 12 feet and a height of 30 feet is found as follows:

$$V = \pi r^2 h \div 3$$

$$V = \frac{3.1416 \times 12 \times 12 \times 30}{3}$$

$$V = \frac{13,571.712}{3}$$

$$V = 4,523.904 \text{ cubic feet}$$

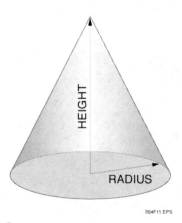

204F11.EPS

Figure 11 ◆ Cone.

A No. 3 taper standard lathe center has a tungsten carbide tip with a base diameter of 0.460 inch. If 0.75 inch of the tip is exposed, the cubic inches of carbide in the exposed tip is found as follows:

Diameter ÷ 2 = radius

0.460 ÷ 2 = 0.23

radius = 0.23

$V = \pi r^2 h \div 3$

$V = \dfrac{3.1416 \times 0.23 \times 0.23 \times 0.75}{3}$

$V = \dfrac{0.1246429}{3}$

V = 0.0415476 cubic inches

7.0.0 ◆ SOLVING CIRCUMFERENCE PROBLEMS

A circle is a closed, curved line on which every point is the same distance from the center. The distance around a circle is called its circumference. The diameter is the length of a straight line drawn from one point on a circle, through its center, to another point on the circle. A straight line drawn from the center of a circle to any point on the circle is the radius (*Figure 12*).

Review Questions

Section 6.5.0

Use the value 3.1416 for pi.

1. An ice cream cone is 2 inches wide at the opening and 4 inches long. After the scoop on top has been eaten down to level with the top, how much ice cream is left?

   a. 4.1888 cu in
   b. 8.8 cu in
   c. 12.5664 cu in
   d. 16.264 cu in

2. A conical inside support is to be machined for a coil. The inside radius for the bottom coil wrap is given as 18 inches. The height of the cone is to be 12 inches. What is the cone's volume?

   a. 3,141.6 cu in
   b. 3,888 cu in
   c. 4,071.5136 cu in
   d. 8,143.272 cu in

3. You are to set up a pump to fill a conical tank with liquid. The diameter of the cone's base is 20 feet, and the height is 10 feet. How much liquid will you need to pump?

   a. 1,047.2 cu ft
   b. 2,104.2 cu ft
   c. 3,141.6 cu ft
   d. 4,188.8 cu ft

4. The cone of a custom concentric reducer is to go from a pinhole on one end, to 6 inches inside diameter on the large end. It is to hold a little over 339 cubic inches of fluid. How long is it?

   a. 36 in
   b. 14 in
   c. 16 in
   d. 20 in

5. The nose cone for a missile is 2 feet in diameter at the base, and 3 feet high. What is its volume in cubic inches?

   a. 458.88 cu in
   b. 1,356.06 cu in
   c. 1,809.3146 cu in
   d. 5,428.6848 cu in

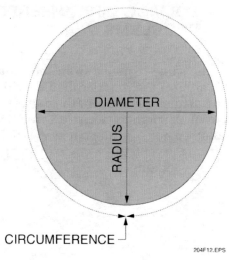

DIAMETER

RADIUS

CIRCUMFERENCE

204F12.EPS

Figure 12 ◆ Parts of a circle.

The circumference of all circles is approximately 3.1416 times the diameter. The Greek letter π, or pi, is used to represent this value. The circumference of any circle can be found by multiplying its diameter by pi, and the diameter can be found by dividing the circumference by pi. For example, if a pulley is 20 inches in diameter, 20 × 3.1416 = 62.832. Its circumference is 62.832 inches.

If the circumference of a pipe is 37.70 mm and the problem is to find its radius, it is found as follows:

$$C \div \pi = D$$

$$37.70 \div 3.1416 = 12$$

$$12 \div 2 = 6$$

Radius = 6

Review Questions

Section 7.0.0

Use the value 3.1416 for pi.

1. What is the circumference of a circle 12 inches in diameter?

    a. 18.8496 inches
    b. 37.6992 inches
    c. 113.36 inches
    d. 216 inches

2. You have to draw a cut line on a piece of 4½-inch outside diameter pipe. You do not have a wrap handy, so you pick up a piece of flexible banding. How long is the minimum amount of banding you need?

    a. 7.0686 inches
    b. 9 inches
    c. 14.1372 inches
    d. 28.2744 inches

3. How much pipe will be required for one full turn of a coil, if the radius of the coil is 15 feet?

    a. 30 feet
    b. 47.124 feet
    c. 94.248 feet
    d. 1,413.0788 feet

4. You have a pipe that is 56½ inches in circumference. What is its diameter?

    a. 6 inches
    b. 9 inches
    c. 12 inches
    d. 18 inches

5. How long a strap will you need to go all the way around a piece of 4½-inch-diameter pipe?

    a. 7.0686 inches
    b. 9 inches
    c. 14.1372 inches
    d. 28.2744 inches

8.0.0 ◆ PYTHAGOREAN THEOREM

The simplest triangles are right triangles. A right triangle has one 90-degree, or right, angle. This angle is usually indicated with a small box drawn in the angle. The right triangle is very important to pipefitters because it is used to determine the components of a piping offset.

The sides of a right triangle have been named for reference. The side opposite the right angle is always called the hypotenuse, and the two sides adjacent to, or connected to, the right angle are called the legs. If one of the other angles is labeled angle A, the leg of the triangle that is not connected to angle A is called its opposite side. The remaining leg that is connected to angle A is called the adjacent side.

The sides of the piping offset have also been named for reference. These sides are called the set, run, and travel. The set is the distance, measured center to center, that the pipeline is to be offset. The run is the total lineal distance required for the offset. The travel is the center-to-center measurement of the offset piping. The angle of the fittings is the number of degrees the piping changes direction. *Figure 13* shows a right triangle and a piping offset.

The Pythagorean theorem states that the square of the hypotenuse is equal to the sums of the squares of the other two sides. For example, in triangle abc, with c being the hypotenuse and a and b being the two legs, the Pythagorean theorem states that $a^2 + b^2 = c^2$. The following steps refer to triangle abc, in which one leg is 3 inches long and the other leg is 4 inches long. *Figure 14* shows triangle abc. Follow these steps to find the length of the hypotenuse, using the Pythagorean theorem:

Step 1   Insert the known values into the Pythagorean theorem.

$$a^2 + b^2 = c^2$$

$$3^2 + 4^2 = c^2$$

Step 2   Square the known values in the formula.

$$9 + 16 = c^2$$

Step 3   Add the squared values in the formula.

$$25 = c^2$$

Step 4   Take the square root of both sides of the equation to determine the value for the unknown side.

$$\sqrt{25} = \sqrt{c^2}$$

$$5 = c$$

The length of the hypotenuse is 5 inches.

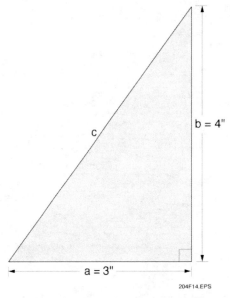

Figure 14 ◆ Triangle abc.

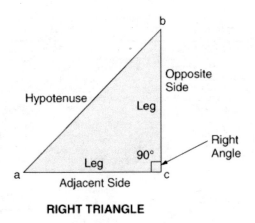

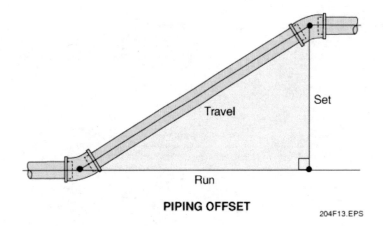

Figure 13 ◆ Right angle and piping offset.

The Pythagorean theorem can also be used to find the length of one of the legs if the other leg and the hypotenuse are known. To do this, start by isolating the unknown value on one side of the equation. As with any equation, isolate the unknown value to one side by doing the same operation to both sides of the equation. The following steps refer to triangle abc, in which one leg is 6 inches long and the hypotenuse is 10 inches long (*Figure 15*). Follow these steps to find the length of one of the legs of a triangle, using the Pythagorean theorem:

Step 1   Write the Pythagorean theorem, and isolate the unknown side.

$$a^2 + b^2 = c^2$$

$$a^2 = c^2 - b^2$$

Step 2   Place the known values into the equation.

$$a^2 = 10^2 - 6^2$$

Step 3   Square the known values in the equation.

$$a^2 = 100 - 36$$

Step 4   Perform the subtraction to the right side of the equation.

$$a^2 = 64$$

Step 5   Take the square root of both sides of the equation to determine the value of the unknown side.

$$\sqrt{a^2} = \sqrt{64}$$

$$a = 8$$

The length of the unknown leg is 8 inches.

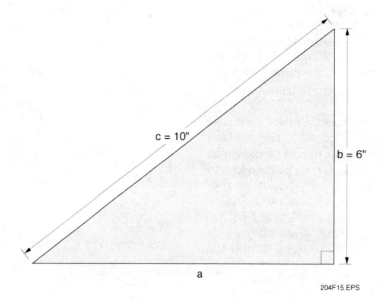

c = 10"

b = 6"

a

204F15.EPS

Figure 15 ◆ Triangle abc.

Section 8.0.0

1. A triangle has one side adjacent to the right angle that is 3 feet long. The other adjacent side to the right angle is 4 feet long. How long is the third side?

    a. 5 feet
    b. 7 feet
    c. 12 feet
    d. 25 feet

2. A pipe runs between two 45-degree ells. The travel is slightly over 7 feet ¾ inches and the set is 5 feet. What is the run? Round off to even feet.

    a. 2 feet
    b. 5 feet
    c. 7 feet
    d. 12 feet

3. You are fitting up a length of welded pipe. You come to a point where you are to connect to another system. The set is 10 feet and the run is 6 feet. What is the travel?

    a. 8 ft
    b. 11.6619 ft
    c. 16 ft
    d. 36.3376 ft

4. A line of water pipe comes to the base of a tank. The connection for the tank will run up a leg and to the top of the tank, 100 feet above. The angle of the leg is such that the run is 20 feet. What length of pipe will you need?

    a. 99.876 feet
    b. 100 feet
    c. 101.98 feet
    d. 105.1416 feet

5. For a pipeline ditch, you need to know how long the horizontal distance is to an object. The set is 25 feet. The travel is 75 feet. What is the run?

    a. 50 feet
    b. 62.1416 feet
    c. 70.711 feet
    d. 81.99 feet

Summary

A pipefitter uses mathematics to install and route piping systems, to determine the volume and capacity of pipes and tanks, and to calculate other requirements such as support pads. We use formulas to express the relationships that exist between quantities that we can measure, so that our measurements let us know what we need to do. Mathematics is an essential tool that you will use in many aspects of your job.

Notes

Trade Terms Introduced in This Module

Adjacent side: The side of a right triangle that is next to the reference angle.

Apex: The point at which the lines of a figure converge.

Arithmetic numbers: Numbers that have definite numerical values, such as 4, 6.3, and ⅝.

Circle: A continuous curved line that encloses a space, with every point on the line the same distance from the center of the circle.

Circumference: The distance around a circle.

Cubic: The designation of a given unit representing volume.

Cylinder: A shape created by a circle moving in a straight line through space perpendicular to the surface of the circle.

Exponent: A number or symbol placed to the right and above another number, symbol, or expression, denoting the power to which the latter is to be raised.

Factors: The numbers that can be multiplied together to produce a given product.

Formula: An equation that states a rule.

Hypotenuse: The longest side of a right triangle. It is always located opposite the right angle.

Literal numbers: Letters that represent arithmetic numbers, such as x, y, and h. Also known as algebraic numbers.

Opposite side: The side of a right triangle that is directly across from the reference angle.

Perpendicular: At a right angle to the plane of a line or surface.

Pi: A number that represents the ratio of the circumference to the diameter of a circle (π). Pi is approximately 3.1416.

Pyramid: A shape with a multi-sided base, and sides that converge at a point.

Radius: A straight line from the center of a circle to a point on the edge of the circle.

Rectangular: Description of a shape having parallel sides and four right angles.

Run: The horizontal distance from one pipe to another.

Set: The vertical distance from the line of flow of a pipe and the line of flow of the pipe to which it is attached.

Solid: A figure enclosing a volume.

Sphere: A shape whose surface is everywhere the same distance from a central point.

Travel: The diagonal distance from one pipe to another.

Volume: The amount of space occupied by an object.

Resources & Acknowledgments

Additional Resources

This module is intended to be a thorough resource for task training. The following reference works are suggested for further study. These are optional materials for continued education rather than for task training.

Pipe Fitter's Math Guide, 1989. Johnny Hamilton. Clinton, NC: Construction Trade Press.

The Pipefitters' Blue Book. W.V. Graves. Webster, TX: Graves Publisher.

The Pipe Fitter's and Pipe Welder's Handbook. Thomas W. Frankland. Milwaukee, WI: The Bruce Publishing Company.

Figure Credits

Cianbro Corporation, Module opener

NCCER CURRICULA — USER UPDATE

NCCER makes every effort to keep its textbooks up-to-date and free of technical errors. We appreciate your help in this process. If you find an error, a typographical mistake, or an inaccuracy in NCCER's curricula, please fill out this form (or a photocopy), or complete the online form at **www.nccer.org/olf**. Be sure to include the exact module ID number, page number, a detailed description, and your recommended correction. Your input will be brought to the attention of the Authoring Team. Thank you for your assistance.

Instructors – If you have an idea for improving this textbook, or have found that additional materials were necessary to teach this module effectively, please let us know so that we may present your suggestions to the Authoring Team.

NCCER Product Development and Revision

13614 Progress Blvd., Alachua, FL 32615

Email: curriculum@nccer.org
Online: www.nccer.org/olf

❏ Trainee Guide   ❏ Lesson Plans   ❏ Exam   ❏ PowerPoints   Other _____

Craft / Level: _____ Copyright Date: _____

Module ID Number / Title: _____

Section Number(s): _____

Description: _____

_____

_____

_____

Recommended Correction: _____

_____

_____

_____

_____

Your Name: _____

Address: _____

_____

Email: _____ Phone: _____

08205-06

Threaded Pipe Fabrication

08205-06
Threaded Pipe Fabrication

Topics to be presented in this module include:

Overview

In many smaller diameter mechanical and industrial applications, pipe and fittings are assembled with threads. The assemblies require an understanding of how threaded connections work, and of the dimensions of threaded piping. The different fittings used in threaded fitups are described, as well as the best methods for achieving accurate, properly aligned, tight assemblies. You will also learn the specific symbols used to represent threaded fittings in drawings.

Objectives

When you have completed this module, you will be able to do the following:

1. Identify and explain the materials used in threaded piping systems.
2. Identify and explain pipe fittings.
3. Read and interpret screwed fitting joint drawings.
4. Identify and explain types of threads.
5. Determine pipe lengths between fittings.
6. Thread and assemble piping and valves.
7. Calculate offsets.

Trade Terms

Branch	Reducer
Bushing	Run
Close nipple	Takeout
Elbow	Tee
Flange	Teflon® tape
Galling	Union
Malleable iron	Wye
Nipple	
National Pipe Thread (NPT)	

Required Trainee Materials

1. Pencil and paper
2. Appropriate personal protective equipment

Prerequisites

Before you begin this module, it is recommended that you successfully complete *Core Curriculum*; *Pipefitting Level One*; and *Pipefitting Level Two*, Modules 08201-06 through 08204-06.

This course map shows all of the modules in the second level of the *Pipefitting* curriculum. The suggested training order begins at the bottom and proceeds up. Skill levels increase as you advance on the course map. The local Training Program Sponsor may adjust the training order.

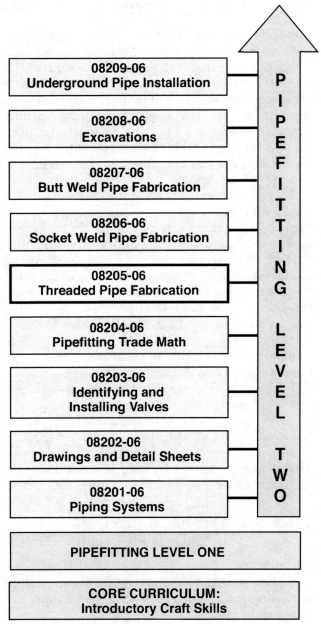

205CMAP.EPS

1.0.0 ◆ INTRODUCTION

The ability to install threaded pipe in accordance with job requirements and specifications is a skill every pipefitter must develop. Threaded connections are relatively inexpensive to fabricate and are a common way to join pipe. Throughout your pipefitting career, you will encounter various threaded piping systems using a variety of materials. The purpose of this module is to introduce the pipefitter trainee to the processes and procedures used to thread pipe.

There are several advantages to a threaded piping system. These systems require no specialized in-shop fabrication and can be easily fabricated on the job site using pipe and various fittings. Portable threading equipment can be used almost anywhere on the job site. If a piping system must be installed near flammable liquids or gases, threaded pipe is much safer because it avoids the fire hazards of welding. Finally, threaded pipe can be cleaned on the inside before it is installed, reducing the possibility of metal particles becoming entrapped and damaging valves or strainers.

Threaded pipe also has its disadvantages. Threaded pipe joints are more prone to leak than other types of joints. The strength of the pipe is reduced by the threading process because the wall thickness of the pipe is reduced at the threads. Threaded pipe cannot be used if erosion, crevice corrosion, extreme shock, or vibration is anticipated.

2.0.0 ◆ MATERIALS USED IN THREADED PIPING SYSTEMS

Various types of materials can be used in threaded piping systems. These materials are classified according to the type of material, size of pipe, and schedule or wall thickness of pipe.

Carbon steel pipe is widely used in piping systems because it is ductile, durable, machinable, and less expensive than most other types of pipe.

Galvanized pipe is made by dipping carbon steel pipe into a mixture of hot zinc. The zinc coats the pipe, both externally and internally, and protects the pipe from corrosion. Galvanized pipe is used to carry air, water, and other fluids.

Stainless steel is an expensive material that can also be used in threaded systems; however, stainless steel pipe is used less frequently because of the difficulty in correctly cutting the threads and making up leak-proof joints. Stainless steel pipe is good for special applications that require the maximum resistance to corrosion, such as processing systems, instrument lines, and heat-transfer equipment. Stainless steel can also be kept sterile to prevent contamination of food, drugs, and dairy products.

Aluminum pipe is used in some specialized applications. Aluminum has the advantage of being relatively soft and easy to machine, but it can be damaged more easily than steel pipe. Aluminum, as used in pipe systems, is almost always an alloy, with widely varying levels of ductility and strength. Other alloys, such as Monel® or titanium steel, are occasionally used for special-purpose piping.

2.1.0 Pipe Sizes

Pipe is listed in inches by its nominal size. For sizes of pipe up to and including 12 inches, nominal size is an approximation of the inside diameter. From 14 inches on, nominal size reflects the outside diameter of the pipe. *Figure 1* shows inside and outside diameters of pipe.

There are times when the nominal size of a pipe and its actual inside or outside diameter differ greatly, but nominal size is used to describe the pipe. The important thing to remember about size is that the sizes of different types of materials varies. A ¾-inch brass tube fitting will not fit a ¾-inch copper pipe. Brass is, however, interchangeable with steel pipes.

2.2.0 Schedules and Wall Thicknesses

Wall thickness can be described in two ways. The first is by schedule. As the schedule numbers increase, the wall thickness gets larger; therefore, the pipe is stronger and can withstand more pressure. Schedule numbers for pipe range from 5 to 160, but no pipe smaller than Schedule 40 should be threaded. It is important to remember that the schedule number refers only to the wall thickness

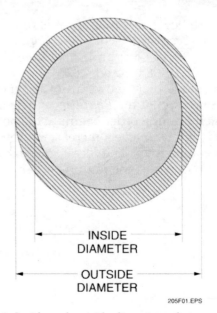

INSIDE
DIAMETER

OUTSIDE
DIAMETER

205F01.EPS

Figure 1 ◆ Inside and outside diameters of pipe.

of a pipe of a given nominal size. A ¾-inch Schedule 40 pipe will not have the same wall thickness as a 1-inch Schedule 40 pipe. Another way to describe pipe wall thickness is by manufacturer's weight. From smallest to largest, there are three classifications in common use today:

- STD – Standard wall
- XS – Extra-strong wall
- XXS – Double extra-strong wall

Schedule numbers and wall thicknesses are somewhat interchangeable. Schedule 40 galvanized pipe and standard wall galvanized pipe have the same wall thickness for all sizes up to and including 10-inch nominal size. Schedule 80 and extra-strong wall galvanized pipe have the same wall thickness through 8-inch nominal size. Common wall thicknesses of carbon steel pipe are Schedule 40, 80, and 160, and double extra-strong. Common wall thicknesses of stainless steel pipe range from Schedule 5 to 160. *Table 1* shows commercial pipe dimensions.

3.0.0 ◆ PIPE FITTINGS

Piping systems consist of pipes, pumps, valves, and other parts, including fittings. Pipe fittings come in various sizes, materials, strengths, and designs to match the various piping systems and can be either plain fittings or banded fittings. The banded fittings are malleable iron and cast fittings, and the plain fittings are forged steel. The malleable iron fittings are banded to provide extra strength. Pipe fittings can be defined as fittings attached to pipe to change the direction of fluid flow, connect a branch line to a main line, close off the end of a line, or join together two pipes of the same or different sizes. Pipe fittings can be placed into different groups according to their use:

- Elbows, offsets, and return bends
- Branch connections
- Caps and plugs
- Line connections

3.1.0 Elbows, Offsets, and Return Bends

Elbows, offsets, and return bends are fittings used for changing the direction of fluid flow in a piping system. An elbow, also known as an ell, L, 90, or 45, is a fitting that forms an angle between different connecting pipes. This angle is always 90 degrees unless another angle is stated. Elbows come in several different angles. The 11¼-, 5⅝-, and 30-degree elbows are rarely used and hard to obtain. The 90- and 45-degree elbows are the most common. *Figure 2* shows 5⅝-, 11¼-, 22½-, 30-, 45-, and 90-degree elbows.

An offset changes the direction of flow for a short distance. The offset moves the pipe run to one side and then returns it to the same direction. An offset can be made by using two fittings, one at each end of the offset. *Figure 3* shows an example of an offset.

Table 1 Commercial Pipe Dimensions

Nominal Size	Outside Diameter	Inside Diameter			
		Standard, Schedule 40	Extra-Strong, Schedule 80	Schedule 160	Double Extra-Strong
⅛	0.405	0.269	0.215	–	–
¼	0.540	0.364	0.302	–	–
⅜	0.675	0.493	0.423	–	–
½	0.840	0.622	0.546	0.466	0.252
¾	1.050	0.824	0.742	0.614	0.434
1	1.315	1.049	0.957	0.815	0.599
1¼	1.660	1.380	1.278	1.160	0.896
1½	1.900	1.610	1.500	1.338	1.100
2	2.375	2.067	1.939	1.689	1.503
2½	2.875	2.469	2.323	1.885	1.771
3	3.500	3.068	2.900	2.625	2.300
3½	4.000	3.548	3.364	–	–
4	4.500	4.026	3.826	3.438	3.152
5	5.562	5.047	4.813	4.313	4.063
6	6.625	6.065	5.761	5.187	4.897
8	8.625	7.981	7.625	6.813	6.875

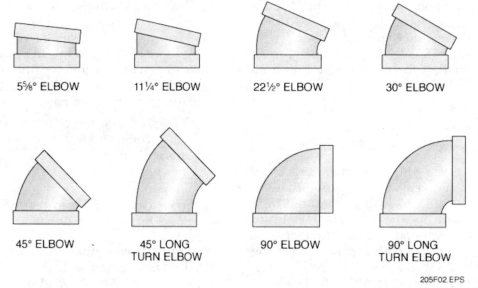

5⅝° ELBOW   11¼° ELBOW   22½° ELBOW   30° ELBOW

45° ELBOW   45° LONG TURN ELBOW   90° ELBOW   90° LONG TURN ELBOW

205F02.EPS

Figure 2 ◆ Elbows.

A return bend is a U-shaped fitting that sends the fluid back in the same direction from which it came. Return bends are often used in boilers, radiators, and systems in which the pipe must pass through the same area several times. Return bends are manufactured in various sizes and come in close, open, and wide patterns. Return bends can be made with tapered threads to give a slant to the pipe. *Figure 4* shows examples of return bends.

3.2.0 Branch Connections

Branch connections are fittings that divide the flow and send the fluid in two different directions. Tees are the most widely used of the branch connections. They are used to make a 90-degree branch in the main pipe. Wyes are another common branch connection that have a side opening set at a 45-degree angle. *Figure 5* shows some common branch connections.

3.3.0 Caps and Plugs

Fittings used for closing off lines are caps and plugs. These have the same pressure ratings as other fittings. A cap fitting screws onto the pipe end, and a plug screws into the pipe or fitting end. Caps can either be plain caps with round heads or caps that have a square projection on the body to allow for tightening with a wrench. Plugs can either be solid square-head plugs or countersink plugs, which have a square depression in which a square key can be inserted for tightening. A bull plug is a long shaft that is threaded on one end that may or may not have a hex head. Bull plugs are used to plug an insulated pipe line so that the head will extend out of the insulation for easy removal. *Figure 6* shows caps and plugs.

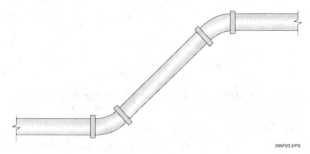

205F03.EPS

Figure 3 ◆ Offset.

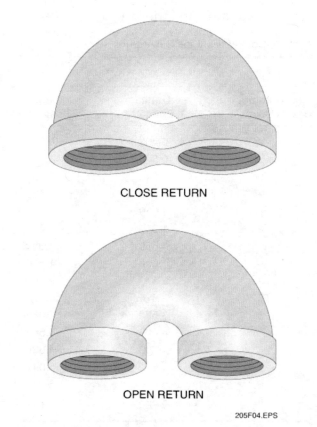

CLOSE RETURN

OPEN RETURN

205F04.EPS

Figure 4 ◆ Return bends.

CROSS

REDUCING TEE

STRAIGHT TEE

205F05.EPS

Figure 5 ◆ Branch connections.

3.4.0 Line Connections

Couplings and unions join pipes of the same size, and bushings, reducers, and reducing couplings join pipes of different sizes.

3.4.1 Couplings

Usually, all lengths of pipe less than 8 inches in diameter are provided by the manufacturer with one straight coupling attached. A coupling is a sleeve that is used to connect two straight pieces of pipe. Both ends of a straight coupling are tapped with right-hand threads. *Figure 7* shows couplings.

3.4.2 Unions

Unions are used either to join two ends of pipe that cannot be turned or to permit the disconnecting of pipes at some future time without cutting them. A union usually consists of three parts, including two sleeves to be threaded on the ends of the pipes and a threaded coupling ring to draw the sleeves together. The male side of the union goes in the direction of the flow. A special type of union is the dielectric union, used to prevent the breakdown of metals caused by directly connecting two different metals, such as bronze valves and cast iron pipe. The dielectric union has an insulating ring between two sleeves that match the metal they are connected to. *Figure 8* shows unions.

3.4.3 Bushings

The function of a bushing is to connect the male end of a pipe to the female end of a fitting of a larger size. It consists of a hollow plug with male and female threads to suit the different diameters. Bushings are usually made with a hexagon top but can also be made without the head. Those without a head are known as face bushings or flush bushings and can be screwed into the fitting to form a neat, flush finish. Bushings are used in low-pressure systems because they restrict the flow in the system. *Figure 9* shows bushings.

3.4.4 Reducers

A reducer is used to join the male ends of two different sizes of pipe. A reducer may be a separate fitting or may be a built-in part of another fitting. Most directional change and branch connection fittings are available with reduced outlet lines. Malleable iron bell reducers are used in low-pressure systems, and swages are used in high-pressure systems. Both of these are made in concentric and eccentric types. A reducing coupling is a concentric high-pressure fitting that restricts flows. Reducing tees are also made in malleable iron or forged steel. *Figure 10* shows reducers.

3.5.0 Nipples

Nipples are short pieces of pipe threaded on both ends and used to make close connections between fittings. They are made in all sizes and types of pipe and are stocked in various lengths classified as close, shoulder, short, and long. The longest nipples are 12 inches. Pipe longer than 12 inches

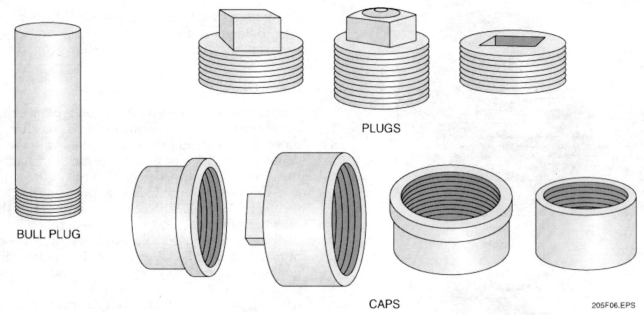

Figure 6 ◆ Caps and plugs.

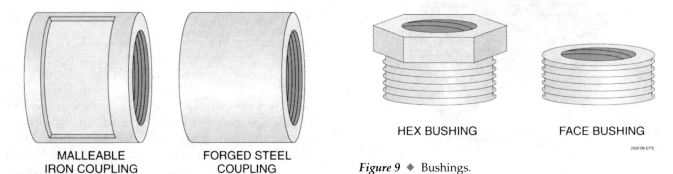

Figure 7 ◆ Couplings.

Figure 9 ◆ Bushings.

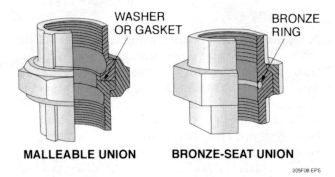

Figure 8 ◆ Unions.

is known as cut pipe instead of as nipples. A close nipple is threaded end to end, and a shoulder nipple has a short section of unthreaded pipe in the center known as the shoulder. *Figure 11* shows types of nipples.

3.6.0 Flanges

The screw-on flange is the weakest type of flange because the threads cut into the pipe. It is normally used only for low-pressure lines and pipes. Screw threads hold the flange to the pipe. As in all flanges, the bolt holes should be carefully aligned when making up the fitting. *Figure 12* shows a screw-on flange.

Each type of flange is available in different face styles, and each flange face style uses a certain gasket. A set of two flanges of the same face style, called companion flanges, is used with the proper gasket. The two face styles of threaded flanges are the flat-face flange and the raised-face flange. Typically, you should never use two different face styles at one joint, but this may be necessary when connecting to manufactured equipment. If you must connect a flat-face flange to a raised-face flange, you must confirm that the flat-face flange is not cast iron. The pressure of the bolts can crack or break the cast iron flange. If the flat-face flange is cast iron, change the raised-face flange to a flat-face flange with engineering approval.

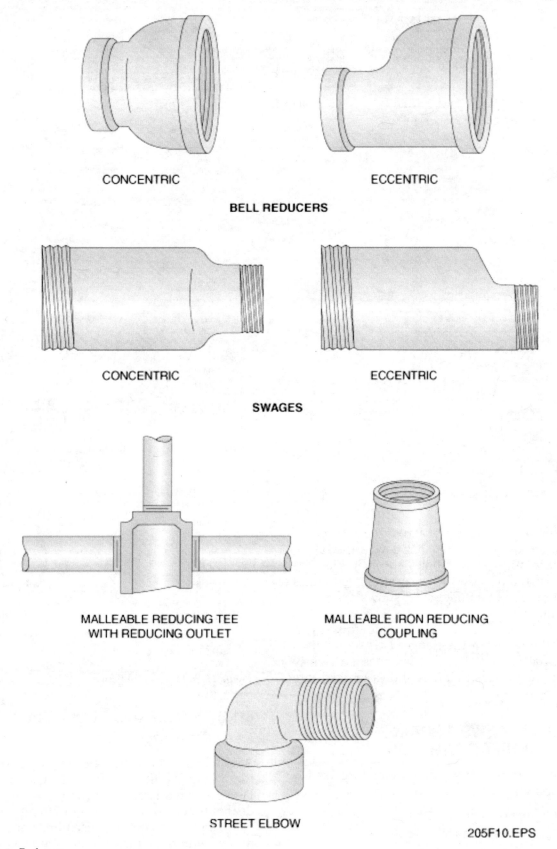

CONCENTRIC

ECCENTRIC

BELL REDUCERS

CONCENTRIC

ECCENTRIC

SWAGES

MALLEABLE REDUCING TEE
WITH REDUCING OUTLET

MALLEABLE IRON REDUCING
COUPLING

STREET ELBOW

205F10.EPS

Figure 10 ◆ Reducers.

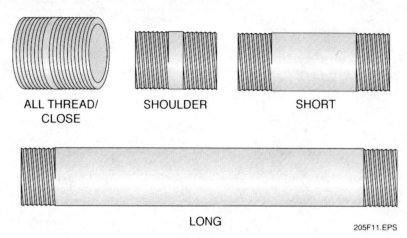

ALL THREAD/ CLOSE SHOULDER SHORT

LONG

205F11.EPS

Figure 11 ◆ Types of nipples

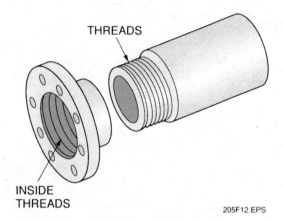

THREADS

INSIDE THREADS

205F12.EPS

Figure 12 ◆ Screw-on flange.

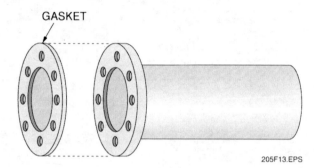

GASKET

205F13.EPS

Figure 13 ◆ Flat-face flange and gasket.

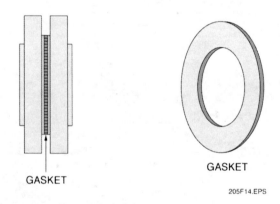

GASKET

GASKET

205F14.EPS

Figure 14 ◆ Raised-face flange.

The flat-face flange is flat across the entire face. This flange uses a gasket that covers the entire flange face and has bolt holes cut in it. Flat-face flanges are mainly used when joining two flanges of dissimilar metals. *Figure 13* shows a flat-face flange and gasket.

The raised-face flange has a wide, raised rim around the center of the flange. This flange uses a wide, flat ring gasket that is squeezed tightly between the flanges. *Figure 14* shows a raised-face flange.

4.0.0 ◆ SCREWED FITTING JOINT DRAWINGS

The following three types of drawings are used to show piping systems:

- Double-line drawings
- Single-line drawings
- Isometric drawings

These drawings can be either plan or elevation drawings. Plan drawings are drawn from the top of the system looking down on it. Elevation drawings are drawn from the side of the system.

4.1.0 Double- and Single-Line Drawings

A double-line drawing is a clear picture of the piping system. It is the easiest drawing to understand. However, this type of drawing requires a great deal of time to draw and is not often used.

A single-line drawing uses lines and symbols to represent the piping system. Information, such as pipeline numbers and lengths, can be added to this type of drawing. *Figure 15* shows double-line and single-line drawings of the same piping system. This drawing is an elevation, section, or side view of the system and shows the system height.

Figure 16 shows a plan drawing, or top view, of the same system.

4.2.0 Isometric Drawings

Piping systems must often be shown from both the elevation (side) and the plan (top) views. This can be done in an isometric drawing. *Figure 17* shows an isometric drawing of the same system with a bill of materials in the system.

4.3.0 Piping Symbols

Piping symbols indicate the types of joints and fittings used in the system. Fittings made with screwed joints are shown by standard symbols. *Figure 18* shows screwed joint symbols.

4.4.0 Line Numbers

Each line in a piping system is numbered by a standard four-part line number. Each part of the line number gives the pipefitter useful information. The first part of the number is the pipe diameter in inches. The second part of the number indicates the fluid carried by the pipe, in abbreviations. Some common fluid abbreviations are CW, for cold water; DP, for drain pipe; HW, for hot water; and S, for steam.

The third part of the line number is the pipe number. This number identifies a specific line within the piping system. The last part of the line number is the specifications book reference number. This number tells which page in the specifications book contains further information about the system. Line numbers are specific to engineering companies and may vary from job to job. *Figure 19* shows a sample line number.

4.5.0 Specifications Book

The specifications book contains information about each piping system. *Table 2* shows a sample page from the specifications book.

This page gives information for the piping system shown in *Figure 17*. Note the following seven items on the sample specification sheet.

- Service
- Pressure rating of the pipe
- Maximum pressure of the system
- Temperature limit
- Specifications book reference number
- Type of pipe, fittings, and gaskets
- Type of valves

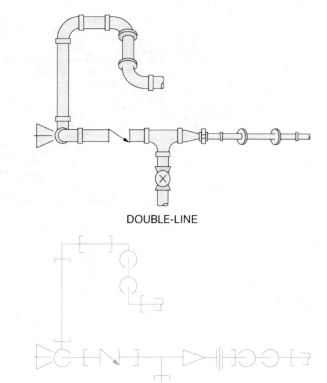

DOUBLE-LINE

SINGLE-LINE

205F15.EPS

Figure 15 ◆ Double-line and single-line drawings: elevation view.

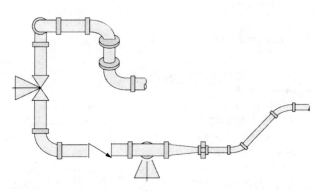

DOUBLE-LINE

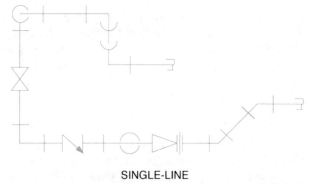

SINGLE-LINE

205F16.EPS

Figure 16 ◆ Plan drawing, top view.

BILL OF MATERIALS			
P.M.	REQ'D	SIZE	DESCRIPTIONS
1		1½"	PIPE SCH/40 ASTM-A-120 GR.B T.C. GALV.
2		¾"	PIPE SCH/40 ASTM-A-120 GR.B T.C. GALV.
3	5	1½'	90° ELL SCREWED ASTM-A-197 GALV.
4	1	1½"	TEE SCREWED ASTM-A-197 GALV.
5	2	¾"	45° ELL SCREWED ASTM-A-197 GALV.
6	1	1½" × ¾"	CONCENTRIC SWAGE SCREWED ASTM-A-197 GALV.
7	1	¾"	UNION SCREWED ASTM-A-197 GALV.
8	2	1½"	GATE VA. SW BRONZE ASTM-B62
9	1	1½"	CHECK VA. SWING BRONZE ASTM-B62

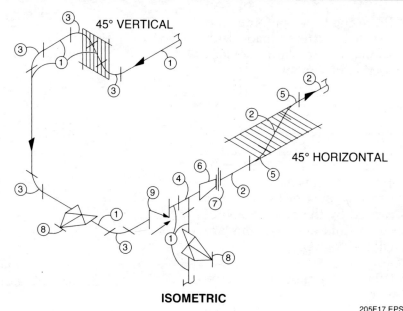

205F17.EPS

Figure 17 ◆ Isometric drawing.

5.0.0 ◆ THREADS

By definition, a pipe thread is an inclined plane wrapped around a cylinder. The thread angle is the angle formed by two inclined faces of the thread. The thread angle varies depending on the specific thread standard used. The pitch refers to the distance between adjacent screw threads when measured from center line to center line. Pitch also refers to threads per inch. The crest of the thread is the outermost part of the thread, or the top of the thread, when viewing it from the side. The root of the thread is the innermost part of the thread, or the bottom, if viewed from the side.

5.1.0 Types of Threads

The National Pipe Thread (NPT) is the standard thread for pipe connections. However, you may also see British Standard Pipe Parallel (BSPP) and British Standard Pipe Tapered (BSPT) threads. British Standard Pipe threads are also known as

Table 2 Sample Piping Specification Sheet

Piping Specification	
Service: Water	Class: ANSI 150#
Design Pressure: 150 psig	Corr. Allow: 0.030"
Max. Pressure: 200 psig	Temp Limit: 200°F
Item: Size	**General Description**
Pipe: ½" to 2"	Sched. 40-ASTM-A-120 Threaded and Capped (T&C) Galv. Grade B
Fittings: ½" to 2"	150# Screwed ASTM-A-197 Galv.
Valves: ½" to 2"	150# Screwed Bronze ASTM-B-62

205T02.EPS

Whitworth threads. The National Pipe Thread has a thread angle of 60 degrees and a slight flat at the crest and the root. While National Pipe Threads can be either straight or tapered, the type normally used has a tapered internal and external thread. The taper used is ¹⁄₁₆-inch per inch of threads. *Figure 20* shows the taper used in cutting pipe threads.

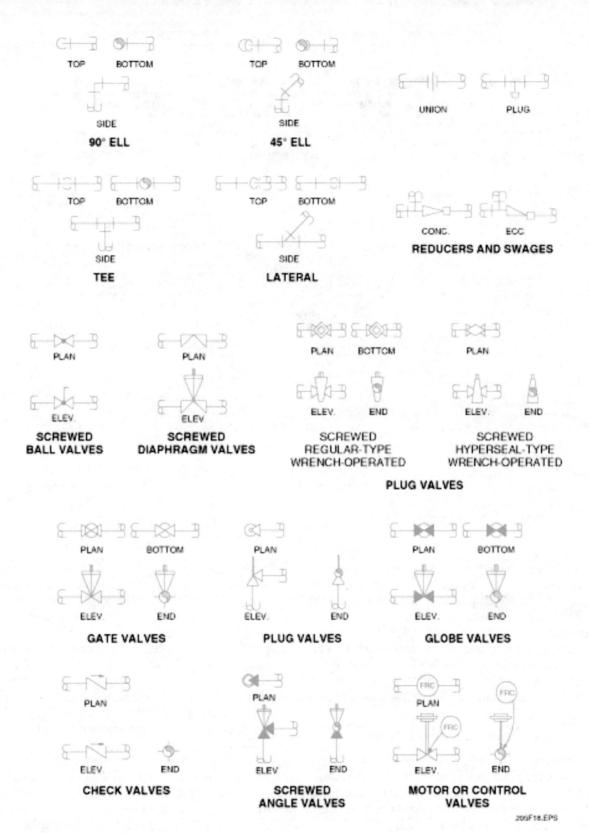

Figure 18 ◆ Screwed joint symbols.

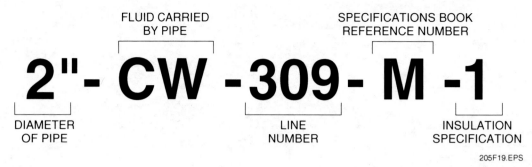

Figure 19 ◆ Line number.

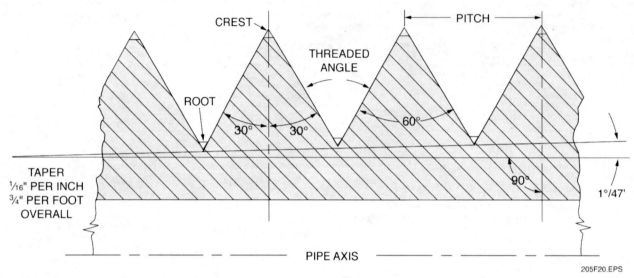

Figure 20 ◆ Taper used in cutting pipe threads.

In addition to the widely used taper pipe thread, there are several other types of standard pipe threads for specific applications. All National Pipe Threads are designated by specifying in sequence the nominal size of the pipe, the number of threads per inch, and the symbols for thread series. For example, ⅜ - 18 - NPT, are threads for a ⅜-inch nominal size pipe having 18 threads per inch and are National Pipe Threads.

5.2.0 Taper Thread Connection

In order to have a strong, leakproof threaded joint, the threads must be clean and smoothly cut. They must have the correct pitch, lead, taper, and form and the threads must be correctly sized. If these conditions are not met, the seal will leak.

The joining or makeup of a taper threaded pipe connection is performed in two distinct operations, known as hand engagement and wrench makeup. Hand engagement refers to how far the fitting can be tightened by hand. Wrench makeup is the additional turning of the fitting with a wrench to completely make up the joint. Hand engagement on properly cut threads is normally between three and four complete revolutions. Hand engagement is a good way to check the quality of threads after cutting. Wrench makeup is usually about three turns for a total of approximately seven revolutions for complete makeup. A greater number of turns is required for larger pipe. *Figure 21* shows hand engagement and wrench makeup.

The standard for taper thread pipe makeup gives basic thread dimensions and the number of rotations for hand engagement and wrench makeup. The taper thread makeup can vary plus or minus one turn from the standard thread makeup allowances. *Table 3* shows the standard for taper pipe thread makeup. Note that these thread makeups are for perfect threads cut under perfect conditions, which seldom occurs.

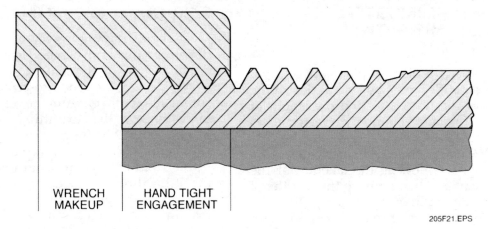

Figure 21 ◆ Hand engagement and wrench makeup.

205F21.EPS

Table 3 Standard for Taper Pipe Thread Makeup

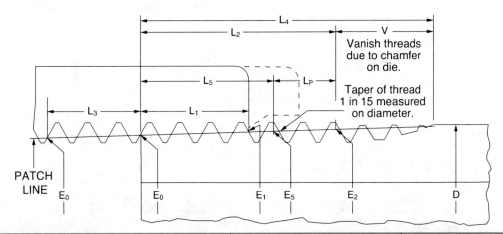

Nominal Pipe Size	Diameter of Pipe (D)	Threads (inches) (n)	Pitch of Thread	Pitch Diameter at Beginning of External Thread	Hand Tight Engagement Length (L1)			Effective Thread, External Length (L1)		
					Inch	Threads	Diameter (E1)	Inch	Threads	Diameter (E2)
1	2	3	4	5	6	7	8	9	10	11
½	0.840	14	0.07143	0.75843	0.320	4.48	0.77843	0.5337	7.47	0.79179
¾	1.050	14	0.07143	0.96768	0.339	4.75	0.98887	0.5457	7.64	1.00179
1	1.315	11.5	0.08696	1.21363	0.400	4.60	1.23863	0.6828	7.85	1.25630
1¼	1.660	11.5	0.08696	1.55713	0.420	4.83	1.58338	0.7068	8.13	1.60130
1½	1.900	11.5	0.08696	1.79609	0.420	4.83	1.82234	0.7235	8.32	1.84130
2	2.375	11.5	0.08696	2.26902	0.436	5.01	2.29627	0.7565	8.70	2.31630
2½	2.875	8	0.12500	2.71953	0.682	5.46	2.76216	1.1375	9.10	2.79062
3	3.500	8	0.12500	3.34062	0.766	6.13	3.38850	1.2000	9.60	3.41562
3½	4.000	8	0.12500	3.83750	0.821	6.57	3.88881	1.2500	10.00	3.91562
4	4.500	8	0.12500	4.33438	0.844	6.75	4.38712	1.3000	10.40	4.41562
5	5.563	8	0.12500	5.39073	0.937	7.50	5.44929	1.4063	11.25	5.47862
6	6.625	8	0.12500	6.44609	0.958	7.66	6.50597	1.5125	12.10	6.54062
8	8.625	8	0.12500	8.43359	1.063	8.50	8.50003	1.7125	13.70	8.54062
10	10.750	8	0.12500	10.54531	1.210	9.68	10.62094	1.9250	15.40	10.66562
12	12.750	8	0.12500	12.53281	1.360	10.88	12.61781	2.1250	17.00	12.66562

1. The basic dimensions of the American National Standard Taper Pipe Thread are given in inches to four or five decimal places. While this implies a greater degree of precision than is ordinarily attained, these dimensions are the basic gauge dimensions and are so expressed for the purpose of eliminating errors in computations.
2. The length L_5 from the end of the pipe determines the plane beyond which the thread form is incomplete at the crest. The next two threads are complete at the root. At this plane the cone formed by the crests of the thread intersects the cylinder, forming the external surface of the pipe. $L_5 = L_2 - L_P$

205T03.EPS

6.0.0 ◆ DETERMINING PIPE LENGTHS BETWEEN FITTINGS

In order to replace a piece of pipe between two fittings, the pipefitter must be able to determine the takeout of the fittings to find the length of pipe needed. Takeout, commonly referred to as makeup, take-up, or takeoff, refers to the distance from the end of the pipe to the center of the fitting (*Figure 22*). Knowing the takeout for the fittings is essential when figuring the lengths of straight pipe. Piping drawings usually show the dimensions of the overall run of pipe, but it is up to the pipefitter to determine the lengths of the individual straight pipe within the run. Two of the most common methods for determining pipe lengths are the center-to-center method and the face-to-face method.

6.1.0 Center-to-Center Method

Most pipe drawings give the center-to-center dimensions for a run of pipe. *Figure 23* shows a pipe drawing showing the center-to-center dimension.

When using the center-to-center method, measurements are taken from the center of one fitting to the center of the next fitting. The actual length of pipe between the fittings is determined by subtracting the center-to-face dimension of each fitting from the overall length and then adding the fitting thread engagement to obtain the actual length that the pipe needs to be. *Figure 24* shows the dimensions of a common 90-degree elbow.

The thread engagement varies with the nominal size of the pipe. Thread engagement should always be measured for accuracy. Manufacturers of fittings do give dimensions of screwed fittings in makeup charts. *Table 4* shows an example of a makeup chart.

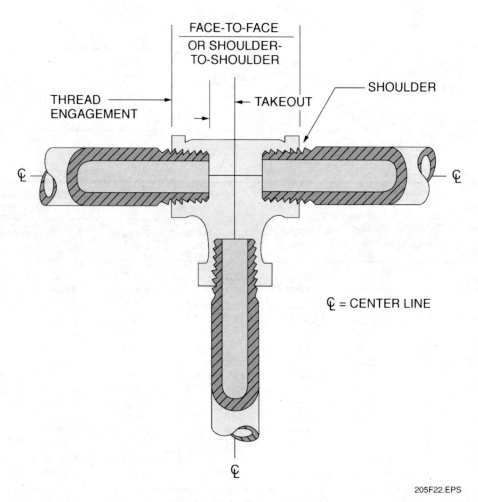

Figure 22 ◆ Takeout conventions.

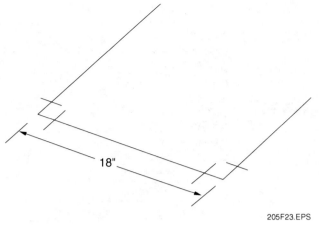

Figure 23 ◆ Center-to-center pipe drawing.

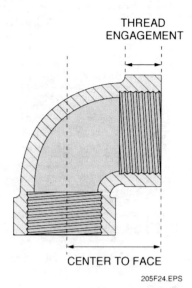

Figure 24 ◆ Dimensions of common 90-degree elbow.

Figure 25 shows the center-to-center method for determining the length of a pipe between two fittings.

Follow these steps to determine the length of a pipe between two fittings, using the center-to-center method:

Step 1 Find the center of each of the two fittings.

NOTE

This can be done by using the manufacturer's makeup chart or by measuring the fitting, using a rule.

Step 2 Measure the distance between the centers of the two fittings.

Step 3 Measure the fitting to determine the center-to-face length of each of the fittings.

Step 4 Subtract the center-to-face length of both fittings from the center-to-center measurement taken in step 2.

Step 5 Determine the thread engagement of the fittings.

NOTE

This can be done by using the manufacturer's makeup chart or by actually measuring the threads, using a rule. Larger fittings allow you to insert the rule inside the fitting to measure the thread engagement.

Step 6 Add the thread engagement of both fittings to the length found in step 4 to find the length of pipe needed.

Table 4 Example of Makeup Chart

Nominal Size	Actual OD	Threads Per Inch	Length Hand Tight	Hand Turns	Makeup Turns	Length of Effective Thread
⅛"	0.405	27	0.18	4.5	2.5	2.6
¼"	0.54	18	0.2	4	3	0.401
⅜"	0.675	18	0.24	4.5	3	0.408
½"	0.84	14	0.32	4.5	3	0.534
¾"	1.058	14	0.34	4.5	3	0.546
1"	1.315	11.5	0.4	4.5	3.25	0.682
1¼"	1.66	11.5	0.42	4.5	3.25	0.707
1½"	1.9	11.5	0.42	4.5	3.25	0.724
2"	2.375	11.5	0.436	5	3	0.756
2½"	2.875	8	0.682	5.5	3	1.136
3"	3.5	8	0.766	6	3	1.2

205T04.EPS

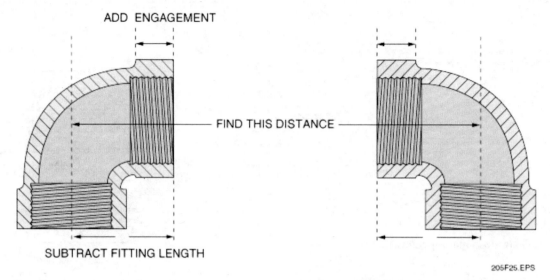

ADD ENGAGEMENT

FIND THIS DISTANCE

SUBTRACT FITTING LENGTH

205F25.EPS

Figure 25 ◆ Center-to-center method.

6.2.0 Center-to-Face Method

The center-to-face method is used when the drawing gives the length from the center of a fitting, such as a 90-degree ell or a tee, to the face of a flange. To find the length of pipe needed, subtract the laying lengths of the fitting and the flange and the thread engagement. *Figure 26* shows the center-to-face method.

Follow these steps to determine the length of pipe between two fittings, using the center-to-face method:

Step 1 Find the center of the fitting.

 NOTE
This can be done by using the manufacturer's dimensions chart or by measuring the fitting using a rule. If you are using a standard dimensions chart not supplied by the fitting manufacturer, you should physically measure the fitting.

Step 2 Measure the distance from the center of the fitting to the face of the flange.

Step 3 Measure the fitting to determine the laying length of the fitting.

Step 4 Measure the flange to determine the laying length of the flange.

 NOTE
Thickness of flange minus socket depth equals takeoff measurement.

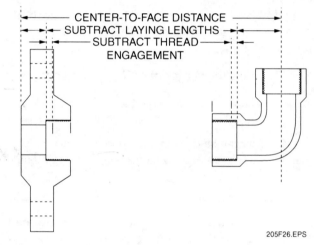

CENTER-TO-FACE DISTANCE
SUBTRACT LAYING LENGTHS
SUBTRACT THREAD
ENGAGEMENT

205F26.EPS

Figure 26 ◆ Center-to-face method.

Step 5 Subtract the laying length of the fitting and of the flange from the center-to-face measurement taken in Step 2.

 NOTE
This gives you the distance from the back of the thread on the fitting to the back of the socket on the flange.

Step 6 Subtract the thread engagement from the measurement found in Step 5.

 NOTE
This is for the thread engagement of each threaded fitting. The length found in step 6 is the final length of the pipe between the fitting and the flange.

6.3.0 Face-to-Face Method

The face-to-face method is used to determine the length of a pipe that has a flange on each end. The measurement is taken from the face of one flange to the face of the other flange. To find the length of pipe needed, subtract the laying lengths of both flanges and the ⅛-inch gap required at each weld. *Figure 27* shows the face-to-face method.

Follow these steps to determine the length of pipe between two flanges, using the face-to-face method:

Step 1   Measure the distance between the faces of the two flanges.

Step 2   Measure the flanges to determine the laying length of each of the flanges.

Step 3   Subtract the laying lengths of the flanges from the face-to-face measurement taken in Step 1.

NOTE

This gives you the distance from the back of the socket on one flange to the back of the socket on the other flange.

Step 4   Subtract ¼ inch from the measurement found in Step 3.

NOTE

This is for the ⅛ inch that the ends of the pipe have to be recessed from the backs of the sockets of each flange. The length found in step 4 is the final length of the pipe between the two flanges.

6.4.0 Calculating Offsets

An offset is a lateral move that moves the pipe out of its original line to another line that is parallel with the original line. Offsets are used when it is necessary to change the position of a pipeline in order to avoid an obstruction, such as a wall or a tank. *Figure 28* shows a simple offset around a wall.

When two lines are to be connected using elbows having an angle other than 90 degrees, the pipefitter must use special calculations to determine the distance of the pipe between the two elbows. All piping offsets are based on the right triangle. The three basic sides of an offset are the set, run, and travel. The set and run are joined by a 90-degree angle, and the travel connects the

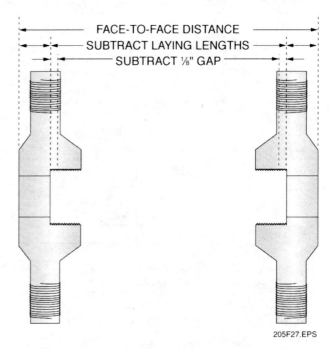

205F27.EPS

Figure 27 ◆ Face-to-face method.

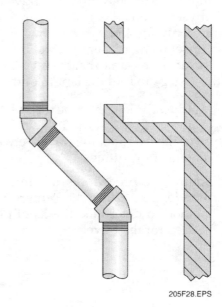

205F28.EPS

Figure 28 ◆ Simple offset around wall.

span between their end points. *Figure 29* shows the components of an offset.

The set is the distance, measured center-to-center, that the pipeline is offset. The run is the total linear distance required for the offset. The travel is the center-to-center measurement of the offset piping. The angle of the fittings determines the number of degrees that the piping changes direction. There are several methods used to calculate offsets. Since many of these require the use of advanced math, they are explained in full detail in a later module. Some of the basic methods for calculating offsets are explained below.

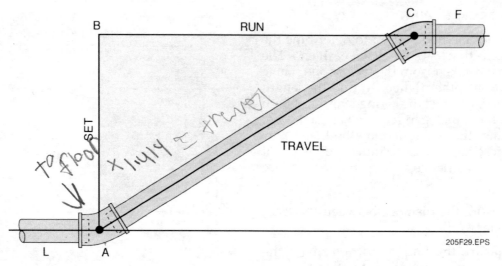

Figure 29 ◆ Offset components.

6.4.1 Calculating 45-Degree Offsets

When using 45-degree elbows to make an off-set, use the following rule: multiply the length of the set times 1.414 to determine the travel, or the length between the centers of the elbows. This rule applies only for 45-degree elbows. *Figure 30* shows an offset using 45-degree elbows.

For this exercise, assume that the set is 20 inches. Follow these steps to find the length of the travel:

Step 1 Multiply the length of the set by 1.414. For example, 20 × 1.414 = 28.28 inches. This gives the length of the travel between the centers of the fittings.

Step 2 Follow the procedures for determining pipe lengths using the center-to-center method to determine the length of pipe needed for the travel.

6.4.2 Calculating Offsets When Set and Elbow Angle Are Known

Offsets can be fabricated using elbows of various angles. Fittings of 60, 30, and 22½ degrees are normally used to offset a pipeline. When these various fittings are used, the distance between the centers can easily be found when the set is known by using the elbow constants listed in *Table 5*.

Table 5 Elbow Constants

Elbow Angle	Travel (AC)	Run (BC)
60 degrees	1.15	0.58
45 degrees	1.41	1.00
30 degrees	2.00	1.73
22½ degrees	2.61	2.41

205T05.EPS

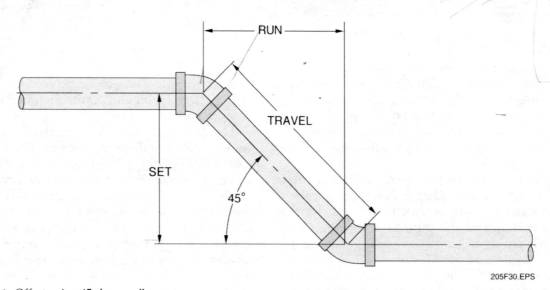

Figure 30 ◆ Offset using 45-degree elbows.

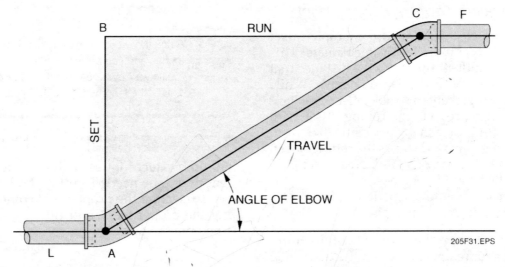

Figure 31 ◆ Elbow angle.

To find the length of the travel or the run, multiply the length of the set by the constant for the elbow used. *Figure 31* shows the elbow angle.

For this exercise, assume the set is 20 inches and the angle of the elbow is 30 degrees.

Travel (AC) = set (AB) × constant for travel (AC)

or

Travel = 20 × 2.00 = 40 inches

Run (BC) = set (AB) × constant for run (BC)

or

Run = 20 × 1.73 = 34.6 inches

6.4.3 *Calculating Offsets Using Multipliers*

This method of calculating offsets involves the use of trigonometry and is based on the fact that the run, set, and travel of an offset represent the three sides of a triangle. The trigonometry involved in this method will be explained in a later module.

For this exercise, simply understand that the multipliers used to calculate offsets are natural trigonometric functions. *Table 6* shows the multipliers used to calculate offsets.

For this exercise, assume that you are trying to determine the run, or developed length, of the offset. You know that the travel of the offset is 15 inches and the angles of the elbows are 22½ degrees. Follow these steps to use the multipliers to calculate offsets.

Step 1 Find the side of the triangle needed in the first column.

Step 2 Find the side that is known in the second column.

Step 3 Continue across in that row to the column headed with the angle of fitting to be used, and read the multiplier.

Step 4 Multiply this number by the length of the known side. The answer is the length of the needed side.

Table 6 Multipliers Used to Calculate Offsets

To Find Side	When Known Side Is	Multiply Side	Using 22½° Elbows	Using 30° Elbows	Using 45° Elbows	Using 60° Elbows
T	S	S	2.613	2.000	1.414	1.155
S	T	T	0.383	0.500	0.707	0.866
R	S	S	2.414	1.732	1.000	0.577
S	R	R	0.414	0.577	1.000	1.732
T	R	R	1.082	1.155	1.414	2.000
R	T	T	0.924	0.866	0.707	0.500
R = run S = set T = travel						

205T06.EPS

6.4.4 Calculating Travel of Rolling Offset

The rolling offset is similar to the simple offset except there is one more factor to calculate. This is the roll of the offset. The roll is the distance that one end of the travel is displaced laterally. While a triangle best represents a simple offset, a rectangular box best represents the rolling offset with the travel moving diagonally across the box from corner to corner. *Figure 32* shows the rolling offset.

Assume that the roll of a 60-degree offset is 15 inches and the height is 8 inches.

Follow these steps to calculate the travel of a rolling offset, using a framing square:

Step 1　Lay out the roll on one side of the framing square and the set on the other side of the square. *Figure 33* shows laying out a rolling offset using a framing square.

Step 2　Measure the distance between these two points on the framing square. In this example, the measurement is 17 inches.

Step 3　Multiply the measurement by the constant for the angle of the fitting from *Table 7*.

Table 7 Constants for Common Angles

Angle	Constant
22½ degrees	2.613
30 degrees	2.000
45 degrees	1.414
60 degrees	1.154

205T07.EPS

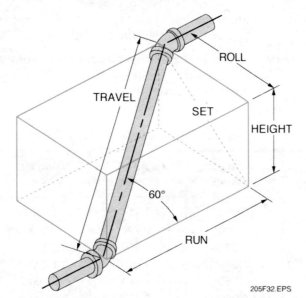

Figure 32 ◆ Rolling offset.

205F32.EPS

7.0.0 ◆ ASSEMBLY TECHNIQUES

Assembly techniques include determining the length of pipe needed, cutting and threading the pipe, selecting the proper joint compound, selecting the proper tools, and fitting the screwed fitting onto the pipe.

7.1.0 Threading Pipe

As you learned in previous modules, pipe can be threaded using manual threaders and power threading machines. This section provides a refresher of the steps you must perform to thread pipe.

7.1.1 Threading Pipe Using a Manual Pipe Threader

Figure 34 shows a manual pipe threader with threading dies. Follow these steps to thread pipe using a manual pipe threader:

Step 1　Identify the material being threaded, such as pipe, tubing, or conduit, and brass, copper, or steel.

Step 2　Identify the diameter of the material being threaded.

Step 3　Select the die head best suited for the threading process.

Step 4　Insert the die head into the pipe threader handle.

Step 5　Inspect the die and die handle to ensure that all parts are clean and in good condition.

Step 6　Secure the pipe in a vise.

Step 7　Make sure that the end of the pipe is square, clean, and deburred.

Step 8　Position your body so that you are balanced, facing the work area, and within easy reach of the work area.

Step 9　Position the die flush on the end of the pipe.

Step 10　Press the die against the end of the pipe, and turn the die handle clockwise to start the threading process.

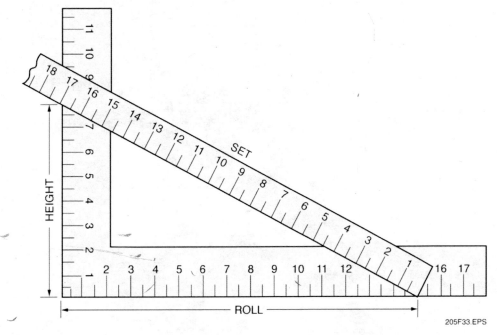

Figure 33 ◆ Laying out rolling offset using framing square.

Step 11 Turn the die handle two full rotations around the pipe. Add cutting oil as necessary to keep a thin coat on the threads while cutting.

Step 12 Stop the work, and back the die off the pipe. This allows you to make sure the threads are properly started and allows the threads to be cleaned from the cutting die.

Step 13 Clean the cutting threads from the die if necessary.

Step 14 Rethread the die onto the pipe, and continue the process until the threading is complete. The threading is complete when two full threads extend past the back edge of the die. If the threading is correct, you should be able to screw a fitting three and a half revolutions by hand onto the new threads.

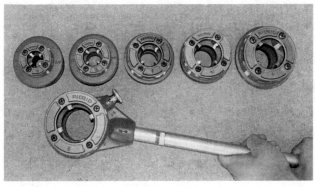

205F34.EPS

Figure 34 ◆ Manual pipe threader.

> **CAUTION**
>
> Do not thread the pipe too far. This can cause an improper fit between the fitting and the pipe, which will result in leaks.

Step 15 Back the die off the pipe.

Step 16 Inspect the new threads, and remove any debris from the threads using a rag.

> **WARNING!**
>
> Freshly cut threads are very sharp. Use care in handling to avoid cuts.

7.1.2 Threading Pipe Using a Threading Machine

Figure 35 shows a powered pipe threading machine. Be sure to check the cutting oil level and prime the cutting oil pump before threading pipe, using the threading machine. Follow these steps to thread pipe:

Step 1 Load the pipe into the threading machine.

Step 2 Place pipe stands under the pipe as needed.

Step 3 Cut and ream the pipe to the required length.

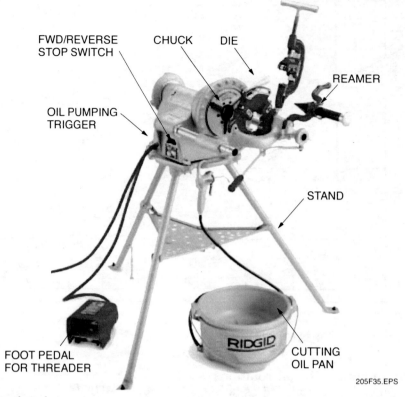

Figure 35 ◆ Powered pipe threader.

Step 4 Make sure that the proper dies are in the die head.

Step 5 Loosen the clamp lever.

Step 6 Move the size bar to select the proper die setting for the size of pipe being threaded.

Step 7 Lock the clamp lever.

Step 8 Swing the die head down to the working position.

Step 9 Close the throw-out lever.

Step 10 Lower the lubrication arm, and direct the oil supply onto the die.

Step 11 Turn the machine switch to the forward position.

Step 12 Step on the foot switch to start the machine. When threading stainless steel or alloys, bump the foot pedal slowly and add water to prevent galling.

Step 13 Turn the carriage to bring the die against the end of the pipe.

Step 14 Apply light pressure on the handwheel to start the die.

CAUTION

Make sure that the die is flooded with oil at all times while the die is cutting to prevent overheating the die and the pipe. Overheating can damage the die and the pipe threads.

NOTE

The dies feed onto the pipe automatically as they follow the newly cut threads.

Step 15 Release the handwheel once the dies have started to thread the pipe.

Step 16 Open the throw-out lever as soon as two full threads extend from the back of the dies.

Step 17 Release the foot switch to stop the machine.

Step 18 Turn the carriage handwheel to back the die off the pipe.

Step 19 Swing the die and the oil spout up and out of the way.

Step 20 Screw a fitting onto the end of the pipe to check the threads. If the threads are correct, you should be able to screw a fitting three and a half revolutions by hand onto the new threads. If the threads are not deep enough, adjust the die with the clamp lever, and repeat the threading operation.

Step 21 Turn the machine switch to the OFF position.

Step 22 Open the chuck, and remove the pipe.

7.2.0 Pipe Joint Compounds

A good grade of pipe joint compound must be applied to the male threads joint to lubricate the joint to prevent galling of the threads and to ease the makeup of the joint. You must be careful about what compound you use to avoid product contamination or dangerous combinations of materials. Depending on the type of liquid being transferred through the pipes, sometimes no compound is permitted. Always check with the engineering specifications and requirements before applying any type of pipe joint compound. Some compounds that can be used include Teflon® tape and pipe dope (*Figure 36*).

7.2.1 Teflon® Tape

Teflon® tape is a special tape that is wrapped around threads to prevent metal-to-metal contact and to create a leakproof joint. Follow these steps to use Teflon® tape:

> **WARNING!**
> Never use Teflon® tape on pipe that is to be welded or pipe that carries steam or other high-temperature service. When Teflon® tape is heated, it emits a highly toxic gas that could be fatal.

Step 1 Remove all excess cutting oil from the threads to improve the grip of the tape on the threads.

> **NOTE**
> Use ½-inch wide Teflon® tape for pipe that is ¾-inch nominal size and smaller. Use ¾-inch wide tape for pipe that is 1-inch nominal size or larger.

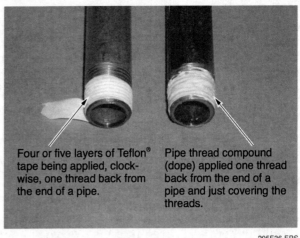

Four or five layers of Teflon® tape being applied, clockwise, one thread back from the end of a pipe.

Pipe thread compound (dope) applied one thread back from the end of a pipe and just covering the threads.

205F36.EPS

Figure 36 ◆ Teflon® tape and pipe joining compound.

Step 2 Start the tape from the end of the thread, leaving the first full thread bare to prevent the tape from overhanging the end of the pipe.

Step 3 Wrap the tape around the pipe in the direction that the joint is to be assembled. Tape should be wrapped around right-hand threads in a clockwise direction and around left-hand threads in a counterclockwise direction.

Step 4 Continue to wrap the tape around the joint, overlapping the edges of each wrap until all remaining threads have been covered.

Step 5 Press the tape against the threads to seal it to the threads and prevent the tape from slipping off the threads once you start to make up the joint.

7.2.2 Liquid Teflon®

Liquid Teflon® is a Teflon®-based pipe joint compound that you paint onto the male threads, using a small brush. It can be used on all types of materials and can withstand temperatures up to 400°F.

7.2.3 Pipe Dope

Pipe dope is manufactured for sealing threaded pipe joints. If manufactured pipe dope is not available, a mixture of graphite paste and water, petroleum jelly and zinc dust, or white or red lead will work as a joint compound. Make sure that the pipe dope is suitable for the type of pipe you are using. Dopes are made specifically for stainless steel pipe and for carbon steel threads. Try to avoid dopes that claim to be suitable for both stainless steel and carbon steel because they tend to fail more often than others. Pipe dope is relatively easy to apply. It is simply brushed onto the male threads to be fitted. Do not apply pipe dope to the female threads, and do not apply dope to the first thread of the pipe end. This is to keep the pipe dope out of the pipeline and avoid contaminating the liquid inside the pipeline.

7.3.0 Selecting Wrenches

When you assemble threaded pipe joints, first tighten the fitting by hand, and then make up the joint completely using a pipe wrench. The pipe should either be secured in a vise, or you should use two pipe wrenches: one to hold the pipe and the other to turn the fitting. In selecting and using the pipe wrench, use the following suggestions as a guide:

- Pipe wrenches should be in good working condition and adjusted properly to avoid damaging the pipe.
- Use a strap wrench when working with brass, chrome-plated, or other specially finished pipe to avoid scratching or marring the surface.
- Select a wrench that is the proper size for the pipe with which you are working. Wrenches that are too short make your job much harder. Wrenches that are too long strip the threads, crack the fitting, or crush the pipeline.
- Threads should never bottom out of the fitting or adjacent pipe.
- Never use a power drive or threading machine to tighten fittings onto pipes.
- Be sure to install unions in accordance with flow requirements.
- Never use a cheater bar on the end of a wrench or strike the wrench handle with a hammer.

7.4.0 Fitting Screwed Pipe and Fittings

Proper fitting of pipe and fittings is a skill that must be learned through experience. It is crucial to fabricating a quality piping system. Poor fitting practices can result in leaks and other defects in the system. When tightening fittings onto a piece of pipe, the pipefitter must be aware of the end position that the fitting must face. If the fitting is turned past the desired end position and then turned back, the fitting will leak. Make sure the union is placed in the direction of flow. Follow these steps to fit screwed pipe and fittings:

Step 1   Determine the type, size, and schedule of pipe being used in the system.

Step 2   Determine the length of pipe needed.

Step 3   Cut the pipe to the desired length.

Step 4   Ream the pipe to remove all internal burrs.

Step 5   Thread the pipe, taking precautions not to cut the threads too deep.

Step 6   Select the proper size, shape, and type of fittings needed.

Step 7   Clean the pipe and fittings thoroughly, inside and out. All sand, dirt, and oil must be removed from the inside of the pipe and fittings to avoid contaminating the system or clogging the line. The fittings and pipe ends may be cleaned with a clean rag soaked in nonflammable solvent.

Step 8   Check the threads on the pipe and the fitting to ensure that they are properly cut and not damaged.

Step 9   Apply joint compound or Teflon® tape to the pipe threads.

Step 10   Start the fitting on the pipe by hand.

Step 11   Tighten the fitting slowly, using a pipe wrench.

CAUTION

Tightening the fitting quickly causes excessive heat due to friction, which could cause the threads or the fitting to expand. Do not allow the threads to bottom out into the fitting. This can damage the threads and the fitting.

8.0.0 ◆ INSTALLING THREADED VALVES

The threaded joint is a common method for joining pipe to smaller valves. In order to have a strong, leakproof threaded joint, the threads must be clean and smoothly cut. They must have the correct pitch, lead, taper, and form, and the threads must be correctly sized. If these conditions are not met, the seal will leak.

The joining, or makeup, of a taper-threaded pipe connection is performed in two distinct operations, known as hand engagement and wrench makeup. Hand engagement on properly cut threads is normally between three and four complete revolutions. Hand engagement is a good way to check the quality of threads after cutting. Wrench makeup is usually about three turns for a total of approximately seven revolutions for complete makeup. More turns are required for larger pipe.

Special precautions must be taken when installing valves with threaded ends to avoid damaging the valve. The steps are the same as those for installing threaded fittings. Follow these guidelines when installing valves with threaded ends:

- Never place the valve in any type of vise. Always secure the pipe in a vise, and screw the valve onto the pipe.
- Always grip the valve with the pipe wrench on the side of the body of the valve closest to the pipe.

- Use a strap wrench on valves with brass bodies to avoid excessive marring of the valve.
- Lubricate the male threads of the pipe. Do not lubricate the female threads inside the valve, and do not lubricate the thread nearest the end of the pipe.
- Do not run the male end of the pipe all the way into the valve. This will damage the seat and cause the valve to leak. Always leave three threads showing outside the valve. *Figure 37* shows proper thread makeup.

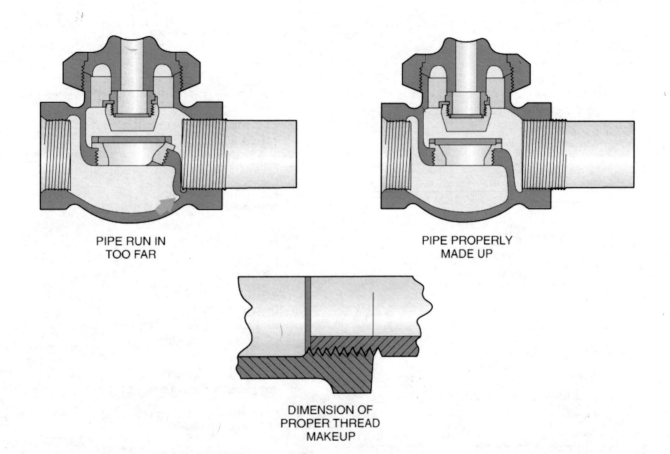

PIPE RUN IN
TOO FAR

PIPE PROPERLY
MADE UP

DIMENSION OF
PROPER THREAD
MAKEUP

DIMENSIONS, IN INCHES									
DIMENSIONS GIVEN DO NOT ALLOW FOR VARIATIONS IN TAPPING OR THREADING									
SIZE	⅛	¼	⅜	½	¾	1	1¼	1½	2	2½
A	¼	⅜	⅜	½	⁹⁄₁₆	¹¹⁄₁₆	¹¹⁄₁₆	¹¹⁄₁₆	¾	¹⁵⁄₁₆
SIZE	3	3½	4	5	6	8	10	12	14	
A	1	1¹⁄₁₆	1⅛	1¼	1⁵⁄₁₆	1⁷⁄₁₆	1⅝	1¾		

205F37.EPS

Figure 37 ◆ Proper thread makeup for a valve.

1. Galvanized pipe is pipe that has been _____.
   a. painted with an aluminum coating
   b. sprayed with Galvan
   c. dipped in a hot zinc mixture
   d. threaded using a hand threading tool

2. One advantage of stainless steel is that it is _____.
   a. inexpensive
   b. easy to thread
   c. resistant to corrosion
   d. difficult to make leak-proof

3. The wall thickness of Schedule 40 and Schedule 80 pipe is the same; it is the outside diameter that is different.
   a. True
   b. False

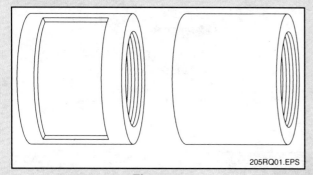

Figure 1

4. The fitting shown in *Figure 1* is a _____.
   a. cap
   b. coupling
   c. nipple
   d. bushing

5. The fitting that is used to connect the male end of a pipe to the female end of a larger fitting is called a _____.
   a. coupling
   b. flange
   c. bushing
   d. reducer

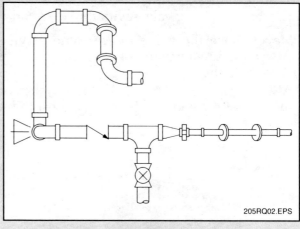

Figure 2

6. The piping drawing depicted in *Figure 2* is called a(n) _____ drawing.
   a. double-line
   b. elevation
   c. isometric
   d. single-line

7. In the piping system line number 2"-CW-309-M-1, the letters CW represent the _____.
   a. pipe diameter
   b. thread direction
   c. fluid carried by the pipe
   d. material the pipe is made of

8. In the thread designation ⅜–18-NPT, the 18 represents the number of threads per inch.
   a. True
   b. False

9. Given a properly threaded pipe, a threaded fitting should be able to be hand-threaded _____ full rotations.
   a. 1 to 2
   b. 2 to 3
   c. 3 to 4
   d. 4 to 5

10. The term *takeout* refers to the _____.
    a. space between fittings
    b. space between the end of the threaded pipe fitting and the center of the fitting
    c. distance between the threaded ends of the pipes connected to the straight-through run of the tee
    d. shoulder-to-shoulder dimension of the fitting

11. The method used to determine the length of a pipe that has a flange on each end is the _____ method.
    a. shoulder-to-shoulder
    b. face-to-face
    c. center-to-center
    d. center-to-face

12. The long side of a straight offset is known as the travel.
    a. True
    b. False

13. In calculating a 45-degree offset, the set is multiplied by _____ to determine the travel.
    a. 0.75
    b. 1.414
    c. 2
    d. 3.1416

14. Teflon® tape must be used on piping that will be used to carry steam.
    a. True
    b. False

15. When threading a pipe to a valve, you would place the valve securely in a vise.
    a. True
    b. False

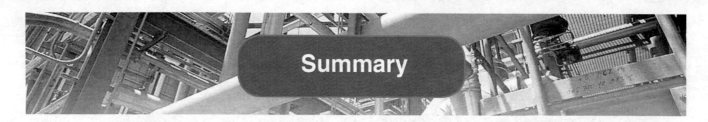

Summary

Threaded pipe is a versatile, widely used method of joining. Pipe must be threaded according to national standards. The tapered pipe thread, or NPT, is the most common thread used on pipe and fittings for industrial and commercial low-pressure applications. The threaded pipe system requires a sealant and proper assembly to ensure a tight, leakproof joint.

Takeout refers to the dimensions of the fittings within any run of a pipeline. In threaded piping systems, takeout also refers to the amount of thread engagement that must be taken into consideration for figuring cut lengths of pipe. Much of the cutting, reaming, and threading of screwed pipe systems is done in the field by pipefitters working from piping drawings.

The ability to install threaded pipe in accordance with the job requirements is a skill that every pipefitter must develop. This lesson has introduced you to the basics of installing screwed pipe. All you have to do is put these basics to work for you to become an experienced, skilled worker in the pipefitting trade.

Notes

Branch: The outlet or inlet of a fitting not in line with the run but which may make any angle.

Bushing: A pipe fitting that connects a pipe with a fitting of a larger nominal size. A bushing is a hollow plug with internal and external threads to suit the different diameters.

Close nipple: A nipple that is about twice the length of a standard thread and threaded from end to end with no shoulder.

Elbow: A fitting that makes an angle between adjacent pipes. An elbow is always 90 degrees unless another angle is stated; also known as an ell.

Flange: A fitting on the end of a pipe that is shaped like a rim and allows the pipe to be bolted to another flanged pipe.

Galling: Deformity of the threads in which the threads are stripped.

Malleable iron: Metallic fitting material that is generally used for air and water and has a pressure rating of 125 to 150 psi.

Nipple: A short piece of pipe threaded on both ends and less than 12 inches long. Any pipe over 12 inches is referred to as cut pipe.

National Pipe Thread (NPT): The United States standard for pipe threads. This thread has a $\frac{1}{16}$-inch taper per inch from back to front.

Reducer: A pipe fitting with female threads on each end, with one end being one or more sizes smaller than the other end.

Run: A length of pipe made up of more than one length of pipe.

Takeout: The dimension that the fittings take out of a center-to-center measurement. Also known as takeup, takeoff, or makeup.

Tee: A fitting that has one side outlet 90 degrees to the run.

Teflon® tape: Tape made of Teflon® that is wrapped around the male threads of a pipe before it is screwed into a fitting to serve as a lubricant and sealant.

Union: A fitting used to connect pipes.

Wye: A fitting that has a side outlet of 45 degrees to the run; also referred to as a lateral.

Resources & Acknowledgments

Additional Resources

This module is intended to be a thorough resource for task training. The following reference works are suggested for further study. These are optional materials for continued education rather than for task training.

Audel Mechanical Trades Pocket Manual. Thomas B. Davis, Carl A. Nelson. New York, NY: Macmillan & Company.

IPT's Pipe Trades Handbook. Robert A. Lee. Clinton, NC: Construction Trades Press.

Figure Credits

Mueller/B&K Industries, Inc., 205F05

Reprinted from *ASME B1.20.1-1983*, by permission of the American Society of Mechanical Engineers. All rights reserved., 205T03

Ridge Tool Co. (RIDGID®), 205F35

Topaz Publications, Inc., 205F34, 205F36

Cianbro Corporation, Module opener

NCCER CURRICULA — USER UPDATE

NCCER makes every effort to keep its textbooks up-to-date and free of technical errors. We appreciate your help in this process. If you find an error, a typographical mistake, or an in-accuracy in NCCER's curricula, please fill out this form (or a photocopy), or complete the on-line form at **www.nccer.org/olf**. Be sure to include the exact module ID number, page number, a detailed description, and your recommended correction. Your input will be brought to the attention of the Authoring Team. Thank you for your assistance.

Instructors – If you have an idea for improving this textbook, or have found that additional materials were necessary to teach this module effectively, please let us know so that we may present your suggestions to the Authoring Team.

NCCER Product Development and Revision

13614 Progress Blvd., Alachua, FL 32615

Email: curriculum@nccer.org
Online: www.nccer.org/olf

❑ Trainee Guide ❑ Lesson Plans ❑ Exam ❑ PowerPoints Other _____

Craft / Level: _____ Copyright Date: _____

Module ID Number / Title: _____

Section Number(s): _____

Description: _____

_____

_____

_____

Recommended Correction: _____

_____

_____

_____

_____

Your Name: _____

Address: _____

_____

Email: _____ Phone: _____

08206-06

Socket Weld Pipe Fabrication

08206-06
Socket Weld Pipe Fabrication

Topics to be presented in this module include:

Overview

Socket weld piping is quick and relatively easy to fit properly. Since it is welded together at the end, remember to measure twice, cut once, as it is better to do the fit only once. The fitter establishes the correct alignment between all of the parts, including the expansion gap inside the socket. The welder tack-welds the assembly for the fitter where the fitter wants tacks, and the fitter aligns the openings and pipes correctly. You will learn more drawing symbols here also, so that you will know what kind of connections you are to make, and you will learn how to get the exact lengths of pipe you need to make the system work.

Objectives

When you have completed this module, you will be able to do the following:

1. Identify and explain types of socket weld piping materials.
2. Identify and explain socket weld fittings.
3. Read and interpret socket weld piping drawings.
4. Determine pipe lengths between socket weld fittings.
5. Fabricate socket weld fittings to pipe.

Trade Terms

Concentric reducer	Insert
Cross	Tack-weld
Eccentric reducer	Straight tee
Fillet weld	Swage
Flange	

Required Trainee Materials

1. Pencil and paper
2. Appropriate personal protective equipment

Prerequisites

Before you begin this module, it is recommended that you successfully complete *Core Curriculum*; *Pipefitting Level One*; and *Pipefitting Level Two*, Modules 08201-06 through 08205-06.

This course map shows all of the modules in the second level of the *Pipefitting* curriculum. The suggested training order begins at the bottom and proceeds up. Skill levels increase as you advance on the course map. The local Training Program Sponsor may adjust the training order.

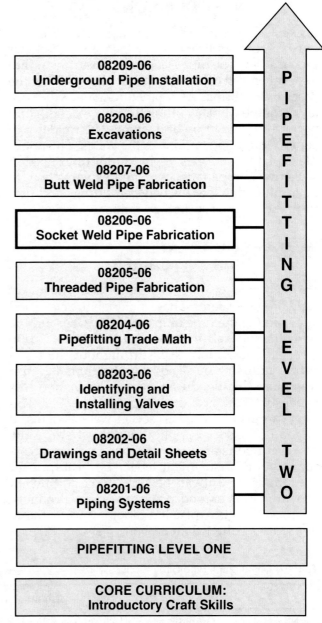

205CMAP.EPS

1.0.0 ◆ INTRODUCTION

Socket weld piping systems are used in a wide variety of piping applications in which small diameter pipe is used. The pipe is joined by inserting it into a socket or well on the fitting and fillet-welding the pipe to the fitting. As a pipefitter, you will work ahead of the welder, cutting the pipe to length and performing the fit-up of the pipe. The welder will come behind you and make the welds. This module explains the types of pipe and fittings used in socket weld pipe fabrication and explains how to determine pipe lengths between fittings and perform the fit-up of socket weld pipe.

2.0.0 ◆ SOCKET WELD PIPE MATERIALS

Various types of materials can be used in socket weld piping systems. These materials are classified according to the type of material, the size, and the schedule, or wall thickness, of the pipe. Carbon steel pipe is widely used throughout the pipefitting industry in various applications because it is more ductile, durable, machinable, and less expensive than most other types of pipe. Stainless steel can also be used in socket weld systems. It is good for special applications that require the maximum resistance to corrosion, such as processing systems, instrument lines, and heat-transfer equipment, and it can be kept sterile to prevent contamination of food, drugs, and dairy products. Stainless steel is expensive, though, and therefore is used less often than carbon steel pipe. Galvanized pipe is rarely used in socket weld piping systems because the heat from welding removes the corrosion-resistant coating of the pipe. Welding galvanized pipe also emits a poisonous gas, and special precautions must be taken during welding.

2.1.0 Pipe Sizes

Socket welds are only used with pipe up to 3 inches nominal size, and sizes above 2 inches are rarely used. Common wall thicknesses of carbon steel pipe used in socket weld systems are Schedule 40, 80, and 160. *Table 1* lists commercial pipe dimensions.

3.0.0 ◆ SOCKET WELD PIPE FITTINGS

Socket weld fittings are manufactured in four classes or pressure ratings: 2,000, 3,000, 6,000, and 9,000 pounds. The rating is forged into the side of each fitting. The most common fittings used in socket weld systems are 3,000- and 6,000-pound fittings. The 3,000-pound fittings are used in compressed air systems and water systems, and 6,000-pound fittings are used in high-pressure systems, such as steam.

Socket weld fittings have deep sockets that the pipe slips into and aligns itself in. The fillet weld is then made on the outer surface of the pipe and fitting. This eliminates the need for special clamps for alignment prior to the final weld. *Figure 1* shows several types of socket weld fittings in a pipe run.

Table 1 Commercial Pipe Dimensions

| Nominal Pipe Size | Outside Diameter | Nominal Wall Thickness | | | | | |
		STD	Schedule 40	XS	Schedule 80	Schedule 160	XXS
⅛	0.405	0.068	0.068	0.095	0.095	–	–
¼	0.540	0.088	0.088	0.119	0.119	–	–
⅜	0.675	0.091	0.091	0.126	0.126	–	–
½	0.840	0.109	0.109	0.147	0.147	0.188	0.294
¾	1.050	0.113	0.113	0.154	0.154	0.219	0.308
1	1.315	0.133	0.133	0.179	0.179	0.250	0.358
1¼	1.660	0.140	0.140	0.191	0.191	0.250	0.382
1½	1.900	0.145	0.145	0.200	0.200	0.281	0.400
2	2.375	0.154	0.154	0.218	0.218	0.344	0.436
2½	2.875	0.203	0.203	0.276	0.276	0.375	0.552
3	3.5	0.216	0.216	0.300	0.300	0.438	0.600

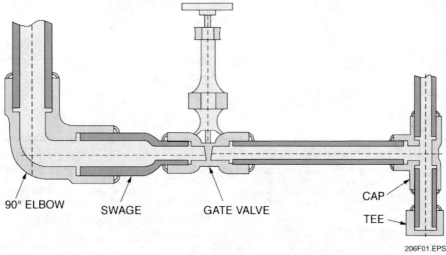

Figure 1 ◆ Socket weld fittings in pipe run.

90° ELBOW   SWAGE   GATE VALVE   CAP   TEE

206F01.EPS

3.1.0 Common Socket Weld Pipe Fittings

The most common socket weld pipe fittings are the 90-degree elbow, 45-degree elbow, cross, tee, and coupling. The 90- and 45-degree elbows are used to change the direction of flow in a piping run. The cross and tee are used to connect branch lines, and the coupling is used to join straight runs of pipe.

The dimensions of the more common socket weld fittings are found in standard dimension charts. These fitting dimensions should always be checked against the chart because many fittings made by foreign manufacturers deviate from the standard dimensions. *Figure 2* shows a standard dimension chart.

3.2.0 Miscellaneous Socket Weld Fittings

Several other socket weld fittings are also available for use in socket weld piping systems. These include unions, caps, and reducers.

3.2.1 Unions

Unions are used either to join two ends of pipe which cannot be turned or to allow pipes to be disconnected at some future time without cutting the pipes. A union usually consists of three parts: two sleeves to be threaded on the ends of the pipes and a threaded coupling ring to draw the sleeves together. *Figure 3* shows socket weld unions. When welding stainless steel unions, let them cool completely before taking them apart. Otherwise, galling will occur.

3.2.2 Caps

A cap is a fitting used to close off the end of a line. Socket weld caps are plain caps with round heads. *Figure 4* shows a socket weld cap.

3.2.3 Reducers

Reducers include swages, reducing couplings, inserts, and reducing tees. The swage used in a socket weld system can be either concentric or eccentric and does not have an internal socket. The swage has enough of a projection on the ends to allow it to be fitted into the socket of another fitting and fillet-welded to that fitting.

A reducing coupling has a socket on each end to allow the pipe to be inserted into the fitting. An insert is a bushing that is inserted into the socket end of a fitting, such as a straight coupling or straight tee, and welded into place, creating a reduction in line size. A reducing tee consists of a socket weld tee with the outlet smaller than the pipe run size. *Figure 5* shows types of reducers.

3.2.4 Special Branch Fittings

Special branch fittings include sockolets, elbolets, and latrolets. These fittings offer an alternative method for branching from a main run of pipe. The ends of these fittings are curved to match the circumference of the straight run of pipe, and installation of these fittings requires cutting a hole into the side of the main run of pipe. The sockolet makes a 90-degree branch from a run of pipe. It can either be full size or reducing. The elbolet makes a reducing branch on an elbow, and the latrolet makes a 45-degree reducing branch from a straight run of pipe. *Figure 6* shows special socket weld branch fittings.

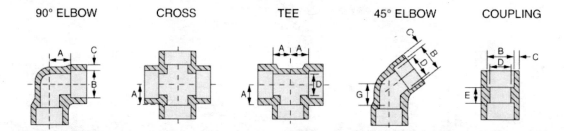

90° ELBOW   CROSS   TEE   45° ELBOW   COUPLING

Pipe Size	Depth of Socket, Min.	Center to Bottom of Socket		Center to Bottom of Socket for 45° Ells		Bore Diameter of Socket, Min.	Socket Wall Thickness, Min.			Bore Diameter of Fitting			Coupling Distance Between Bottoms of Sockets	Pipe Size
		Sched. 40 and 80	Sched. 160	Sched. 40 and 80	Sched. 160		Sched. 40	Sched. 80	Sched. 160	Sched. 40	Sched. 80	Sched. 160		
		A		G		B	C			D			E	
⅛	⅛	7/16		5/16		0.420	0.125	0.125		0.269	0.215		¼	⅛
¼	⅜	7/16		5/16		0.555	0.125	0.149		0.364	0.302		¼	¼
⅜	⅜	17/32		5/16		0.690	0.125	0.158		0.493	0.423		¼	⅜
½	⅜	⅝	¾	7/16	½	0.855	0.136	0.184	0.234	0.622	0.546	0.466	⅜	½
¾	½	¾	⅞	½	9/16	1.065	0.141	0.193	0.273	0.824	0.742	0.614	⅜	¾
1	½	⅞	1 1/16	9/16	11/16	1.330	0.166	0.224	0.313	1.049	0.957	0.815	½	1
1¼	½	1 1/16	1¼	11/16	13/16	1.675	0.175	0.239	0.313	1.380	1.278	1.160	½	1¼
1½	½	1¼	1½	13/16	1	1.915	0.181	0.250	0.351	1.610	1.500	1.338	½	1½
2	⅝	1½	1⅝	1	1⅛	2.406	0.193	0.273	0.429	2.067	1.939	1.689	¾	2
2½	⅝	1⅝	2¼	1⅛	1¼	2.906	0.254	0.345	0.469	2.469	2.323	2.125	¾	2½
3	⅝	2¼	2½	1¼	1⅜	3.535	0.270	0.375	0.546	3.068	2.900	2.626	¾	3

206F02.EPS

Figure 2 ◆ Standard dimension chart.

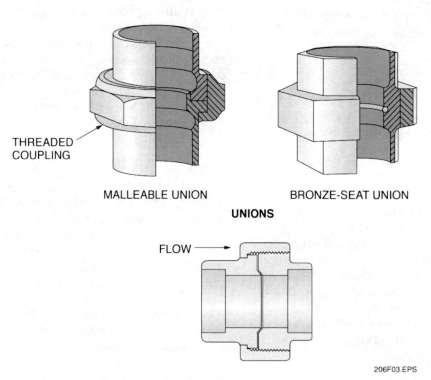

THREADED COUPLING

MALLEABLE UNION   BRONZE-SEAT UNION

UNIONS

FLOW →

206F03.EPS

Figure 3 ◆ Socket weld unions.

Figure 4 ◆ Socket weld cap.

3.3.0 Socket Weld Flanges

Socket weld flanges are available in four classes, or pressure ratings. These are 150-pound, 300-pound, 600-pound, and 1,500-pound. Always check the engineering specifications to determine what class or rating flange to use in a specific piping system. *Figure 7* shows socket weld flanges and their dimensions.

When installing flanges, you must align the bolt holes before welding. The method used to align the bolt holes is the two-hole method, commonly referred to as two-holing. With the two-hole method, the top two holes of the flange are level. One hole is to the left of top dead center, and the other hole is to the right of top dead center. *Figure 8* shows the two-hole method of leveling a flange.

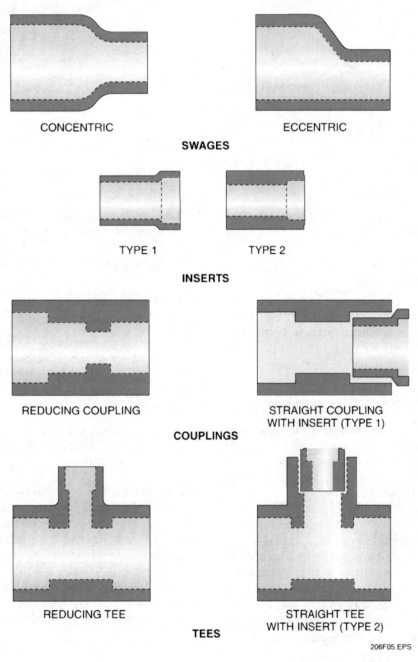

CONCENTRIC   ECCENTRIC

SWAGES

TYPE 1   TYPE 2

INSERTS

REDUCING COUPLING   STRAIGHT COUPLING WITH INSERT (TYPE 1)

COUPLINGS

REDUCING TEE   STRAIGHT TEE WITH INSERT (TYPE 2)

TEES

206F05.EPS

Figure 5 ◆ Types of reducers.

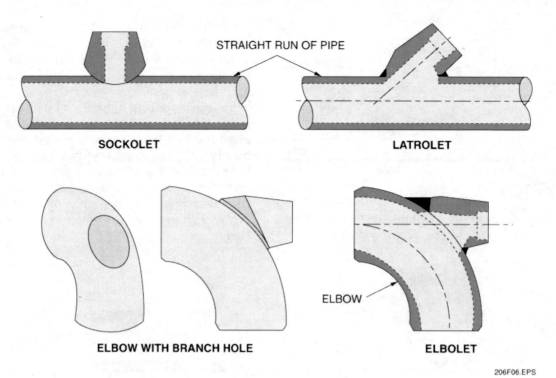

SOCKOLET

LATROLET

STRAIGHT RUN OF PIPE

ELBOW WITH BRANCH HOLE

ELBOW

ELBOLET

206F06.EPS

Figure 6 ◆ Special socket weld branch fittings.

4.0.0 ◆ SOCKET WELD PIPING DRAWINGS

Remember the three types of drawings used to show piping systems:

- Double-line
- Single-line
- Isometric

These drawings can be either plan or elevation drawings. Plan drawings are drawn from the top of the system looking down on it. Elevation drawings are drawn from the side of the system.

4.1.0 Double- and Single-Line Drawings

A double-line drawing of a socket weld system is the easiest to understand. Looking at *Figure 9*, you can see that the double-line drawing shows the sockets on the fittings, just as they were shown on the threaded pipe drawings.

The single-line drawing in *Figure 9* shows the same system in socket weld symbols. You can see that the sockets are shown as openings around the single line of the pipe as it enters the fitting. This is the way that socket weld fittings are represented in single-line drawings.

Figure 10 shows a plan drawing, or top view, of the same system. If you compare the two drawings, you can see the way that the pipe and fittings are assembled in three dimensions.

4.2.0 Isometric Drawings

Piping systems must often be shown from both the elevation, or side, and the plan, or top, views. This can be done in an isometric drawing. *Figure 11* shows an isometric drawing of the same system with a bill of material for the system.

4.3.0 Piping Symbols

Fittings made with socket weld joints are shown by standard symbols. *Figure 12* shows socket weld fitting symbols. Observe the shapes of the connections, showing that these valves and fittings are socket-weld, rather than butt-weld or threaded.

4.4.0 Line Numbers and Specifications Book

The line number in *Figure 13* is an example of a socket-weld pipe line number. It does not tell what kind of connection is to be used. That information is contained in the specification, shown in *Table 2*. The general description on *Table 2* describes the pipe as ASTM-A-197 SW CS. The ASTM-A-197 describes the material standard of the pipe; SW tells the pipefitter that the connections are socket-weld; and the CS stands for carbon steel.

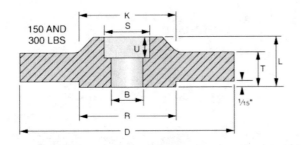
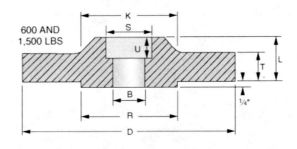

Figure 7 ◆ Socket weld flanges and dimensions.

Pipe Size	150-LB STANDARD				300-LB STANDARD				600-LB STANDARD				Pipe Size
	D	T	L	K	D	T	L	K	D	T	L	K	
¼	3½	7/16	5/8	1 3/16	3¾	9/16	7/8	1½	9/16	3¾	7/8	1½	¼
⅜	3½	7/16	5/8	1 3/16	3¾	9/16	7/8	1½	9/16	3¾	7/8	1½	⅜
½	3½	7/16	5/8	1 3/16	3¾	9/16	7/8	1½	9/16	3¾	7/8	1½	½
¾	3⅞	½	5/8	1½	4⅝	5/8	1	1⅞	5/8	4⅝	1	1⅞	¾
1	4¼	9/16	11/16	1 15/16	4⅞	11/16	1 1/16	2⅛	11/16	4⅞	1 1/16	2⅛	1
1¼	4⅝	5/8	13/16	2 5/16	5¼	¾	1 1/16	2½	13/16	5¼	1⅛	2½	1¼
1½	5	11/16	7/8	2 9/16	6⅛	13/16	1 3/16	2¾	7/8	6⅛	1¼	2¾	1½
2	6	¾	1	3 1/16	6½	7/8	7/8	3 15/16	1	6½	1 7/16	3 5/16	2
2½	7	7/8	1⅛	3 9/16	7½	1	1	3 15/16	1⅛	7½	1⅝	3 15/16	2½
3	7½	15/16	1 3/16	4¼	8¼	1⅛	1 1/16	4⅝	1¼	8¼	1 13/16	4⅝	3

Pipe Size	1,500-LB STANDARD				DIMENSIONS COMMON TO 150-, 300-, 600-, AND 1,500-LB FLANGES				Pipe Size
	D	T	L	K	S	B	R	U	
¼	-	-	-	-	0.58	0.36	1⅜	⅜	¼
⅜	-	-	-	-	0.71	0.49	1⅜	⅜	⅜
½	4¾	7/8	1¼	1½	0.88	0.62	1⅜	⅜	½
¾	5⅛	1	1⅜	1¾	1.09	0.82	1 11/16	7/16	¾
1	5⅞	1⅛	1⅝	2 1/16	1.36	1.05	2	½	1
1¼	6¼	1⅛	1⅝	2½	1.70	1.38	2½	9/16	1¼
1½	7	1¼	1¾	2¾	1.95	1.61	2⅞	5/8	1½
2	8½	1½	2¼	4⅛	2.44	2.07	3⅝	11/16	2
2½	9⅝	1⅝	2½	4⅞	2.94	2.47	4⅛	¾	2½
3	-	-	-	-	3.57	3.07	5	13/16	3

206F07.EPS

Table 2 Sample Piping Specification Sheet

Piping Specification	
Service: Water	Class: ANSI 150#
Design Pressure: 150 psig	Corr. Allow: 0.030"
Max Pressure: 200 psig	Temp Limit: 200°F
Item: Size	**General Description**
Pipe: ½" to 2"	ASTM-A-197 SW CS
Fittings: ½" to 2"	150# SW ASTM-A-197 CS
Valves: ½" to 2"	150# SWCS ASTM-A-105

206T02.EPS

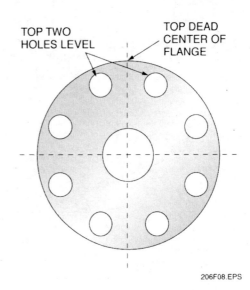

Figure 8 ◆ Two-hole method of leveling flange.

206F08.EPS

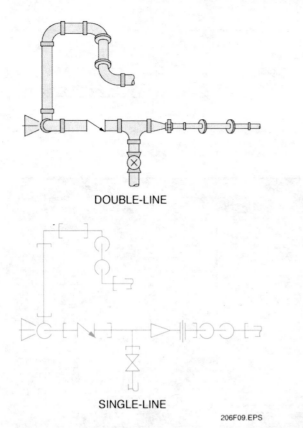

DOUBLE-LINE

SINGLE-LINE

206F09.EPS

Figure 9 ◆ Double- and single-line drawings: elevation view.

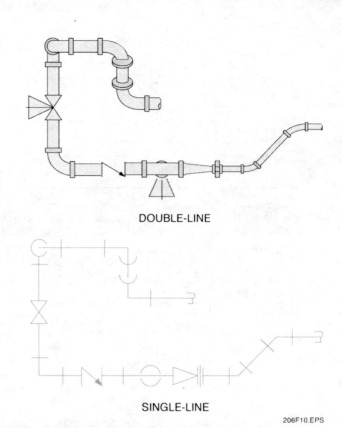

DOUBLE-LINE

SINGLE-LINE

206F10.EPS

Figure 10 ◆ Plan drawing, top view.

BILL OF MATERIALS			
P.M.	REQ'D	SIZE	DESCRIPTIONS
1		1½"	PIPE SCH/40 ASTM-A-120 GR. B CS
2		¾"	PIPE SCH/40 ASTM-A-120 GR. B CS
3	5	1½'	90° ELL SCREWED ASTM-A-197 CS
4	1	1½"	TEE SCREWED ASTM-A-197 CS
5	2	¾"	45° ELL SCREWED ASTM-A-197 CS
6	1	1½" × ¾"	CONC. SWAGE SW ASTM-A-197 CS
7	1	¾"	UNION SCREWED ASTM-A-197 CS
8	2	1½"	GATE VA. SW CARBON STEEL ASTM-A-105
9	1	1½"	CHECK VA. SWING ASTM-A-105

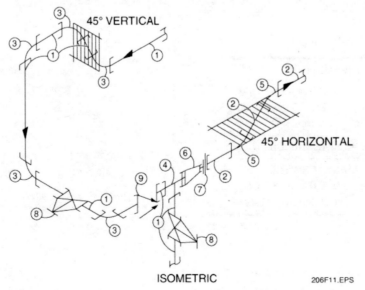

45° VERTICAL

45° HORIZONTAL

ISOMETRIC

206F11.EPS

Figure 11 ◆ Isometric drawing.

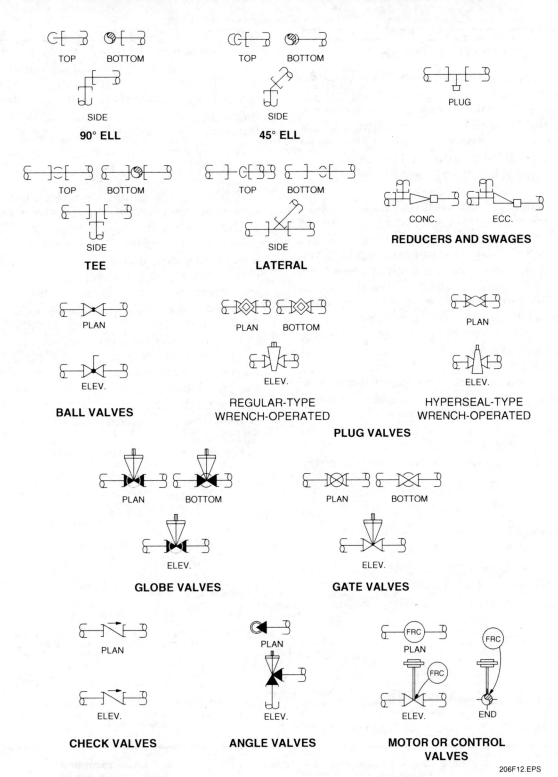

Figure 12 ◆ Socket weld fitting symbols.

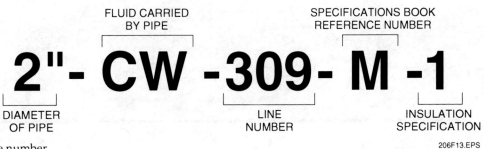

Figure 13 ◆ Line number.

206F13.EPS

5.0.0 ◆ DETERMINING PIPE LENGTHS BETWEEN FITTINGS

As you have learned, knowing the takeout for the fittings is essential when figuring the lengths of straight pipe. The takeout of a socket weld fitting refers to the laying length of the fitting, or the distance from the back of the socket to the center of the fitting. *Table 3* lists the center-to-end dimensions and laying lengths of common socket weld fittings.

The required gap between the end of the pipe and the back of the socket must also be considered when determining pipe lengths between fittings. There must be a ¹⁄₁₆-inch minimum gap between the end of the pipe and the back of the socket to allow for thermal expansion when the pipe is welded. The gap is usually set at ⅛ inch, with the exception of high-pressure steam applications. *Figure 14* shows the proper placement of the pipe in the fitting.

There are two ways normally used for setting the gap between the back of the socket and the end of the pipe: measure and move, and shimming. In the first case, you push the pipe as far in as it will go, scribe a mark where the pipe enters the fitting, and then slide the pipe out so that the mark is the required gap distance from the face of the fitting. In shimming, the appropriate gap is held by a piece of material having the correct thickness, such as a piece of matchstick or similar material. There is a spring ring insert, called a Gap-A-Let, that is placed inside the socket and is designed to maintain the minimum required gap by seating the end of the pipe against it.

5.1.0 Center-to-Center Method

When using the center-to-center method, measurements are taken from the center of one fitting to the center of the next fitting. *Figure 15* shows a socket-weld piping drawing showing the center-to-center dimension. The actual length of pipe between the fittings is determined by subtracting the laying length of each of the fittings and the required gap for each weld on each fitting. *Figure 16* shows the center-to-center method for determining the length of a pipe between two fittings.

The only difference in applying the center-to-center method in socket-weld pipe is that the calculation of length of pipe between fittings must include the required gap for the pipe in the socket. The measurement of the takeout of the fitting is the distance from the gap to the center of the fitting.

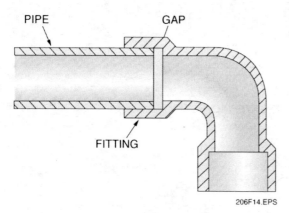

Figure 14 ◆ Proper placement of pipe in fitting.

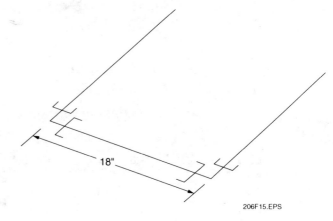

Figure 15 ◆ Center-to-center piping drawing.

Table 3 Center-to-End Dimensions and Laying Lengths

Size of Fittings	Center-to-End Dimensions (inches)						Laying Lengths (inches)					
	½	¾	1	1¼	1½	2	½	¾	1	1¼	1½	2
2,000-lb	1⅛	1⁵⁄₁₆	1½	1¾	2	2⅜	⅝	¾	⅞	1¹⁄₁₆	1¼	1½
3,000-lb	1⅛	1⁵⁄₁₆	1½	1¾	2	2⅜	⅝	¾	⅞	1¹⁄₁₆	1¼	1½
6,000-lb	1⁵⁄₁₆	1½	1¾	2	2⅜	2½	¾	⅞	1	1¼	1½	1⅝

206T03.EPS

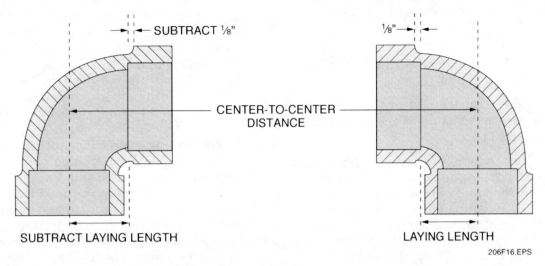

Figure 16 ◆ Center-to-center method as applied to socket-weld piping.

5.2.0 Center-To-Face Method

The center-to-face method is used when the drawing gives the length from the center of a fitting, such as a 90-degree ell or a tee, to the face of a flange. To find the length of pipe needed, subtract the laying lengths of the fitting and the flange and the gap required at each weld. *Figure 17* shows the center-to-face method as applied to socket-weld fittings and flanges.

5.3.0 Face-to-Face Method

As you learned in the previous module, to find the length of pipe needed, subtract the laying lengths of both flanges and the ⅛-inch gap required at each weld. *Figure 18* shows the face-to-face method in socket-weld flanges.

6.0.0 ◆ FABRICATING SOCKET WELD FITTINGS TO PIPE

When performing socket weld pipe fabrication, you must be sure to cut the end of the pipe square, even though the end of the pipe will be inside the fitting socket. This ensures that the socket covers the proper amount of the pipe end at all points around the pipe. The end of the pipe does not have to be beveled, but all large internal and external burrs must be removed from the pipe.

The two duties that a pipefitter must perform to fabricate socket weld pipe fittings to pipe include preparing the pipe and fittings for alignment and aligning fittings to be welded. As a pipefitter, you will not be expected to actually perform the weld. The welder works with you, tack-welding several fits. Then, as you cut and prepare pipe for more fits, the welder goes back and welds out the fittings that have been tacked.

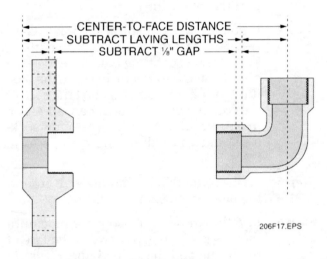

Figure 17 ◆ Center-to-face method.

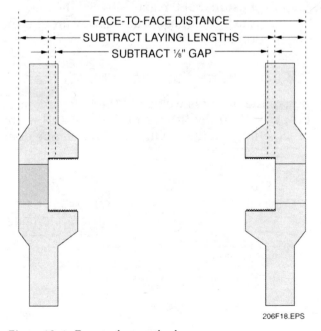

Figure 18 ◆ Face-to-face method.

6.1.0 Preparing Pipe and Fittings for Alignment

Follow these steps to prepare pipe and fittings for alignment:

Step 1 Ensure that the end of the pipe is cut square and that there are no internal or external burrs.

Step 2 Clean the end of the pipe and the end of the fitting to remove any scale, rust, grease, paint, or other contamination or surface coating material that could reach the weld area. Make sure to clean the pipe and fitting to at least one inch from the end of the fitting after the fit.

Step 3 Secure the pipe in a pipe vise and support the pipe with a pipe jack.

Step 4 If you are using a shim, such as a Gap-A-Let, place it in the socket of the fitting. Place the socket fitting onto the end of the pipe. Push the socket fitting onto the pipe until the pipe is seated against the back of the socket. If you are using a Gap-A-Let, make sure that the pipe is evenly socketed, and you will not need to perform Steps 5, 6, and 7.

Step 5 Scribe or mark a line all the way around the pipe right next to the fitting.

Step 6 Pull the fitting away from the pipe until there is a ⅛-inch gap between the face of the fitting and the scribed line. *Figure 19* shows proper placement of the fitting.

Step 7 Ensure that the gap between the line and the fitting is equal all the way around the pipe.

6.2.0 Aligning Fittings to be Welded

Before the fitting is welded to the pipe, you must properly align the fittings to the pipe and to the other fittings in the piping run. This can be done using squares or levels. Many of the alignment procedures require you to use both squares and levels. Squares provide more accurate alignments in the field and in other locations where the working conditions are not perfect. Perfect working conditions include being indoors with smooth, level floors and having access to accurate, adjustable jacks and stands. There is also a higher risk of error in using levels because levels get dropped, have weld splattered on them, and are abused due to normal wear and tear. If a level is dropped or abused, it will no longer read true. Therefore, a square gives a more accurate reading of level. When aligning fittings to be welded, it is critical that you properly support the fabricated spool so that the fittings are not disturbed before or during weld-out. This section explains the procedures for aligning fittings to be welded using squares.

6.2.1 Aligning 90-Degree Elbow to Pipe

Follow these steps to align a 90-degree elbow to pipe. This procedure assumes there is no other fitting attached to the pipe at this time.

Step 1 Measure and cut a 2-inch carbon steel pipe 3 feet long.

Step 2 Follow the steps to prepare the pipe end and position the elbow on the pipe. The elbow should be positioned on the pipe so that the elbow turns straight up.

Step 3 Have the welder tack-weld the elbow to the pipe in the throat of the fitting at the 12-o'clock position (*Figure 20*).

Step 4 Place a square over the fitting so that the long leg of the square is aligned with the pipe (*Figure 21*).

Step 5 Adjust the fitting on the pipe until the distance between the pipe and the long leg of the square is the same for the entire length of the square, as shown in *Figure 22*.

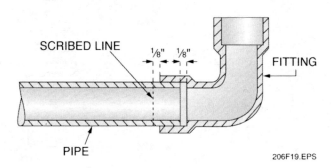

Figure 19 ◆ Proper placement of fitting.

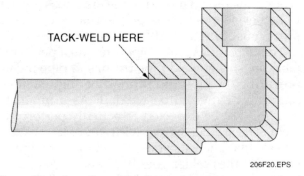

Figure 20 ◆ Location of the first tack-weld.

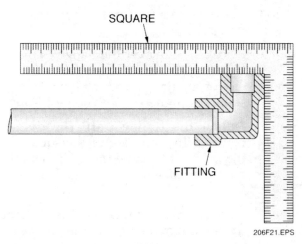

Figure 21 ◆ Position of the square.

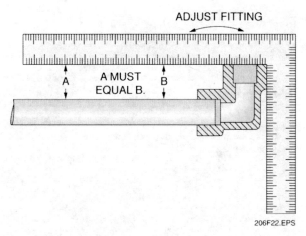

Figure 22 ◆ Adjusting fitting.

NOTE

The elbow can also be leveled from front to back using a torpedo level, as long as the pipe is level in the pipe vise. Always remove the level before the fitting is tack-welded.

Step 6   Have the welder tack-weld the fitting to the pipe at the 6-o'clock position, align the fitting from side to side, and have the welder tack the 3-o'clock and 9-o'clock positions.

6.2.2 Squaring Pipe into 90-Degree Elbow

In this procedure, you will add a 3-foot length of pipe into the elbow fabricated in the previous section. To perform this procedure, the pipe and elbow should be secured in the pipe vise with the elbow turned straight up. The two pipes connected by the elbow must form a true 90-degree angle. Prepare and tack the pipe to the elbow as before. Align the pipe and fitting using the following steps:

Step 1   Hold a square against the pipe so that the long leg of the square is aligned with the pipe in the vise, as shown in *Figure 23*.

Step 2   Adjust the pipe in the elbow until the distance between the pipe and the long leg of the square is equal for the entire length of the square (*Figure 24*).

Step 3   Have the welder tack-weld the fitting to the pipe opposite to the first tack-weld position. Align the fitting from side to side, and have the welder tack the other sides.

6.2.3 Aligning Flange to Pipe

To align a flange to pipe, you must align the flange horizontally, using the two-hole method, and vertically, using a square. For this exercise, you will

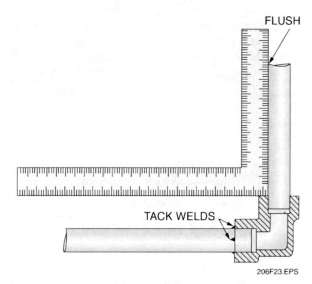

Figure 23 ◆ Position of the square to the pipe.

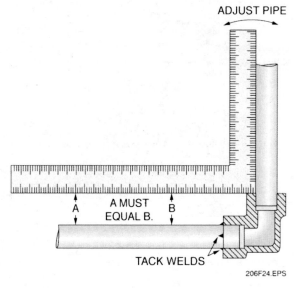

Figure 24 ◆ Adjusting the pipe.

fit a flange to the pipe that is still in the pipe vise from the previous section. Follow the same steps as earlier to prepare the pipe and flange and position the new flange on the pipe.

> **NOTE**
>
> A standard flange has a bore (inside diameter) measurement that is the same as Schedule 40 or standard weight pipe. If you are ordering flanges or taking them from stock and you do not specify the bore, you will receive standard bore. If you are using Schedule 80 or 120 pipe, you will not be able to make a smooth interior alignment because the inside diameters will be different. For example, at 2-inch nominal pipe size, the Schedule 40 bore would be 2.07 inches, the Schedule 80 bore would be 1.94 inches, and the Schedule 160 bore would be 1.69 inches. In the case of the Schedule 160 pipe, the pipe inside diameter is ⅓ of an inch smaller than that of the standard flange. Be sure to consider the bore difference when ordering and choosing flanges for either socket weld or butt weld applications.

Step 1 Plumb the vertical pipe in the pipe vise from side to side (*Figure 25*).

Step 2 Use the two-hole method to align the flange to the pipe (*Figure 26*).

Step 3 Have the welder tack-weld the sides of the flange to the pipe at the 3-o'clock and the 9-o'clock positions.

Step 4 Remove the two-hole pins from the flange.

Step 5 Place a square over the flange so that the short leg of the square is flush with the face of the flange and the long leg of the square is aligned over the pipe (*Figure 27*).

Step 6 Adjust the flange until the distance between the long leg of the square and the pipe is the same the entire length of the square. *Figure 28* shows adjusting the flange.

Step 7 Have the welder tack-weld the top and the bottom of the pipe to the flange.

206F26.EPS

Figure 26 ◆ Two-hole method of aligning a flange.

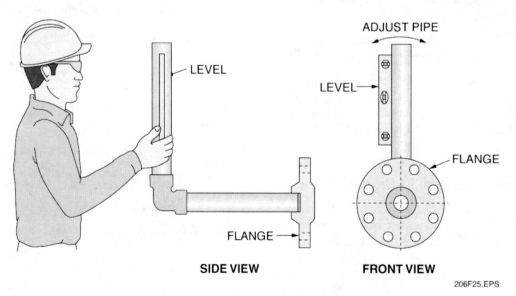

206F25.EPS

Figure 25 ◆ Plumbing a vertical pipe.

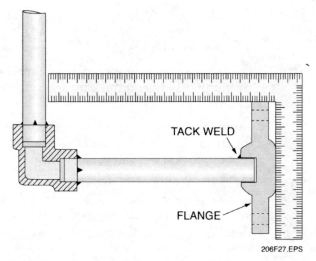

Figure 27 ◆ Positioning the square on the flange.

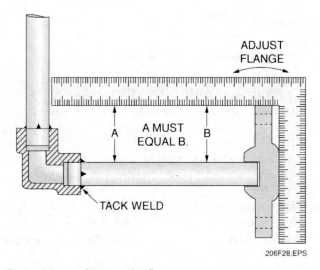

Figure 28 ◆ Adjusting the flange.

6.2.4 Aligning 45-Degree Elbow to Pipe Using Levels

A 45-degree elbow can be aligned to pipe, using two squares or a torpedo level with a 45-degree vial. Follow the same initial procedure as in fitting a 90-degree elbow to a pipe, and have the first tack made at the 12-o'clock position on the pipe. Then use the following steps:

Step 1 Place a torpedo level with a 45-degree vial on the face of the fitting and adjust the fitting until its face is at a true 45-degree angle (*Figure 29*).

Step 2 Remove the level from the fitting, and have the welder tack-weld the bottom of the fitting to the pipe. Check the alignment, and have the welder place the other two tacks.

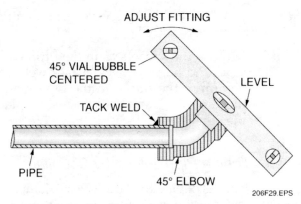

Figure 29 ◆ Checking a fitting for a true 45-degree angle.

6.2.5 Aligning 45-Degree Elbow Using Squares

Two squares can also be used to align a 45-degree elbow to the end of a straight pipe. Follow the same initial procedure as in fitting a 90-degree elbow to a pipe, and have the first tack made at the 12-o'clock position on the pipe. Then use the following steps to align the elbow:

Step 1 Place the short leg of the square on the top center of the straight pipe so that the long leg of the square is vertically in line with the throat of the elbow.

Step 2 Place the long leg of the second square on the face of the elbow (*Figure 30*).

Step 3 Adjust the elbow so that the same inch marks on the short and long legs of the second square touch the inner edge of the first square (see *Figure 30*).

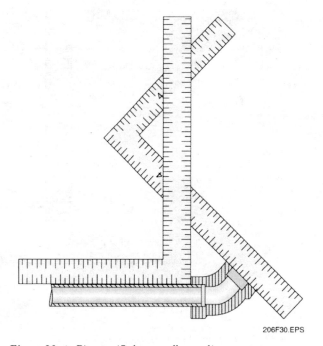

Figure 30 ◆ Pipe to 45-degree elbow alignment.

Step 4 Have the welder tack-weld the bottom of the pipe to the elbow.

Step 5 Remove the squares from the fitting and pipe, and have the welder tack-weld the bottom of the fitting to the pipe. Check the alignment, and have the welder place the other two tacks.

6.2.6 Aligning Pipe Joined by Couplings

Straight lengths of pipe the same size are joined by socket weld couplings. To fit a coupling, prepare the pipe and coupling as you have been doing up to now. Have the coupling tacked to the pipe at the 12-o'clock position. Align the coupling to the first pipe using the square, as you did with the flange, and have the welder complete the tacks (*Figure 31*).

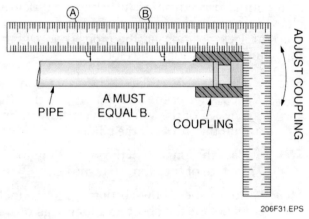

Figure 31 ◆ Positioning the square on the coupling.

Step 1 Place the second length of pipe on jack stands in line with the first pipe and align it for the first tack-weld.

Step 2 Position a square on each pipe so that the coupling is between the long legs of each square (*Figure 32*).

Step 3 Adjust the second pipe up or down until the distance is the same between the long legs of the squares at every point.

Step 4 Have the welder tack-weld the coupling to the pipe at the 6-o'clock position. Rotate the pipes 90 degrees in the jack stands, and repeat the alignment with the squares.

Step 5 Have the welder tack-weld the coupling to the pipe at the new 12- and 6-o'clock positions.

6.3.0 Installing Welded Valves

Welded valves are available in various steel alloys and are used mainly for high-pressure, high-temperature systems that do not require frequent dismantling. The two types of welded joints used are butt weld joints and socket weld joints. Butt weld joints are available in all sizes, while socket weld joints are limited to the smaller sizes, usually less than 2 inches.

Heat from welding can distort the metal of a valve. Before installing a welded valve, all components of the valve that may be damaged by the

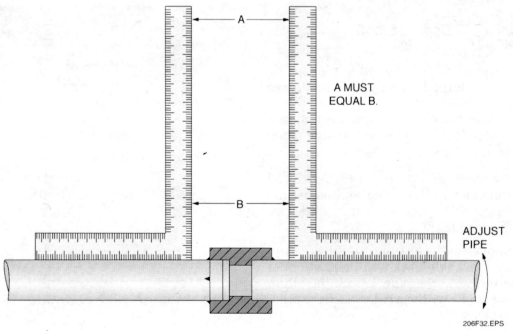

Figure 32 ◆ Position of squares on pipes.

heat must be removed. These components include fiber or rubber packing and Teflon® seats. The valve must also be fully open before it is welded into place. The heat from the weld can warp the valve seats and cause the closing mechanism to mate unevenly with the seat. The welder may have to weld only a section of the joint and then allow the valve to cool before continuing the weld. The pipefitter's responsibilities include preparing the pipe ends and fitting the valve to the pipe.

6.3.1 Fitting Socket Weld Valves

Step 1 Prepare the pipe and valve just as you did with the coupling, and open the valve to its fully open position.

Step 2 Place the socket valve onto the end of the pipe, with the valve stem pointing straight up, ensure proper gap, and have the welder make the first tack-weld (*Figure 33*).

Step 3 Place a square over the valve, with the long leg of the square aligned with the pipe and the short leg of the square flush against the valve end (*Figure 34*).

Step 4 Adjust the valve until the distance between the long leg of the square and the pipe is the same the entire length of the square (*Figure 35*).

Step 5 Have the welder tack-weld the valve to the pipe in the 6-o'clock position.

Step 6 Rotate the framing square on the valve so that the long leg of the square is at the center line of the side of the pipe.

Step 7 Repeat step 4 with the square in this position.

Step 8 Have the welder tack-weld the valve to the pipe at the 3- and 9-o'clock positions.

Step 9 Prepare, insert, and get a first tack-weld on the other pipe in the open side of the valve, and, just as you did with the coupling, align the valve and pipes with two squares (*Figure 36*).

Step 10 Have the welder make the other tacks.

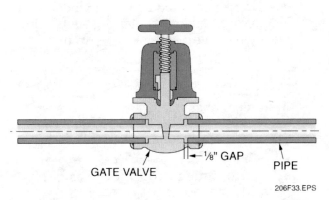

Figure 33 ◆ Proper placement of valve.

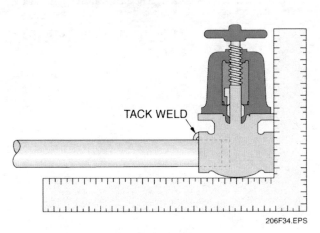

Figure 34 ◆ Positioning a square on valves.

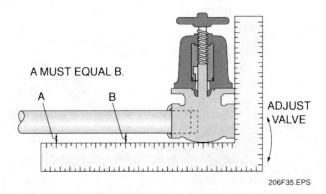

Figure 35 ◆ Adjusting the valve.

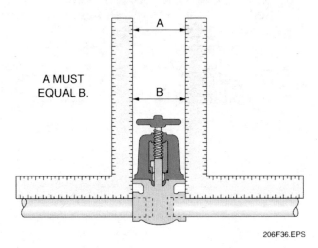

Figure 36 ◆ Position of the squares on the pipe.

1. What pipe material is most commonly used in socket weld applications?
   a. Galvanized steel
   b. Copper
   c. Carbon steel
   d. Aluminum

2. The maximum pipe size used in socket weld applications is _____.
   a. 3 inches
   b. 4½ inches
   c. 6 inches
   d. 10 inches

3. Pressure classes of pipe for socket weld fittings do *not* include _____ pound.
   a. 2,000
   b. 3,000
   c. 4,000
   d. 6,000

4. Which of the following is *not* one of the most common socket weld fittings?
   a. Tee
   b. Ell
   c. Latrolet
   d. Cross

5. In socket weld applications, a swage is a _____.
   a. socketless reducer
   b. flaring tool
   c. special elbow
   d. union

6. The welded coupling that makes a 45-degree reducing branch from a straight run of pipe is a(n) _____.
   a. reducing tee
   b. latrolet
   c. sockolet
   d. elbolet

7. Which of the following is *not* a class of flange?
   a. 150 pound
   b. 300 pound
   c. 1,500 pound
   d. 3,000 pound

8. The purpose of two-holing a flange is to _____.
   a. reduce the number of holes
   b. align the bolt holes
   c. increase the number of holes
   d. set the face at perpendicular to the pipe

9. To find the maximum pressure for a part of the system shown on a drawing, look in the _____.
   a. title block
   b. specifications
   c. pipe markings
   d. schematics

10. The length of the pipe between socket weld ells can be calculated by finding the design length from center to center and subtracting the laying length and the _____ from it.
    a. minimum gap
    b. takeout
    c. pipe diameter
    d. number of ells

11. The laying length of a socket weld fitting is the length _____.
    a. from the fitting face to the end of the pipe inside the socket
    b. of the fillet
    c. from the face of the socket to the very back of the socket
    d. from the center of the ell to the bottom of the socket

12. A Gap-A-Let is used to set a standard gap length from _____.
    a. the back of the socket to the face of the socket.
    b. the back of the socket to the end of the pipe inside
    c. the outside of the pipe to the inside of the socket
    d. one pass of the fillet to the next pass

13. The center-to-center method is used to determine the correct length of cut pipe between two _____.
    a. valves
    b. flanges
    c. fittings
    d. pipe caps

14. The face-to-center method is used to determine the correct length of cut pipe between _____.
    a. valves and fittings
    b. caps and flanges
    c. two fittings
    d. a fitting and a flange

15. The face-to-face method is used to determine the correct length of cut pipe between _____.
    a. valves and fittings
    b. two fittings
    c. two flanges
    d. a fitting and a flange

Summary

Socket welding is a very common method used to join pipe in the field. The proper pipe end preparation and fit-up procedures are critical to producing a quality, leak-proof system. The purpose of this module is to teach you methods of aligning socket fittings to pipe that will be welded to produce quality joints. There are other methods to perform many of these alignment procedures that you will encounter in the field. The most important thing to remember is that regardless of the procedure you use to align the fitting, the end result must be a fit-up that is square and level. Through practice, you will be able to rapidly and properly fabricate socket weld piping systems.

Notes

Concentric reducer: A reducer that maintains the same center line between the two pipes that it joins.

Cross: A fitting with four branches all at right angles to each other.

Eccentric reducer: A reducer that displaces the center line of the smaller of the two joining pipes to one side.

Fillet weld: A weld with a triangular cross section joining two surfaces at right angles to each other.

Flange: A form of attachment between pipes consisting of a ring perpendicular to the line of pipe, with holes drilled through the ring parallel to the pipe line, through which bolts may be used to attach the pipe or fitting to another flanged pipe or fitting.

Insert: A type of reducer that fits into the socket of a fitting to reduce the line size.

Tack-weld: A weld made to hold parts together in proper alignment until the final weld is made.

Straight tee: A fitting that has one side outlet 90 degrees to the run.

Swage: A type of reducer.

Resources & Acknowledgments

Additional Resources

This module is intended to be a thorough resource for task training. The following reference work is suggested for further study. This is optional material for continued education rather than for task training.

The Pipe Fitters Blue Book. W.V. Graves. Webster, TX: W.V. Graves Publishing Company.

Figure Credits

Mathey Dearman, 206F26

Cianbro Corporation, Module Opener

NCCER CURRICULA — USER UPDATE

NCCER makes every effort to keep its textbooks up-to-date and free of technical errors. We appreciate your help in this process. If you find an error, a typographical mistake, or an inaccuracy in NCCER's curricula, please fill out this form (or a photocopy), or complete the online form at **www.nccer.org/olf**. Be sure to include the exact module ID number, page number, a detailed description, and your recommended correction. Your input will be brought to the attention of the Authoring Team. Thank you for your assistance.

Instructors – If you have an idea for improving this textbook, or have found that additional materials were necessary to teach this module effectively, please let us know so that we may present your suggestions to the Authoring Team.

NCCER Product Development and Revision
13614 Progress Blvd., Alachua, FL 32615

Email: curriculum@nccer.org
Online: www.nccer.org/olf

❏ Trainee Guide   ❏ Lesson Plans   ❏ Exam   ❏ PowerPoints   Other _____

Craft / Level: _____   Copyright Date: _____

Module ID Number / Title: _____

Section Number(s): _____

Description: _____

_____

_____

_____

_____

Recommended Correction: _____

_____

_____

_____

_____

Your Name: _____

Address: _____

_____

Email: _____   Phone: _____

08207-06

Butt Weld Pipe Fabrication

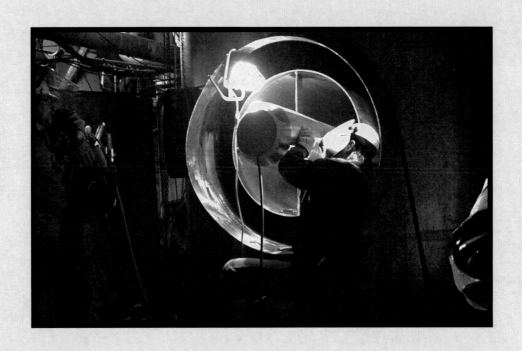

08207-06
Butt Weld Pipe Fabrication

Topics to be presented in this module include:

Overview

Most larger industrial piping in aboveground applications are a combination of butt weld and boltups. The skills in this module are required in working in the oil and chemical industries, in power plants, and in aboveground pipeline work. This is more difficult fitting than socket weld, because the alignment of the pipe ends is so critical to proper flow. You will learn about the tools and jigs you will use to get the alignment correct for the first tack, and how to determine and adjust for small differences in the actual shapes and sizes of pipe and fittings. The drawing symbols for butt weld assemblies are here, and again the fitter must do the fit carefully and exactly, or the welder's job won't be right.

Objectives

When you have completed this module, you will be able to do the following:

1. Identify butt weld piping materials and fittings.
2. Read and interpret butt weld piping drawings.
3. Prepare pipe ends for fit-up.
4. Determine pipe lengths between fittings.
5. Select and install backing rings.
6. Perform alignment procedures for various types of fittings.

Trade Terms

Align	Fit up
Alloy	Full-penetration weld
ASTM International	Lateral
Bevel	Oxide
Burn-through	Root opening
Chamfer	

Required Trainee Materials

1. Pencil and paper
2. Appropriate personal protective equipment

Prerequisites

Before you begin this module, it is recommended that you successfully complete *Core Curriculum*; *Pipefitting Level One*; and *Pipefitting Level Two*, Modules 08201-06 through 08206-06.

This course map shows all of the modules in the second level of the *Pipefitting* curriculum. The suggested training order begins at the bottom and proceeds up. Skill levels increase as you advance on the course map. The local Training Program Sponsor may adjust the training order.

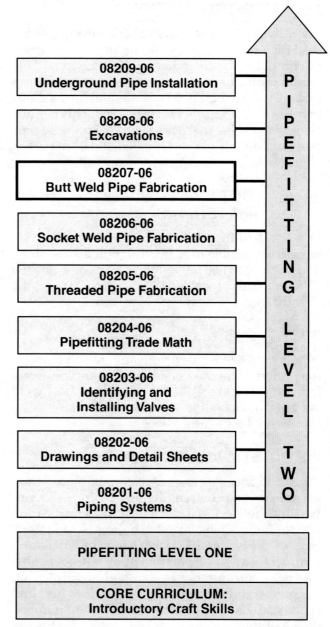

207CMAP.EPS

1.0.0 ◆ INTRODUCTION

The strength, availability, and versatility of welded steel pipe have made it common in a wide variety of commercial and industrial applications. This module introduces the types of pipe and fittings that are used in butt weld systems and the methods of fitting and aligning butt weld pipe. This module also explains that the mating ends of butt weld pipe must be properly cut and beveled and be in perfect alignment with each other for proper fit-up.

2.0.0 ◆ BUTT WELD PIPING MATERIALS

Pipes and fittings used in butt weld piping systems are available in many different alloys, sizes, and strengths. The following sections describe the different sizes of welded piping materials and the classifications of each type of material.

2.1.0 Pipe Sizes

A butt weld can be used with any size pipe, and is the preferred method of joining larger pipe. Common wall thicknesses of pipe used in butt weld systems are Schedule 40, 80, and 160. *Table 1* lists commercial pipe dimensions.

2.2.0 ASTM Identification System

Pipes and fittings are available in many different alloys, sizes, and strengths. All piping is identified by the ASTM International identification system for weldable metals. In the ASTM system, each metal is identified by a series of letters and numbers, called an identification number. This identification number is stamped or printed on the pipe and fittings. When the job drawings and specifications list a pipe by identification number, the pipefitter must use that kind of pipe for the job.

An example of an ASTM identification number is ASTM-A-120. If the metal contains iron, it is considered a ferrous material, and the identification number starts with an A. If the metal does not contain iron, the number begins with a B. After the first letter is a three-digit number that identifies the specific metal within the ferrous/nonferrous group.

The ASTM identification system is only one of several systems used by engineers. The other systems are by the American National Standard Institute (ANSI) and the American Society of Mechanical Engineers (ASME), each of which has its own numbering system. When selecting a pipe for any piping system, always check the drawings and specifications for the job, and then select a pipe with the correct identification number. The two standard types of steel used in welded piping systems are carbon steel and stainless steel.

2.2.1 Carbon Steel

Carbon steel pipe is widely used throughout the pipefitting industry in various applications because it is durable, machinable, and less expensive than most other types of pipe. Often the identification number for carbon steel pipe is followed by either Grade A or Grade B. Grade A refers to mild carbon steel; Grade B refers to high carbon steel. The grade of the pipe also refers to the mechanical properties of the pipe, the tensile strength and the yield strength. The tensile strength is the ultimate amount of stretching the steel can withstand without breaking. The yield strength is the maximum amount of stretching steel can withstand before it becomes permanently deformed or before it loses its ability to return to its original shape. Grade A steel has less carbon content, lower tensile strength, and lower yield strength than Grade B steel. Grade A steel is also more ductile, and therefore is better for cold bending and close coiling applications. Grade B is better for applications where pressure, structural strength, and collapse are factors, and it is easier to machine because of its higher carbon content.

2.2.2 Stainless Steel

Stainless steel is good for special applications that require the maximum resistance to corrosion, such as processing systems, instrument lines, and heat-transfer equipment. Stainless steel can also be kept sterile to prevent contamination of food, drugs, and dairy products.

Stainless steel pipe is made by alloying iron with a certain percentage of chromium and nickel. The alloys increase the resistance to corrosion, strengthen the steel, and make the steel able to withstand extreme temperatures. However, stainless steel is very expensive, so it is only used in critical applications that expose the pipe to high temperature, extreme low temperature, or high corrosion.

3.0.0 ◆ BUTT WELD FITTINGS

As you know, fittings are used in conjunction with straight pipe to change the direction of flow, change the pipe diameter in a system, close off a line, join two straight pipes, or allow a branch line to tie into the main pipe run. Butt weld fittings are sized the same as straight pipe as far as diameter and wall thickness, and end preparation is the same for fittings as it is for straight pipe.

3.1.0 Elbows and Return Bends

An elbow, also known as an ell, a 90, or a 45, is a fitting that forms an angle between connecting pipes. Elbows come in several different angles,

Table 1 Commercial Pipe Dimensions

Nominal Pipe Size	Outside Diam.	Nominal Wall Thickness												
		Sched. 10	Sched. 20	Sched. 30	STD	Sched. 40	Sched. 60	XS	Sched. 80	Sched. 100	Sched. 120	Sched. 140	Sched. 160	XXS
⅛	0.405	-	-	-	0.068	0.068	-	0.095	0.095	-	-	-	-	-
¼	0.540	-	-	-	0.088	0.088	-	0.119	0.119	-	-	-	-	-
⅜	0.675	-	-	-	0.091	0.091	-	0.126	0.126	-	-	-	-	-
½	0.840	-	-	-	0.109	0.109	-	0.147	0.147	-	-	-	0.188	0.294
¾	1.050	-	-	-	0.113	0.113	-	0.154	0.154	-	-	-	0.219	0.308
1	1.315	-	-	-	0.133	0.133	-	0.179	0.179	-	-	-	0.250	0.358
1¼	1.660	-	-	-	0.140	0.140	-	0.191	0.191	-	-	-	0.250	0.382
1½	1.900	-	-	-	0.145	0.145	-	0.200	0.200	-	-	-	0.281	0.400
2	2.375	-	-	-	0.154	0.154	-	0.218	0.218	-	-	-	0.344	0.436
2½	2.875	-	-	-	0.203	0.203	-	0.276	0.276	-	-	-	0.375	0.552
3	3.500	-	-	-	0.216	0.216	-	0.300	0.300	-	-	-	0.438	0.600
3½	4.000	-	-	-	0.226	0.226	-	0.318	0.318	-	-	-	-	-
4	4.500	-	-	-	0.237	0.237	-	0.337	0.337	-	0.438	-	0.531	0.674
5	5.563	-	-	-	0.258	0.258	-	0.375	0.375	-	0.500	-	0.625	0.750
6	6.625	-	-	-	0.280	0.280	-	0.432	0.432	-	0.562	-	0.719	0.864
8	8.625	-	0.250	0.277	0.322	0.322	0.406	0.500	0.500	0.594	0.719	0.812	0.906	0.875
10	10.750	-	0.250	0.307	0.365	0.365	0.500	0.500	0.594	0.719	0.844	1.000	1.125	1.000
12	12.750	-	0.250	0.330	0.375	0.406	0.562	0.500	0.688	0.844	1.000	1.125	1.312	1.000
14 OD	14.000	0.250	0.312	0.375	0.375	0.438	0.594	0.500	0.750	0.938	1.094	1.250	1.406	-
16 OD	16.000	0.250	0.312	0.375	0.375	0.500	0.656	0.500	0.844	1.031	1.219	1.438	1.594	-
18 OD	18.000	0.250	0.312	0.438	0.375	0.562	0.750	0.500	0.938	1.156	1.375	1.562	1.781	-
20 OD	20.000	0.250	0.375	0.500	0.375	0.594	0.812	0.500	1.031	1.281	1.500	1.750	1.969	-
22 OD	22.000	0.250	0.375	0.500	0.375	-	0.875	0.500	1.125	1.375	1.625	1.875	2.125	-
24 OD	24.000	0.250	0.375	0.562	0.375	0.688	0.969	0.500	1.218	1.531	1.812	2.062	2.344	-
26 OD	26.000	0.312	0.500	-	0.375	-	-	0.500	-	-	-	-	-	-
28 OD	28.000	0.312	0.500	0.625	0.375	-	-	0.500	-	-	-	-	-	-
30 OD	30.000	0.312	0.500	0.625	0.375	-	-	0.500	-	-	-	-	-	-
32 OD	32.000	0.312	0.500	0.625	0.375	0.688	-	0.500	-	-	-	-	-	-
34 OD	34.000	0.312	0.500	0.625	0.375	0.688	-	0.500	-	-	-	-	-	-
36 OD	36.000	0.312	0.500	0.625	0.375	0.750	-	0.500	-	-	-	-	-	-
42 OD	42.000	-	-	-	0.375	-	-	0.500	-	-	-	-	-	-

207T01.EPS

but the angle is always assumed to be 90 degrees unless otherwise stated. The 45- and 90-degree elbows are the most common fittings used in butt weld piping and are available in both long radius and short radius. Any other angle will likely be a special order item, or may be cut in the field from the standard 45- or 90-degree elbows. Long radius elbows are considered standard and should always be used unless the drawings specify short radius elbows. Reducing elbows are elbows that have different sized ends, to be used where the run changes sizes. *Figure 1* shows 90- and 45-degree long and short radius elbows.

A return bend is a U-shaped fitting that sends the fluid back in the same direction from which it came. They are also known as 180-degree returns and are also available in long and short radius. Return bends are often used in boilers, radiators, and other systems in which the pipe must pass through the same area several times. Return bends are manufactured in various sizes and come in close, open, and wide patterns (*Figure 2*).

3.2.0 Branch Connections

As you have seen, branch connections are fittings that divide the flow and send the fluid in two or more different directions. The branch connections used in butt weld pipe fabrication include tees, laterals, crosses, and saddles. Tees are the most

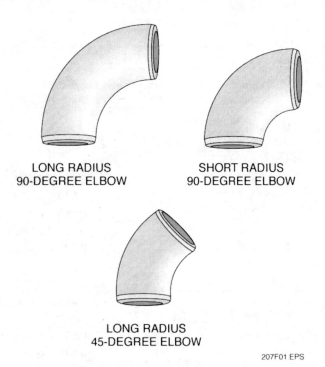

LONG RADIUS
90-DEGREE ELBOW

SHORT RADIUS
90-DEGREE ELBOW

LONG RADIUS
45-DEGREE ELBOW

207F01 EPS

Figure 1 ◆ Elbows.

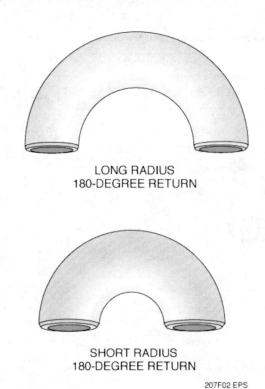

LONG RADIUS
180-DEGREE RETURN

SHORT RADIUS
180-DEGREE RETURN

207F02 EPS

Figure 2 ◆ Return bends.

widely used of the branch connections. They are used for making a 90-degree branch into the main pipe run. They can either be straight tees, meaning that all ends are the same diameter, or reducing tees, meaning that one end of the branch connection is smaller than the others. Tees are sized by listing the dimensions straight through the tee and then listing the size of the branch, such as 3 × 3 × 2.

Laterals are another type of common branch connections that have a side opening set at any degree other than a 90-degree angle. Laterals can also be either straight or reducing and are sized the same way as tees. Crosses are branch connections that serve as an intersection of four pipes. They can also be either straight or reducing. Weldolets are branch connections that are commonly used to run a reduced line off a header.

A saddle is used as a reinforcement pad to a branch connection in the main pipe run. The job specifications govern the use of saddles in branch connections. The saddle can be either one piece or two pieces, with one piece split down the middle. A leakage detection hole, also known as a weep hole or tell-tale hole, is drilled into the side of the saddle. When using a split saddle, each half of the saddle contains a weep hole. If there is an internal weld failure in the branch connection beneath the saddle, the fluid within the pipe will escape through the weep hole, allowing workers to detect the problem. The weep hole can be plugged so that service can continue until a proper repair can be made. *Figure 3* shows branch connections.

3.3.0 Caps

Caps are fittings that are butt-welded onto pipe ends to close off lines. Caps have the same pressure ratings as other fittings (*Figure 4*).

3.4.0 Reducers

Since straight pieces of pipe of the same size can be butt-welded together, couplings are not necessary in butt weld systems. Reducers (*Figure 5*) are used to join pipes of different sizes in butt weld systems in either a straight run of pipe or as part of another fitting, such as an elbow or a tee. Most directional-change and branch-connection fittings are available with reduced outlet lines. Reducers are made in eccentric and concentric types. Use an eccentric reducer when the top or the bottom of the pipe must be kept level. An eccentric reducer displaces the center line of the smaller pipe at the same time that it reduces the size of the pipe. Use a concentric reducer to maintain the same center line between two pipes.

3.5.0 Flanges

A flange is a ring-shaped plate that is attached to the end of a pipe. The flange has holes in it to allow it to be bolted to a mating flange on a pipe, fitting, or valve. Each flange has a specific pressure rating varying from 150 to 2,500 pounds. When you are welding flanges onto the end of a pipe, the face of

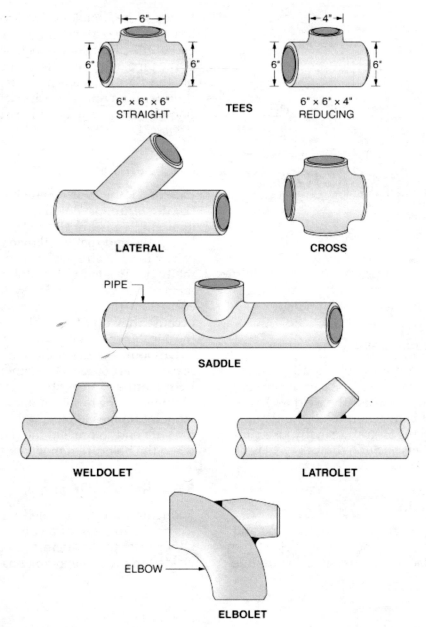

Figure 3 ◆ Branch connections.

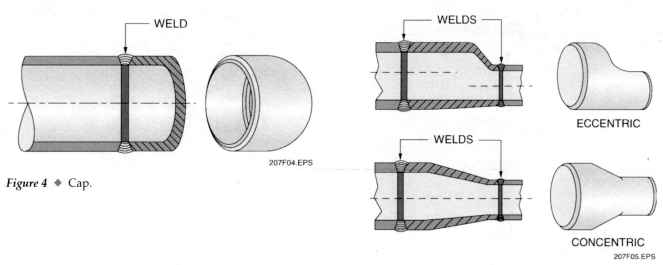

Figure 4 ◆ Cap.

207F04.EPS

Figure 5 ◆ Reducers.

207F05.EPS

Figure 6 ◆ Weld neck flange.

the flange must always be protected from welding arc damage. The most common types of flanges are weld neck, slip-on, and lap-joint flanges.

3.5.1 Weld Neck Flanges

Weld neck flanges are available in all pressure ratings and are commonly used when severe conditions are anticipated because the weld neck provides the strongest connection of all the flanges. Long weld neck flanges are normally used for vessel and tank outlets. *Figure 6* shows a weld neck flange.

3.5.2 Slip-On Flanges

Slip-on flanges slip over the end of the pipe and are fillet-welded to the pipe at the neck of the flange and at the face of the flange, allowing the flange to be double-welded to the pipe. Normal practice is to extend the flange face beyond the end of the pipe about ⅜ inch or the wall thickness of the pipe, whichever is greater, to keep the weld off the flange face. These flanges are normally used with low-pressure applications of 150 to 300 pounds. They may be either straight or reducing (*Figure 7*).

3.5.3 Lap-Joint Flanges

Lap-joint flanges, also known as Van Stone flanges (Figure 8), are used with lap-joint stub ends. Lap-joint stub ends are straight pieces of pipe with a lap on the end that the flange rests against. These can be used at all pressures, and the flange is normally not welded to the stub end, making alignment easier. When used in systems subject to high corrosion, the flanges can usually be salvaged because they do not actually come in contact with the substance flowing through the pipe.

Lap-joint flanges are often used for economic reasons. Carbon steel lap-joint flanges can be used with stainless steel stub ends in stainless steel piping systems to save money. If you are welding lap-joint flanges to pipe, weld a lug to each side of the stub end to prevent the flange from dropping when the bolts are removed.

3.5.4 Reducing Flanges

Reducing flanges are used where it is necessary to change the size of a run of pipe at the flange. Otherwise, the reducing flange is attached and welded in the same manner as any other flange.

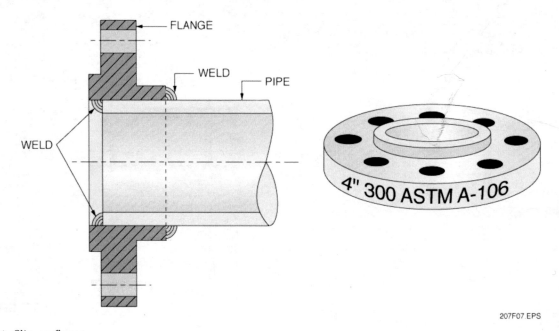

Figure 7 ◆ Slip-on flange.

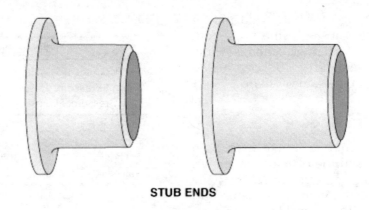

STUB ENDS

4" 300 ASTM A-106

LAP-JOINT FLANGE

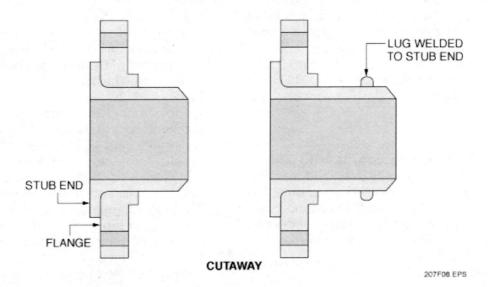

LUG WELDED
TO STUB END

STUB END

FLANGE

CUTAWAY

207F08.EPS

Figure 8 ◆ Lap-joint flange and stub end.

4.0.0 ◆ BUTT WELD PIPING DRAWINGS

As you have seen, the following three types of drawings are used to show piping systems:

- Double-line
- Single-line
- Isometric

Remember, these drawings can be either plan or elevation drawings.

4.1.0 Double- and Single-Line Drawings

Double-line drawings of a butt weld system show the system as it actually appears, with the weld shown as a line. Compare this with drawings of socket weld or threaded pipe, and you will see that the socket weld and threaded fittings are shown with a ring at the socket thicker than the line of the pipe.

A single-line drawing of a butt weld piping system shows the pipe as a single line, with a dot at the weld line. *Figure 9* is an elevation, section, or side view of the system. *Figure 10* shows a plan drawing, or top view, of the same system.

4.2.0 Isometric Drawings

Piping systems must often be shown from both the elevation, or side, and the plan, or top, views. This can be done in an isometric drawing. *Figure 11* shows an isometric drawing of the same butt weld system with a bill of material for the system.

4.3.0 Piping Symbols

Fittings made with butt weld joints are shown by standard symbols. Do not always depend on the symbol shown on the piping drawing to determine what kind of weld connection to make. Always double-check the specifications book to verify that the symbol on the drawing agrees with the specification. Symbols vary from job to job depending on the engineering company's specifications. Always refer to the symbol legend in the piping and instrumentation drawing (P&ID) when reading piping drawings. *Figure 12* shows typical butt weld fitting and valve symbols.

4.4.0 Line Numbers

Line numbers do not necessarily tell you what kind of connections are to be used, and may vary from job to job. *Figure 13* shows a sample line number. Notice that this sample line number does not state the type of material required. It does tell you what page of the specifications book holds the necessary information.

4.5.0 Specifications Book

The specifications book contains information about each pipe line. *Table 2* shows a sample page from a specifications book. Notice that the fittings are identified as 150# BW ASTM-A-197 CS. In this specification, the fitting is described as a 150-pound butt weld carbon steel fitting meeting the ASTM standard *A-197*.

5.0.0 ◆ PREPARING PIPE ENDS FOR FIT-UP

The pipe end and fitting must be prepared before a joint can be fitted up for welding. The two steps in preparing pipe ends include preparing the edges and cleaning the surfaces.

Table 2 Sample Piping Specification Sheet

Piping Specification	
Service: Steam	Class: ANSI 150#
Design Pressure: 150 psi	Corr. Allow: 0.030"
Max Pressure: 200 psi	Temp Limit: 200°F

Item: Size	General Description
Pipe: 4"	ASTM-A-120 CS
Fittings: 4"	150# BW ASTM-A-197 CS
Valves: 4"	150# Flanged Bronze ASTM-B-62

207T02.EPS

5.1.0 Preparing Edges

Pipe edges are prepared by either chamfering or beveling (*Figure 14*). A bevel is a 30-degree angle cut on the pipe end. A chamfer is an angle cut only on the edge of the pipe end. The flat edge of a pipe left after a chamfer is cut is called the land.

A chamfer is sometimes called a bevel, and a land is often called a root face. When using butt weld fittings, the pipe ends must be chamfered. Chamfering helps the welder get good penetration at the joint and produces a strong weld. The chamfer should extend to about 1/16 inch from the inside wall. The land that is left makes a wall that helps fit up the joint.

Pipes and fittings are often prepared in the shop and sent to the field for welding, but many times they must be prepared at the job site. Power bevellers are used to prepare pipe for butt weld fabrication. Bevellers place the correct angle, or bevel, on the end of a pipe before it is welded. There are many different ways to bevel the end of a piece of pipe to be butt-welded. The joint can be beveled mechanically, using grinders, nibblers, or cutters; or it can be beveled thermally, using an oxyacetylene cutting torch. Mechanical joint beveling is most often used on alloy steel, stainless steel, and nonferrous metal piping and is frequently required for the materials that could be affected by the heat of the thermal process. Mechanical beveling is slower than thermal beveling, but has the advantage of high precision with low heat and the absence of oxides commonly left by the thermal methods. The method used to bevel the pipe depends on the type of base material, the ease of use, and the code or procedure specifications.

5.1.1 Beveling Using Grinders

Portable hand grinders can be used to bevel pipe, but great care must be taken to make the bevel at the specified angle and keep the bevel uniform.

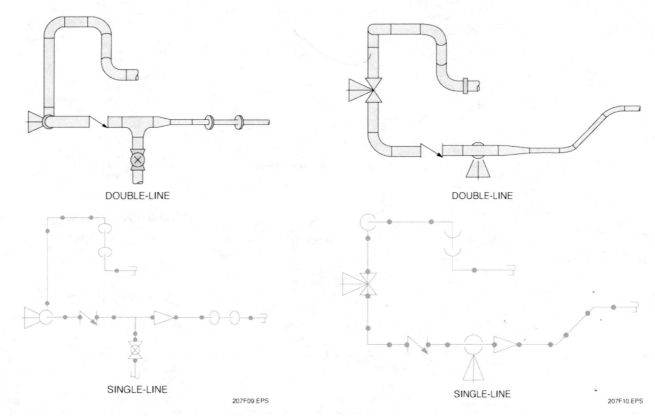

DOUBLE-LINE

DOUBLE-LINE

SINGLE-LINE

SINGLE-LINE

207F09.EPS

207F10.EPS

Figure 9 ◆ Double- and single-line drawings: elevation view.

Figure 10 ◆ Plan drawing, top view.

BILL OF MATERIALS			
P.M.	**REQ'D**	**SIZE**	**DESCRIPTIONS**
1		1½"	PIPE SCH/40 ASTM-A-120 GR. B
2		¾"	PIPE SCH/40 ASTM-A-120 GR. B
3	5	1½'	90° ELL ASTM-A-197 BW
4	1	1½"	TEE ASTM-A-197 BW
5	2	¾"	45° ELL ASTM-A-197 BW
6	1	1½" × ¾"	CONCENTRIC REDUCER ASTM-LAT
7	2	¾"	GATE VA. BW ASTM-A-105
8	1	1½"	CHECK VA. SWING BW-A-105

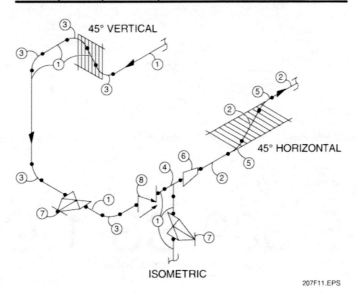

45° VERTICAL

45° HORIZONTAL

ISOMETRIC

207F11.EPS

Figure 11 ◆ Isometric drawing.

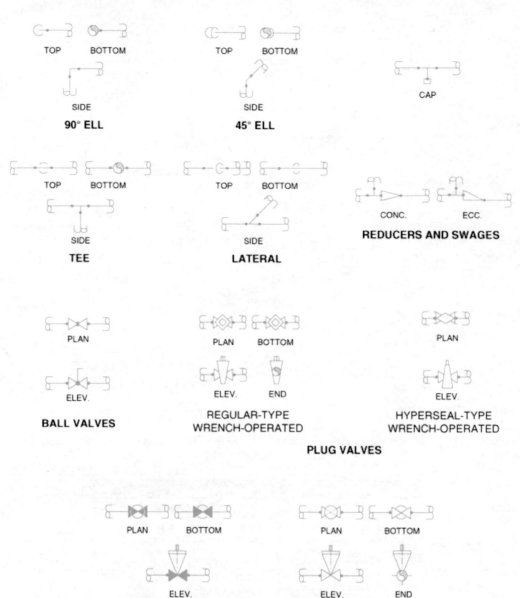

Figure 12 ◆ Butt weld fitting and valve symbols.

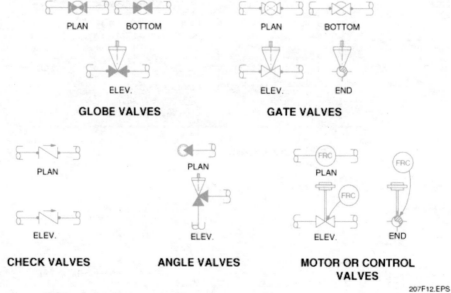

Figure 13 ◆ Line number.

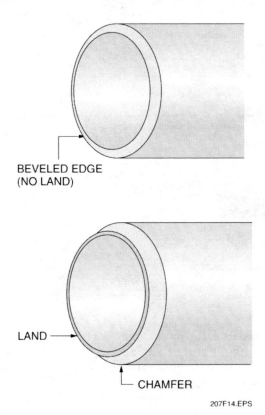

Figure 14 ◆ Bevel and chamfer.

Special grinders are available that can grind the bevel and land of the pipe end more precisely than a portable grinder can. These grinders are mounted inside the pipe and are locked in position to ensure that the joint preparation is square. An electric or air grinder is mounted to an arm that can be adjusted for the bevel angle required or set at 90 degrees to grind the land. The grinder is then rotated to prepare the bevel.

There are also special cutoff machines that are mounted to the outside of the pipe by a ring or a special chain with rollers. An electric- or air-operated machine with a cutoff blade is mounted on the ring and manually or electrically powered around the pipe to cut it off.

5.1.2 Beveling Using Pipe Bevellers

The pipe beveller is a portable, air- or electric-powered beveling tool designed to bevel pipe from very small to very large diameters. The pipe beveller comes in a number of sizes, from small hand-held models to very large diameter bevellers. The beveller can make precision weld preps on all metals, including stainless steel, high-alloy steel, and aluminum pipe or tubes. The beveller has a self-centering mandrel system that holds the beveller on the pipe and ensures accuracy. The beveller is one of the easiest and most accurate tools used to bevel pipe ends. *Figure 15* shows a small-diameter Wachs pipe beveller.

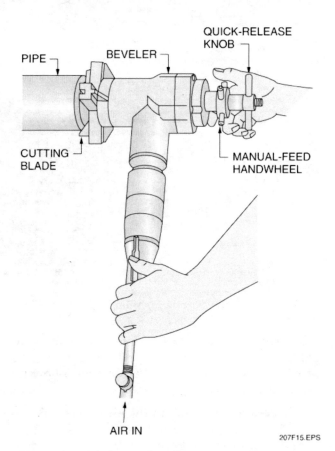

Figure 15 ◆ Wachs pipe beveller.

5.1.3 Thermal Beveling

Joint preparation using an oxyacetylene torch is a quick and easy way to bevel pipe ends; however, the cutting torch leaves an oxide coating on the pipe that must be ground away before welding. The oxyacetylene pipe-beveling machine (*Figure 16*) is fitted over the pipe end and tightened to the pipe. The torch is then adjusted to cut the correct angle. An electric drive motor moves the cutting torch around the pipe end to cut the preset bevel.

There are many different types of oxyacetylene pipe-beveling machines on the market. The best way to learn how to operate the pipe-beveling machine is through on-the-job training with your supervisor or the equipment manufacturer.

5.2.0 Cleaning Surfaces

After the pipe edge has been prepared, it must be cleaned before welding. Portable grinders are often used to clean the pipe edge. Grinders must be thoroughly inspected before they are used.

5.2.1 Inspecting Grinders Before Use

Whether the portable grinder being used is electric or pneumatic, a thorough inspection of the equipment is required before the grinder is used.

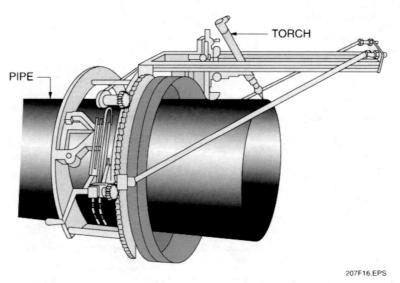

TORCH

PIPE

207F16.EPS

Figure 16 ◆ Oxyacetylene pipe-beveling machine.

If the grinder is found to be defective in any way, tag the defect and do not use the grinder until the problem has been repaired. Perform the following procedures to inspect a grinder:

- Inspect the air inlet and the air line of a pneumatic grinder to ensure that there are no signs of damage that could cause a bad connection or loss of air.
- Inspect the power cord and plug on electric models to ensure that there are no signs of damage.
- Inspect the handle to ensure that it is not loose.

 WARNING!
If the handle is loose, it could cause a loss of control.

- Inspect the grinder housing, wheel, and body for defects.
- Ensure that the trigger switch works properly and does not stick in the ON position.
- Ensure that the safety guard is in good condition and securely attached to the grinder.
- Check the oil level in pneumatic grinders.
- Ensure that the maximum rotating speed of the grinding wheel is higher than the maximum rotating speed of the grinder.
- Ensure that the grinding wheel is compatible with the metal being ground, and is tightened on the arbor.
- Start the grinder and allow it to run for 1 to 2 minutes while checking for visual abnormalities, excessive vibration, extreme temperature changes, or noisy operation.

- Inspect the work area to ensure the safety of yourself and others and to ensure that the heat and sparks generated by the grinder cannot start any fires.

5.2.2 Operating Grinders

When operating grinders, you must pay full attention to the grinder, the work being performed, and the flow of sparks and metal bits coming off the wheel. Always remember that each grinding accessory is designed to be used in only one way. When a flat grinding disc is being used to clean a bevel, put the flat surface of the disc against the work only. Follow these steps to operate a grinder:

 WARNING!
Ensure that you are equipped with the proper personal safety equipment, including gloves, a full-face shield, and safety shoes. Make sure that all loose clothing is tucked in. Do not wear frayed clothing, as it may catch fire.

Step 1 Inspect the grinder.

NOTE
Perform the procedures in the previous section to inspect the grinder.

Step 2 Secure the object to be ground in a vise or clamps to ensure that it does not move.

Step 3 Obtain any hot-work permits required.

Step 4   Attach the grinder to the power source.

Step 5   Position yourself with good footing and balance, and establish a firm hold on the grinder to avoid kickback.

WARNING!

Special precautions must be taken to avoid kickback while grinding. Always use a grinding wheel or a flapper wheel rather than a buffing brush when cleaning a bevel on a pipe end. Remember that the wheel rotates clockwise, and you must position yourself and the wheel on the pipe end correctly to avoid the wheel grabbing the pipe end and kicking back toward you. Always grind with the direction of the bevel, and position the wheel to the bevel so that if it kicks off, it will kick off away from you. The wheel is properly positioned on the bevel when the torque pulls the wheel in the direction that the wheel would leave the pipe end if it grabbed. Reposition the wheel on the pipe end when moving from the top half to the bottom half of the pipe.

Step 6   Pull the trigger to start the grinder.

Step 7   Apply the grinder to the work, and direct the sparks to the ground whenever possible and away from any hazards in the area, such as combustible debris or acetylene tanks.

CAUTION

Protect nearby alloy metals, such as stainless steel tanks, from cross-contamination of other metals being ground in the area.

Step 8   Apply proper force to the grinder on the grinding surface. If you are applying too much force on the grinder, you can hear the motor strain.

Step 9   Stop grinding and inspect the work and the grinding wheel periodically. If the wheel is damaged or too worn, replace it.

Step 10   Turn off and disconnect the grinder from the power source when grinding is complete.

Step 11   Inspect the grinder for any signs of damage.

Step 12   Return the grinder and any accessories to the storage area.

After grinding the bevel, the surfaces around the bevel must be cleaned. A wire brush can be used to remove any rust or corrosion. Wipe the pipe end with a clean rag to remove all grease and oil.

Stainless steel, aluminum, and other alloy pipes are cut with horizontal or vertical bandsaws, abrasive saws, or circular saws. The circular saws used to cut stainless steel or carbon steel are called coldsaws, because they do not produce really high temperatures while cutting, as do the abrasive cutters. The blades on coldsaws are tipped with carbide or a kind of cement. Another type of coldsaw is the guillotine saw, which is basically a powered bow saw, that usually has hydraulic feed mechanisms. The guillotine is mounted on the pipe itself, rather than on a table.

6.0.0 ◆ DETERMINING PIPE LENGTHS BETWEEN FITTINGS

To determine the length of a piece of pipe between two butt weld fittings, the pipefitter must be able to determine the takeout of the fitting and the required welding gap, or root opening. Both the takeout of the fitting and the welding gap must be subtracted from the dimension of the piping run.

6.1.0 Calculating Takeout

The takeout of a butt weld fitting refers to the laying length of the fitting, or the distance from the face of one side of the fitting to the center of the fitting. The takeout of common welding fittings can be found in charts showing general dimensions for welding fittings (*Figure 17*).

The general dimensions for welding pipe fittings are supplied by fitting manufacturers and are also found in many pipefitting guidebooks. However, many times you will have to calculate takeout in the field. Formulas are provided for the following types of fittings:

- Long radius 90-degree elbows
- Short radius 90-degree elbows
- 45-degree elbows

6.1.1 Long Radius 90-Degree Elbows

The takeout of long radius 90-degree elbows is always 1½ times the nominal size of the elbow. This is true of all long radius 90-degree elbows except for ½-inch elbows. For example, to determine the takeout of an 8-inch LR 90, multiply 8 times 1½, and the answer, 12, is the takeout (*Figure 18*).

NOTE

The takeout of an extra long radius fitting is three times the diameter.

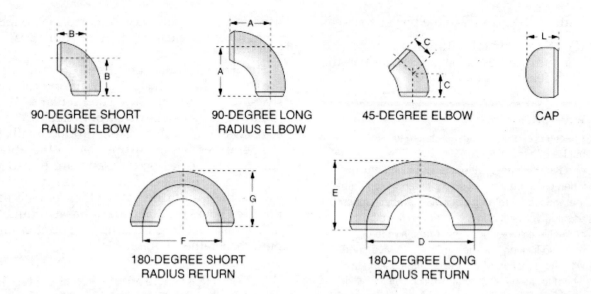

90-DEGREE SHORT
RADIUS ELBOW

90-DEGREE LONG
RADIUS ELBOW

45-DEGREE ELBOW

CAP

180-DEGREE SHORT
RADIUS RETURN

180-DEGREE LONG
RADIUS RETURN

DIMENSIONS (in inches)													
	90° Elbow		90° Elbow	180° Return Elbows				Cap	Wall Thickness				
	Long Radius Center to End A	Short Radius Center to End B	Center to End C	Long Radius		Short Radius							
Pipe Size				Center to Center D	Back to Face E	Center to Center F	Back to Face G	Length L	S	XS	Sched. 160	XXS	Pipe Size
½	1½		⅝	3	1⅞			1	0.109				½
¾	1⅛		⁷⁄₁₆	2½	1¹¹⁄₁₆			1	0.113	0.154			¾
1	1½	1	⅞	3	2³⁄₁₆	2	1⅝	1½	0.133	0.179	0.250	0.358	1
1¼	1⅞	1¼	1	3¾	2¾	2½	2¹⁄₁₆	1½	0.140	0.191	0.250	0.382	1¼
1½	2¼	1½	1⅛	4½	3¼	3	2⁷⁄₁₆	1½	0.145	0.200	0.281	0.400	1½
2	3	2	1⅜	6	4³⁄₁₆	4	3³⁄₁₆	1½	0.154	0.218	0.343	0.436	2
2½	3¾	2½	1¾	7½	5³⁄₁₆	5	3¹⁵⁄₁₆	1½	0.203	0.276	0.375	0.552	2½
3	4½	3	2	9	6¼	6	4¾	2	0.216	0.300	0.438	0.600	3
3½	5¼	3½	2¼	10½	7¼	7	5½	2½	0.226	0.318		0.636	3½
4	6	4	2½	12	8¼	8	6¼	2½	0.237	0.337	0.531	0.674	4
5	7½	5	3⅛	15	10⁵⁄₁₆	10	7¾	3	0.258	0.375	0.625	0.750	5
6	9	6	3¾	18	12⁵⁄₁₆	12	9⁵⁄₁₆	3½	0.280	0.432	0.718	0.864	6
8	12	8	5	24	16⁵⁄₁₆	16	12⁵⁄₁₆	4	0.322	0.500	0.906	0.875	8
10	15	10	6¼	30	20⅜	20	15⅜	5	0.365	0.500	1.125		10
12	18	12	7½	36	24⅜	24	18⅜	6	0.375	0.500	1.312		12
14	21	14	8¾	42	28	28	21	6½	0.375	0.500	1.406		14
16	24	16	10	48	32	32	24	7	0.375	0.500	1.593		16
18	27	18	11¼	54	36	36	27	8	0.375	0.500			18
20	30	20	12½	60	40	40	30	9	0.375	0.500			20
22	33	22	13½	66	44	44	33	10	0.375	0.500			22
24	36	24	15	72	48	48	36	10½	0.375	0.500			24

207F17.EPS

Figure 17 ◆ General dimensions for welding fittings.

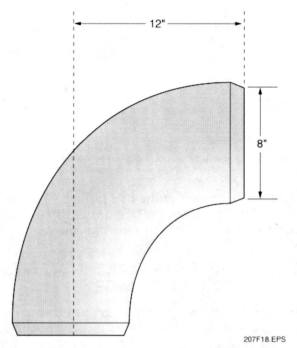

Figure 18 ◆ Takeout of long radius 90-degree elbow.

6.1.2 Short Radius 90-Degree Elbows

The takeout of a short radius 90-degree elbow is always the same as the nominal size of the elbow. For example, a 3-inch SR 90 has a takeout of 3 inches, and a 6-inch SR 90 has a takeout of 6 inches (*Figure 19*).

6.1.3 45-Degree Elbows

Do not divide the takeout of a 90-degree elbow above 4 inches by two to find the takeout of a 45-degree elbow of the same nominal size. To calculate the takeout of a 45-degree elbow, multiply the nominal size of the elbow by ⅝ or 0.625.

If a calculator is not available, the takeout of a 45-degree elbow can be determined using the following method. Follow these steps to find the takeout of a 45-degree elbow:

Step 1 Write down the nominal size of the elbow.

8 inches

Step 2 Divide this number by 2, and write the answer next to the nominal size.

8 inches   4 inches

Step 3 Divide this number by 2, and write this number next to the last number.

8 inches   4 inches   2 inches

Step 4 Divide this number by 2, and write this number next to the last number.

8 inches   4 inches
2 inches   1 inch

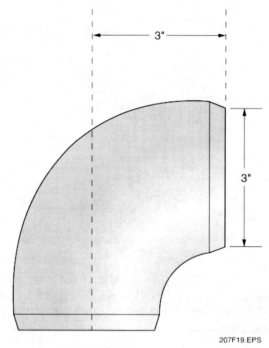

Figure 19 ◆ Takeout of short radius 90-degree elbow.

Step 5 Add the second and fourth numbers that you have written down to find the takeout of the 45-degree elbow.

4 inches + 1 inch = 5 inches

Therefore, the takeout for an 8-inch, 45-degree elbow is 5 inches.

6.2.0 Obtaining Proper Spacing Between Pipes and Fittings

To ensure that the welder has enough space to make a full-penetration weld, the pipefitter must leave a space between the pipe edges to be joined. This space is called the root opening, or gap. The proper gap is crucial for a good fit-up. If the gap is too small, it is difficult to fuse the pipes properly. If the gap is too large, burn-through may result. Always check your job specifications to determine pipe spacing. Almost all pipe welding gaps are between 1/16 and 1/8 inch; however, the welder determines the gap. A simple way to obtain the correct gap between pipes is to bend a welding rod that is the same size as the required gap and place it between the pipes. A scrap piece of steel that is the correct thickness can also be used. *Figure 20* shows pipe spacing using a welding rod.

6.3.0 Calculating Pipe Lengths

Piping drawings usually show the dimensions of the overall run of pipe, but it is up to the pipefitter to determine the lengths of the individual straight pipes within the run. As in the case of threaded

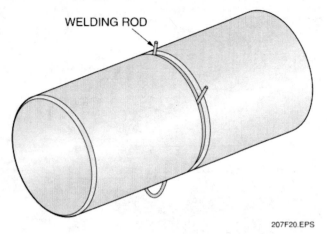

Figure 20 ◆ Pipe spacing using welding rod.

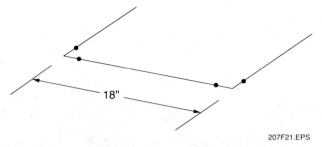

207F21.EPS

Figure 21 ◆ Center-to-center piping drawing.

and socket weld piping, the three methods of determining pipe lengths are the center-to-center method, the center-to-face method, and the face-to-face method.

6.3.1 Center-to-Center Method

Figure 21 shows a piping drawing showing the center-to-center dimension.

As you have learned, when using the center-to-center method, measurements are taken from the center of one fitting to the center of the next fitting (*Figure 22*). The actual length of pipe between butt weld fittings is determined by subtracting the takeout of each of the fittings and the required gap for each weld on each fitting.

In butt weld piping, the center-to-center method is the same as in socket weld, except that the welding gap is added to the takeout of the two fittings. The sum is then subtracted from the center-to-center dimension to determine the length of pipe to be cut.

6.3.2 Center-to-Face Method

In butt weld piping, the length of a run of pipe from a fitting to a flange is determined by the center-to-face method (*Figure 23*). The center-to-face method is the same as in socket weld, except that the welding gap, which is at the welder's discretion (usually ⅛ inch), is added to the takeout of each of the two fittings. The sum is then subtracted from the center-to-face dimension to determine the length of pipe to be cut.

6.3.3 Face-to-Face Method

In butt weld piping, the face-to-face method (*Figure 24*) is used to determine the length of pipe between two flanges. This method is the same as in socket weld, except that the welding gap, which is at the welder's discretion, but usually ⅛ inch, is added to

the takeout of each of the two fittings. The sum is then subtracted from the face-to-face dimension to determine the length of pipe to be cut.

7.0.0 ◆ SELECTING AND INSTALLING BACKING RINGS

Backing rings are generally required in piping systems where severe conditions are anticipated. Job specifications govern the use of backing rings. Other names for backing rings are chill rings and welding rings. Backing rings eliminate weld defects that often form on the inside of the pipe as a direct result of the welding process and that can restrict the flow of material through a system. To ensure a smooth flow across the backing ring, the ends of the ring are usually chamfered about 15 degrees.

Backing rings are usually stocked in sizes ranging from ¾ inch to 36 inches, with larger sizes available on request. There are many types of backing rings available from different manufacturers, but the four common types include Type CCC, Type CC, Type C, and plain rings. *Figure 25* shows the four common types of backing rings.

The Type CCC backing rings have long nubs that serve as spacers at the joint. Type CCC rings are usually used when the high-low misalignment between the pipe and fitting is great. The nubs are usually spot-welded onto the ring during manufacture and can be broken off cleanly after alignment.

The nubs of the Type CC backing rings are shorter than those on the Type CCC backing rings. These rings are typically used when the variation in high-low alignment is not as great. The nubs on Type CC rings can either be chipped off prior to welding or they can be fused with the welding root pass.

The nubs on the Type C welding ring are actually a part of the backing ring and cannot be removed prior to welding. These rings are used in situations in which the high-low alignment is very close.

Plain backing rings do not have nubs, and the gap between the pipe and fitting must be set manually. It is important that the backing ring and the pipe being welded be made of the same material.

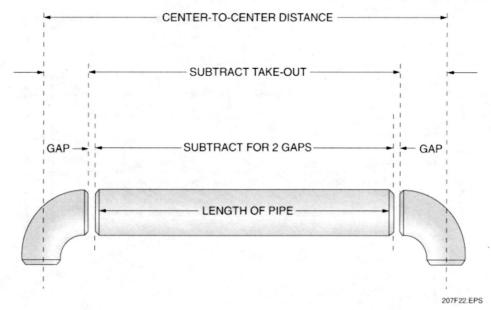

Figure 22 ◆ Center-to-center method.

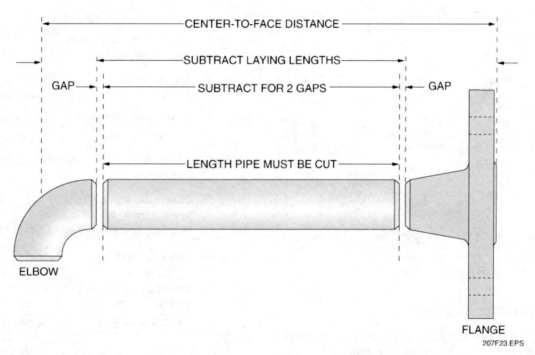

Figure 23 ◆ Center-to-face method.

Follow these steps to install backing rings:

Step 1 Ensure that the pipe ends have been cut square and are properly beveled with a smooth land on the face of the bevel.

Step 2 Compress the backing ring slightly, and insert it into the pipe end until the nubs are flush against the bevel land.

Step 3 Place the other pipe or fitting over the ring, and butt it against the nubs. The pipe is now ready to be aligned and welded.

8.0.0 ◆ USING AND CARING FOR CLAMPS AND ALIGNMENT TOOLS

In butt weld piping, alignment is a very critical task. Unlike socket weld fit-ups, the pipe must be aligned with other pipes or with fittings in the absence of any sort of self-aligning socket or other adaptation. The result has been a set of specialized clamps and tools for alignment. Some are clamps to hold the pieces; some are capable of aligning the pieces as well.

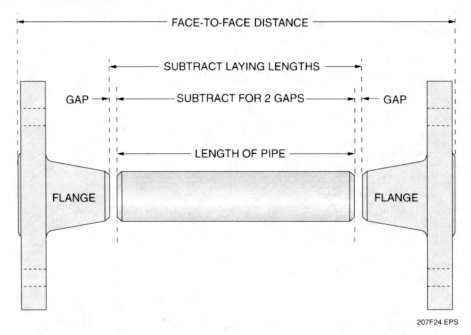

Figure 24 ◆ Face-to-face method.

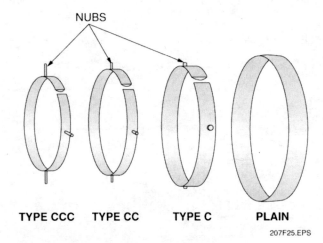

Figure 25 ◆ Common backing rings.

8.1.0 Angle Iron Jigs

Angle iron jigs are fabricated on the job site and are used to hold smaller diameter pipe for welding. They can be fabricated out of angle iron or channel iron that will hold the pipe in line or at the desired angle for welding. *Figure 26* shows some typical angle iron jigs.

Follow these steps to fabricate an angle iron jig:

> **NOTE**
> This procedure explains how to make an angle iron jig from ⅛ by 1½-inch channel iron that is 3 feet, 9 inches long. Channel iron of this size can be used for pipe sizes ranging from 1¼ inch to 3 inches.

Step 1   Cut out 90-degree notches about 9 inches from one end (*Figure 27*).

Step 2   Heat the bottom of the notch, using a torch, until the metal is red-hot.

Step 3   Bend the channel iron into a 90-degree angle.

Step 4   Have the welder weld the sides of the corners.

Step 5   Place an elbow in the jig and saw halfway through the sides of the channel at the bottom face of the elbow (*Figure 28*).

Step 6   Repeat Step 5 with several different sizes of elbows so that the jig can be used for several applications. A used hacksaw blade can be placed in the notch to provide the proper welding gap (*Figure 29*).

8.2.0 Shop-Made Aligning Dogs

Many pipefitters fabricate their own aligning devices. In the pipefitting trade, these devices are known as aligning dogs. *Figure 30* shows an example of an aligning dog.

> **NOTE**
> Aligning dogs are not permitted under some codes.

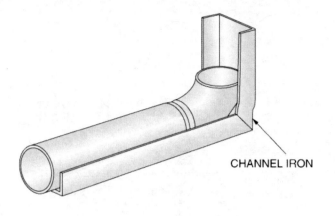

CHANNEL IRON

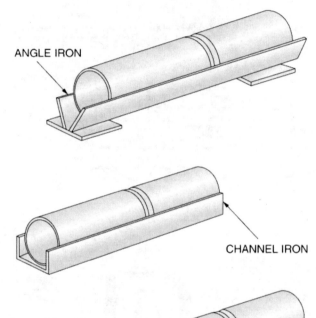

ANGLE IRON

CHANNEL IRON

STEEL RODS

207F26.EPS

Figure 26 ◆ Angle iron jigs.

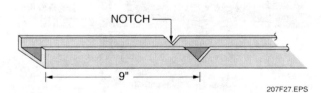

NOTCH

9"

207F27.EPS

Figure 27 ◆ Notches.

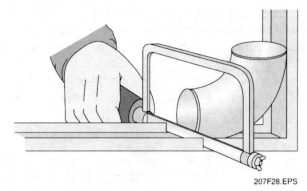

207F28.EPS

Figure 28 ◆ Sawing bottom sides of channel.

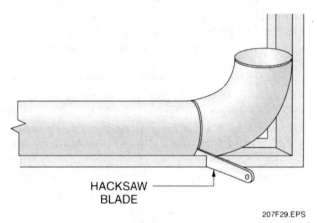

HACKSAW BLADE

207F29.EPS

Figure 29 ◆ Using a jig for alignment.

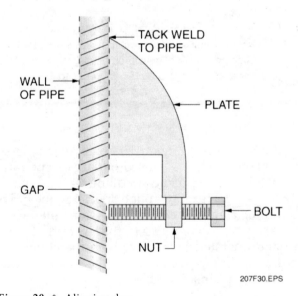

TACK WELD TO PIPE

WALL OF PIPE

PLATE

GAP

BOLT

NUT

207F30.EPS

Figure 30 ◆ Aligning dog.

The aligning dog shown in *Figure 30* is made from plate of the same material as the pipe being welded, to which a nut has been welded. The aligning devices are tack-welded onto one section of the pipe so that the other section can be moved by tightening the bolt. Usually, two such dogs would be used to provide more control over the fit-up. Once the fit-up is complete, the aligning

dogs are removed from the pipe. After the dog is removed from the pipe, the pipe must be ground to remove any welding deposits. Aligning dogs also require periodic grinding to remove welding deposits.

Welding codes in some areas prohibit anything being welded onto the pipe. Make sure you know the code before using any aids that must be welded to the pipe. If codes do not permit the dog to be tacked on, make three dogs with the weld leg parallel top and bottom. Three of these dogs can be used to align the pipe by placing them at three points 120 degrees apart on the pipe and holding the weld leg against the pipe with a chain and a ratchet tensioner.

8.3.0 Lever-Type Clamps

Lever-type clamps are very useful if several straight butt joints must be made in a single piping run. They are usually made for one size of pipe, ranging from 2 to 36 inches in diameter, with larger sizes available through special manufacturers. These clamps cannot be used to align flanges.

Lever-type clamps are usually made from two rings joined by four or more horizontal bars. A lever opens and closes the clamp. When the lever is drawn tight, the horizontal bars butt against the pipe, aligning each section with the other. Fine adjustments are usually possible by loosening or tightening lock nuts attached to the arms that the lever controls (*Figure 31*). Many companies no longer use the lever-type clamp because of worker injuries from pinching fingers in the clamps.

> **NOTE**
>
> Another common type of pipe alignment clamp, called the ultra-clamp, is a screw-type clamp that allows the alignment of the joint to be adjusted with the adjusting screws. It makes contact with the two pieces at only three points, and is made in three sizes, 1 to 2", 2 to 6", and 5 to 12".

8.4.0 Hydraulic Clamps

Hydraulic clamps (*Figure 32*) are similar to lever-type clamps. They are used on larger pipe ranging from 16 to 36 inches in diameter, with larger sizes manufactured through special request. The hydraulic clamp operates in the same manner as the lever-type clamp except that instead of using a hand-operated lever to tighten the clamp, the hydraulic clamp uses a hydraulic jack to tighten the clamp against the pipe.

8.5.0 Chain-Type Clamps

Chain-type welding clamps use the same principle as the angle iron jigs, but have chains that hold the pipe in place. These clamps are made of tough, durable metal that will not be warped by the heat of the welding. Chain-type clamps are generally available for the common fit-ups, such as straight pipe, elbow, T-joint, and flange fit-ups.

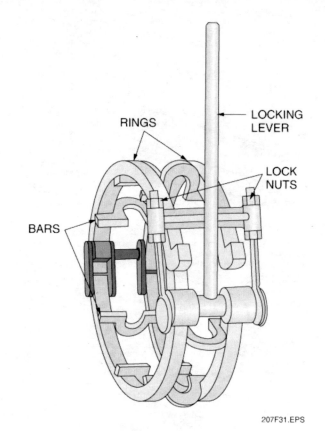

207F31.EPS

Figure 31 ◆ Lever-type clamp.

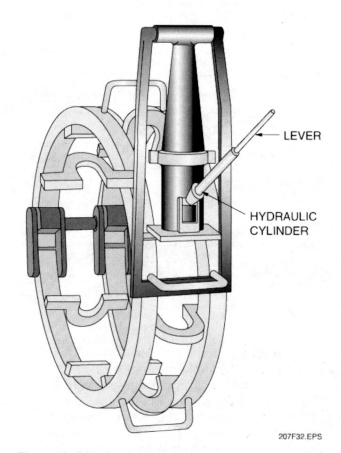

207F32.EPS

Figure 32 ◆ Hydraulic clamp.

8.5.1 Straight Pipe Welding Clamps

Straight pipe clamps are chain-type clamps used to secure two straight pipes together end to end. They consist of straight sections of steel approximately 14 to 18 inches long and two lengths of chain, one near each end, attached to screw locks that tighten the chain once it has been wrapped around the pipe. Straight pipe welding clamps can be used on pipe with ½- to 8-inch outside diameters (*Figure 33*).

Follow these steps to use and care for a straight pipe welding clamp:

NOTE

For this exercise, assume that two previously prepared 4-inch steel pipes must be welded together end to end.

Step 1 Inspect the straight pipe clamp for any obvious damage, such as broken or worn chain links, bent or broken clamp feet, a warped clamp back, bent or broken chain hooks, damaged chain tighteners, and grease, rust, or excessive dirt. Repair or replace any damaged part. Clean and oil any part as needed.

Step 2 Place the pipes on the clamp body so that the pipes line up end to end. Install a backing ring if it is required.

Step 3 Set the proper gap between the pipes.

Step 4 Position the clamp on top of, and evenly spaced over, the two sections of pipe so that the chains drop down the back side of the clamp as you face the pipes.

Step 5 Hold the clamp in place with one hand, and reach under the pipe and grasp the chain with the other hand.

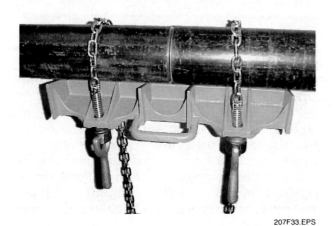

Figure 33 ◆ Straight pipe welding clamp.

207F33.EPS

Step 6 Pull the chain around the pipe and up to the locking notch on the clamp.

Step 7 Secure the chain to the clamp.

Step 8 Rotate the clamp around the pipe to the bottom of the pipe.

Step 9 Turn the screw handle to tighten the chain. Tighten only until the chain is snug. Some readjustment may be needed later.

Step 10 Check the alignment of the unclamped section of pipe, ensuring that it is still aligned with the first section of pipe and is ready to be welded.

Step 11 Wrap the chain around the second section of pipe.

Step 12 Attach the chain to the clamp.

Step 13 Tighten the second chain onto the second section of pipe.

Step 14 Adjust both chains until they are tight.

Step 15 Check the pipe sections again to ensure that they are still aligned with each other and ready for welding.

Step 16 When welding is complete, loosen the chains and remove the clamp.

WARNING!

Handle the pipe and the clamps with care. Welded objects remain hot for several minutes after the welding is completed.

Step 17 Clean the clamp, if necessary, and inspect it for damage. If it is damaged, replace it.

Step 18 Store the straight pipe clamp.

9.0.0 ◆ PERFORMING ALIGNMENT PROCEDURES

Aligning pipe for welding requires great skill. The pipes must be correctly aligned to make a smooth inner wall that does not restrict flow. The pipes must also be aligned to put the completed pipe in the proper position with respect to other pipes and fittings in the piping system. This is complicated by the fact that fittings and pipe may have small but significant differences in inside diameter. If this is detected, it may require the fitter to counterbore the pipe or fitting to allow a closer match. The joint cannot be offset or angled but must be joined evenly (*Figure 34*).

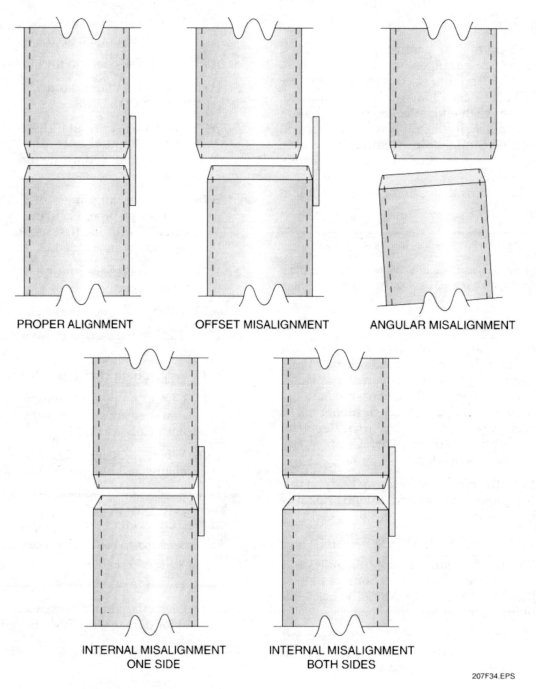

PROPER ALIGNMENT   OFFSET MISALIGNMENT   ANGULAR MISALIGNMENT

INTERNAL MISALIGNMENT
ONE SIDE

INTERNAL MISALIGNMENT
BOTH SIDES

207F34.EPS

Figure 34 ◆ Pipe alignment.

NOTE

During alignment of fitting, compensate for movement of pipe due to weld draw. This should be
taken into account prior to weld completion.

If the walls of joining pipe or fittings being
joined are not exactly the same thickness, you
must ensure that the internal misalignment is the
same all the way around the joining pipes. You
must also ensure that the internal misalignment
is less than the maximum allowable tolerance as
stated in the engineering specifications. If there is
internal misalignment on one side only, either the
pipe or fitting may be out of round and therefore
must be replaced. This happens most often in
large-diameter pipe.

Pipes can be aligned using a level, squares, or a
Hi-Lo gauge. Levels and squares are used to check
end alignment; the primary purpose of a Hi-Lo
gauge is to check for inner wall misalignment. The
name of the gauge comes from the relationship
between the inside wall of one pipe and that of
another pipe. To check for internal misalignment,
Hi-Lo gauges have two prongs, or alignment stops,
that are pulled tightly against the inside wall of
the joint so that one stop is flush with each side of

the joint. The variation between the two stops is read on a scale marked on the gauge. To measure misalignment using a Hi-Lo gauge, insert the prongs of the gauge into the joint gap. Pull up on the gauge until the prongs are snug against both inside surfaces, and read the misalignment on the end of the scale. *Figure 35* shows checking internal misalignment using a Hi-Lo gauge. Major misalignment could be predetermined by measuring pipe circumference before fit-up. The Hi-Lo gauge can also be used to measure gap.

9.1.0 Aligning Straight Pipe

Straight pipe is aligned using framing squares. To align two pieces of straight pipe, two framing squares are needed. If the pipe being joined is less than 10-inch nominal size, support the pipe with 3-legged jack stands. If the pipe is larger than 10-inch nominal size, use 4-legged jack stands. Follow these steps to align straight pipe:

Step 1 Set up four pipe supports in a straight line to hold the two pipes.

> **NOTE**
> For this exercise, assume that two 6-foot sections of 4-inch steel pipe must be welded into one 12-foot section.

Step 2 Place the pipes on the supports so that the pipes line up end to end.

> **NOTE**
> Insert a backing ring between the flange and the pipe if required.

Step 3 Set the proper gap between the pipes. When setting the gap between larger sizes

of pipe, you can tap the jack stands with a hammer opposite to the direction that you are trying to move the pipe. This will walk the pipes together from a maximum distance of about 3 inches (*Figure 36*).

Step 4 Place the framing squares on the top centers of the pipes and slide the squares toward each other. *Figure 37* shows how to position the squares.

Step 5 Adjust the pipe until the squares are flush from bottom to top.

Step 6 Position the squares on the side centers of the pipes and adjust the pipes again until the squares are flush against each other.

Step 7 Recheck the alignment on top of the pipes and adjust as necessary.

Step 8 Remove the squares and have the welder make a light tack-weld at the top and bottom of the joint.

> **WARNING!**
> Handle the pipe with care. Welded objects remain hot for several minutes after the welding is completed.

Step 9 Recheck the alignment on both the top and side. If the pipe needs to be adjusted, tap it with a hammer until it is properly aligned to be welded.

Step 10 Have the welder make light tack-welds on the sides of the joint.

9.2.0 Aligning Pipe to 45-Degree Elbow

Just as when you were working with socket weld pipe, two framing squares can also be used to align a 45-degree elbow to the end of a straight

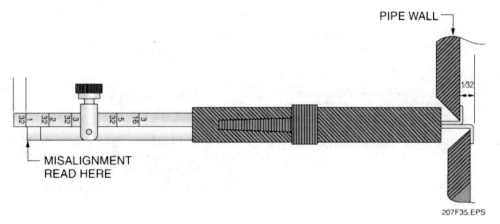

Figure 35 ◆ Checking internal misalignment using a Hi-Lo gauge.

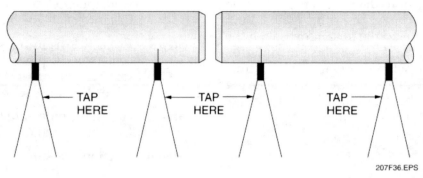

PIPES WILL MOVE
TOGETHER.

207F36.EPS

Figure 36 ◆ Walking pipes together.

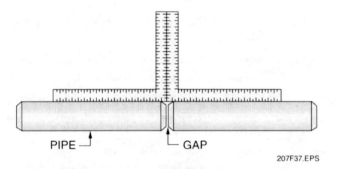

PIPE ⌐ └ GAP

207F37.EPS

Figure 37 ◆ Positioning of squares to align straight pipe.

pipe when using smaller diameter pipe. Follow these steps to align straight pipe to a 45-degree elbow:

Step 1 Set up a 6-foot section of 4-inch steel pipe on two pipe supports.

Step 2 Set up a pipe jack at the end of the pipe to support the elbow.

Step 3 Place the 45-degree elbow on the support so that it lines up with the pipe and turns to the side. Install a backing ring between the pipe and the elbow if required.

Step 4 Set the proper gap between the pipe and the elbow.

Step 5 Hold the elbow in line with the pipe to make a preliminary alignment.

Step 6 Have the welder make a light tack-weld at the 12-o'clock position. Since the elbow is turned to one side, this tack is actually made on the side of the elbow.

Step 7 Rotate the pipe and elbow 180 degrees so that the elbow faces the opposite direction.

Step 8 Set the proper gap between the pipe and the fitting.

Step 9 Have the welder make a light tack-weld at the 12-o'clock position.

Step 10 Rotate the pipe and elbow 90 degrees so that the elbow turns up.

Step 11 Place the tongue of one framing square on the top center of the straight pipe so that the body is vertically in line with the end of the pipe.

Step 12 Place the body of the second framing square on the end of the elbow.

Step 13 Adjust the elbow so that the same inch marks on the tongue and the body of the second square contact the inner edge of the first square (*Figure 38*). When fabricating larger diameter pipe, the framing square is not large enough to use with the elbow. In this case, level the pipe in the jack stands, and use a 24-inch spirit level with a 45-degree vial to set the position of the elbow (*Figure 39*).

Step 14 Have the welder tack-weld the top and bottom of the joint.

Step 15 Check the alignment both ways, and adjust it if necessary by tapping the fitting with a hammer.

Step 16 Have the welder tack-weld the sides of the joint.

9.3.0 Aligning Pipe to 90-Degree Elbow

Two framing squares can also be used to align a 90-degree elbow to the end of a straight pipe. Prepare and put the first tack-weld on the pipe and elbow at the 12-o'clock position on the pipe.

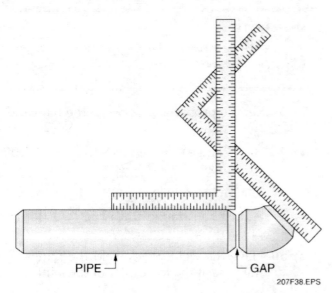

Figure 38 ◆ Pipe to 45-degree elbow alignment.

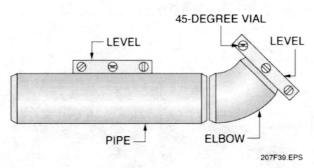

Figure 39 ◆ Aligning 45-degree elbow using level.

Step 1 Place the tongue of one framing square on the top center of the straight pipe so that the body is vertically in line with the end of the pipe.

Step 2 Place the body of the second framing square on the end of the elbow so that the tongue of the square lines up with the body of the other square (*Figure 40*).

Step 3 Adjust the elbow until the squares are flush against each other. When fabricating larger diameter pipe, the framing square is not large enough to use with the elbow. In this case, level the pipe in the jack stands, and use a 24-inch spirit level to set the position of the elbow (*Figure 41*).

Step 4 Have the welder tack-weld the bottom of the joint.

Step 5 Place a square on the sides of the elbow and pipe and check for misalignment from side to side.

Step 6 Have the welder tack-weld the sides of the joint.

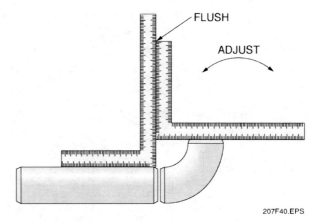

Figure 40 ◆ Pipe to 90-degree elbow alignment.

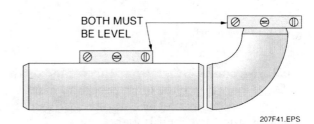

Figure 41 ◆ Aligning 90-degree elbow using level.

9.4.0 Squaring 90-Degree Corner

Long pipe runs with a 90-degree corner are difficult to square with a framing square because the squares are too small. These long runs can be squared using the 3-4-5 method. The 3-4-5 foot method, based on the Pythagorean theorem, uses a triangle with a 3-foot side, a 4-foot side, and a 5-foot side. A triangle with these sides will always have a 90-degree angle at the intersection of the 3- and 4-foot sides. For longer runs of pipe, any multiple of 3-4-5 can be used to make a square corner. For example, the triangle can have sides of 6, 8, and 10 feet; 9, 12, and 15 feet; or 12, 16, and 20 feet. This section explains how to square a pipe into the 90-degree elbow that was fit up in the previous section of this module. The pipe must be welded at a true 90-degree angle to the other pipe connected by the elbow. *Figure 42* shows the 3-4-5 method.

To fit a 90-degree angle using the 3-4-5 method, prepare the pipe and elbow as you did with any elbow, and get the first tacks made on the 3-o'clock and 9-o'clock points. Then use the following steps:

Step 1 Mark the top dead center of the elbow using a center finder and the takeout formula.

Step 2 Measure 3 feet from the center of the elbow down one pipe, and mark the top dead center of the pipe at this point using the center finder.

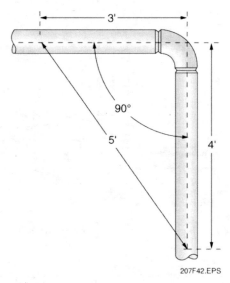

Figure 42 ◆ 3-4-5 method.

Step 3 Measure 4 feet from the center of the elbow down the other pipe, and mark the top dead center of the pipe at this point using the center finder.

Step 4 Measure the distance between these two points and adjust the pipe until this distance is 5 feet.

Step 5 Have the welder tack-weld the joint in the throat of the elbow.

Step 6 Have the welder tack-weld the joint on the outside of the elbow.

9.5.0 Aligning Pipe to Flange

Flanges must be aligned to pipe both horizontally and vertically. The bolt holes must also be aligned with the other flanges on the pipe. If you are welding a flange to a pipe that already has a flange on one end, level the existing flange first using the two-hole method and a torpedo level. Follow these steps to align a flange to the end of a pipe:

Step 1 Set up two adjustable pipe stands to support the pipe.

Step 2 Secure the section of pipe onto the stands with the end to be welded extending at least 12 inches beyond one of the stands.

Step 3 Level the pipe between the two stands.

Step 4 Line up the flange with the end of the pipe.

Step 5 Place a jack stand underneath the flange to help support the flange.

NOTE

Insert a backing ring between the flange and the pipe if required.

Step 6 Set the proper gap between the pipe and the flange.

Step 7 Place flange pins into the top two holes of the flange.

Step 8 Place a torpedo level on the flange pins.

Step 9 Adjust the flange until the two flange pins are level (*Figure 43*).

Step 10 Have the welder tack-weld the flange to the pipe at the top center of the joint.

Step 11 Remove the jack stand from underneath the flange.

Step 12 Remove the flange pins from the flange.

Figure 43 ◆ Aligning flange horizontally.

Step 13 Place a framing square over the flange so that the tongue of the square is flush with the face of the flange and the body of the square is aligned over the pipe (*Figure 44*).

Step 14 Adjust the flange until the distance between the body of the square and the pipe is the same the entire length of the square (*Figure 45*).

Step 15 Have the welder tack-weld the sides and the bottom of the joint.

9.6.0 Aligning Pipe to Tee

Two framing squares can also be used to align a tee to the end of a straight pipe. Prepare the pipe and tee as you would a 90-degree elbow and use the same procedure as you used to align the 90-degree ell (*Figure 46*).

Step 1 Adjust the tee until the squares are flush. When fabricating larger diameter pipe, the framing square is not large enough to use with the tee. In this case, level the pipe in the jack stands, and use a 24-inch spirit level to set the position of the tee. *Figure 47* shows aligning a tee using a level.

Step 2 Have the welder tack-weld the bottom of the joint.

Step 3 Place the square on the sides of the tee and pipe and check for misalignment from side to side.

Step 4 Have the welder tack-weld the sides of the joint.

9.7.0 Fitting Butt Weld Valves

Fitting valves in a butt weld piping system is essentially the same process as fitting other components. For this exercise, assume that a valve is to be butt-welded to the end of a 4-inch pipe that has been beveled and cleaned and is ready to be welded. Follow these steps to fit a butt weld valve to pipe. You must open the valve all the way and remove any parts of the valve that might be damaged by the welding heat. Follow the same procedures to obtain preliminary alignment with the valve stem up, just as you did with the 90-degree elbow.

Step 1 Set the proper gap between the pipe and the valve.

Step 2 Hold the valve in line with the pipe to make a preliminary alignment (*Figure 48*).

Step 3 Have the welder make a light tack-weld at the 12-o'clock position.

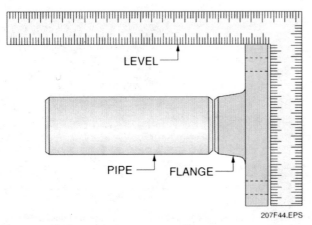

Figure 44 ◆ Positioning square on flange.

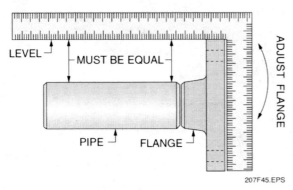

Figure 45 ◆ Adjusting flange.

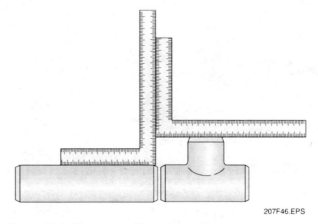

Figure 46 ◆ Pipe to tee alignment.

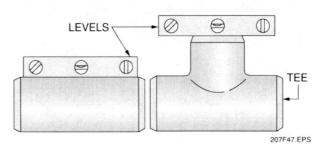

Figure 47 ◆ Aligning tee using level.

Step 4 Place a square over the valve, with the long leg of the square aligned with the pipe and the short leg of the square flush against the valve end.

Step 5 Adjust the valve until the distance between the long leg of the square and the pipe is the same the entire length of the square.

Step 6 Have the welder make a light tack-weld at the 6-o'clock position.

Step 7 Rotate the square on the pipe 90 degrees to check the squareness of the valve from side to side.

Step 8 Adjust the valve until the distance between the long leg of the square and the pipe is the same the entire length of the square.

Step 9 Have the welder tack-weld the valve to the pipe in the 3- and 9-o'clock positions.

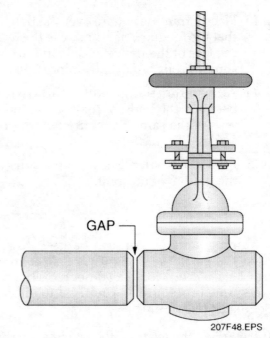

GAP

207F48.EPS

Figure 48 ◆ Preliminary valve alignment.

1. Nominal size and outside diameter are the same from _____ up.

   a. 6 inches
   b. 8 inches
   c. 10 inches
   d. 14 inches

2. All piping is marked and identified by the _____ identification number.

   a. ANSI
   b. EPA
   c. ASTM
   d. ASME

3. Metal pipe that does *not* contain iron will be labeled with a number that starts with _____.

   a. Z
   b. B
   c. A
   d. F

4. A pipe with the ASTM specification *A-120* is _____.

   a. T-6 aluminum
   b. bronze
   c. ferrous
   d. titanium

5. Unless otherwise stated, elbows are always _____.

   a. 30 degrees
   b. 45 degrees
   c. 60 degrees
   d. 90 degrees

6. Unless the drawings call for something else, use _____ -radius elbows.

   a. short
   b. slow
   c. large
   d. long

7. A return bend is a _____ fitting.

   a. 0-degree
   b. 45-degree
   c. 120-degree
   d. 180-degree

8. From a straight run of pipe, a weldolet makes a _____ reducing branch.

   a. 30-degree
   b. 45-degree
   c. 60-degree
   d. 90-degree

9. A reducer that displaces the center line of the smaller of the two joining pipes is called a(n) _____.

   a. concentric reducer
   b. displacing reducer
   c. eccentric reducer
   d. deviating reducer

10. Which of the following is *not* a pressure rating of a flange?

    a. 150 pound
    b. 300 pound
    c. 1,500 pound
    d. 3,000 pound

11. Long weld neck flanges are normally used for _____ outlets.

    a. sewer and stormdrain
    b. vessel and tank
    c. truck
    d. potable water

12. Slip-on flanges normally extend beyond the end of the pipe either ⅜ of an inch or the _____, whichever is greater.

    a. diameter of the pipe
    b. thickness of the pipe wall
    c. thickness of the flange
    d. diameter of the bolt holes.

13. Lap-joint flanges are also known as _____ flanges.

    a. Van Ronk
    b. Van Stone
    c. Van Allen
    d. raised face

14. Which of the following is *not* a type of piping system drawing?
    a. Double-line
    b. Single-line
    c. Plan-line
    d. Isometric

15. A side view of a system of pipe is also called a(n) _____ view.
    a. isometric
    b. schematic
    c. elevation
    d. plan

16. A top view of a system of pipe is called a(n) _____ view.
    a. isometric
    b. schematic
    c. elevation
    d. plan

17. Specific information about components in piping systems is found in the _____.
    a. isometric view
    b. title block
    c. specifications book
    d. detail drawing

18. The third part of a line number is the _____.
    a. ASTM designation
    b. number of fittings
    c. actual pipeline number
    d. high-temperature limit

19. The flat edge of a pipe that is left after a chamfer is cut is called the _____.
    a. bevel
    b. ear
    c. face
    d. land

20. If you find a crack in a grinding wheel, you should _____.
    a. replace it
    b. glue and clamp it
    c. wear it down
    d. use it

21. Stainless steel piping is cut with a(n) _____.
    a. oxyacetylene torch
    b. handsaw
    c. burn table
    d. abrasive wheel saw

22. A guillotine saw is a type of abrasive saw.
    a. True
    b. False

23. The takeout for a fitting is the distance from _____.
    a. one side to the other
    b. one face to the center of the fitting
    c. the outside to the outside
    d. one fitting to the next

24. The takeout for a long-radius elbow is always _____ the nominal size.
    a. twice
    b. one-half
    c. one and one-half
    d. the same as

25. The length of the pipe between a fitting on one end and the flange on the other end is determined using the _____ method.
    a. center-to-face
    b. suzuki
    c. face-to-face
    d. center-to-center

26. Backing rings that are used for the greatest high-low alignment are Type _____.
    a. A
    b. CCC
    c. B
    d. C

27. It is always safe to tack-weld aligning dogs on the pipe.
    a. True
    b. False

28. Hydraulic clamps are used to _____.
    a. hold the grinder against the bevel
    b. realign the inner wall
    c. hold the alignment for welding
    d. support the jackstands

29. A Hi-Lo gauge is used to measure _____.
    a. the distance from the ground
    b. the diameter of the pipe
    c. misalignment of pipes
    d. the center-to-center dimension

30. When setting the gap between larger sizes of pipe, you can walk the ends together by hitting the _____.
    a. other end with a ballpeen hammer
    b. jack stands in the direction you want the pipe to go
    c. pipe on top, to bounce it
    d. jack stands in the opposite direction

Summary

The most important thing to remember about performing butt weld pipe fabrication is that the fit-up must be square and level. It is the pipefitter's job to align pipe and fittings as near perfectly as possible to produce safe systems that will give many years of service and stay maintenance-free. Practice the alignment procedures to perfect your skills and produce quality fit-ups every time.

Notes

Trade Terms
Introduced in This Module

Align: To make straight or to line up evenly.

Alloy: Two or more metals combined to make a new metal.

ASTM International: An organization that publishes standards relating to materials, products, and services. Formerly known as the American Society for Testing and Materials.

Bevel: An angle cut or ground on the end of a piece of solid material.

Burn-through: A hole that is formed in a weld due to improper grinding or welding.

Chamfer: An angle cut or ground only on the edge of a piece of material.

Fit up: To put piping material in position to be welded together.

Full-penetration weld: Complete joint penetration for a joint welded from one side only.

Lateral: A fitting or branch connection that has a side outlet that is any angle other than 90 degrees to the run.

Oxide: A type of corrosion that is formed when oxygen combines with a base metal.

Root opening: The space between the pipes at the beginning of a weld; the gap is usually ⅛ of an inch.

Resources & Acknowledgments

Additional Resources

This module is intended to be a thorough resource for task training. The following reference work is suggested for further study. This is an optional material for continued education rather than for task training.

The Pipe Fitters Blue Book. W.V. Graves. Webster, TX: W.V. Graves Publishing Company.

Figure Credits

H & M Beveling Machine Company, Inc., 207F33

Mathey Dearman, 207F43

Cianbro Corporation, Module Opener

NCCER CURRICULA — USER UPDATE

NCCER makes every effort to keep its textbooks up-to-date and free of technical errors. We appreciate your help in this process. If you find an error, a typographical mistake, or an inaccuracy in NCCER's curricula, please fill out this form (or a photocopy), or complete the online form at **www.nccer.org/olf**. Be sure to include the exact module ID number, page number, a detailed description, and your recommended correction. Your input will be brought to the attention of the Authoring Team. Thank you for your assistance.

Instructors – If you have an idea for improving this textbook, or have found that additional materials were necessary to teach this module effectively, please let us know so that we may present your suggestions to the Authoring Team.

NCCER Product Development and Revision

13614 Progress Blvd., Alachua, FL 32615

Email: curriculum@nccer.org
Online: www.nccer.org/olf

❏ Trainee Guide ❏ Lesson Plans ❏ Exam ❏ PowerPoints Other _____

Craft / Level: _____ Copyright Date: _____

Module ID Number / Title: _____

Section Number(s): _____

Description: _____

_____

_____

_____

_____

Recommended Correction: _____

_____

_____

_____

_____

Your Name: _____

Address: _____

_____

Email: _____ Phone: _____

08208-06

Excavations

08208-06
Excavations

Topics to be presented in this module include:

Overview

The most dangerous parts of pipefitting come when the fitter is working high up in a structure, or in an excavation. In this module, you will learn the hazards and safety equipment and procedures for pipeline excavations. You will learn how to lay out the line from the surveyor's reference points and how to use various kinds of equipment to get the pipeline trenches to hold the pipe where they should be. You will learn how to identify soil types, and the OSHA requirements for trenching safety.

Objectives

When you have completed this module, you will be able to do the following:

1. Identify and explain the use of shoring materials.
2. Identify and explain the use of premanufactured support systems.
3. Install a vertical shore to be used for shoring.
4. Determine the overall fall of a sewer line.
5. Determine and set the grade and elevation of a trench.
6. Explain backfilling procedures.

Trade Terms

Angle of repose	Oxidation
Bedding	Protective system
Bench mark	Raveling
Benching	Screw jack
Carbonation	Sheeting
Catchment basin	Shield
Coffer dam	Shore
Compaction	Shoring
Competent person	Skeleton
Compression strength	Sloping
Cross braces	Spoils
Disturbed soil	String line
Excavation	Subsidence
Finished grade	Support system
Grade	T-bars
Grade pin	Tight sheeting
Grade rod	Trench
Hub	Trench box
Hydration	Uprights
Information stake	Void
Lift	Wales
Monitoring well	

Required Trainee Materials

1. Pencil and paper
2. Appropriate personal protective equipment

Prerequisites

Before you begin this module, it is recommended that you successfully complete *Core Curriculum*; *Pipefitting Level One*; and *Pipefitting Level Two*, Modules 08201-06 through 08207-06.

This course map shows all of the modules in the second level of the *Pipefitting* curriculum. The suggested training order begins at the bottom and proceeds up. Skill levels increase as you advance on the course map. The local Training Program Sponsor may adjust the training order.

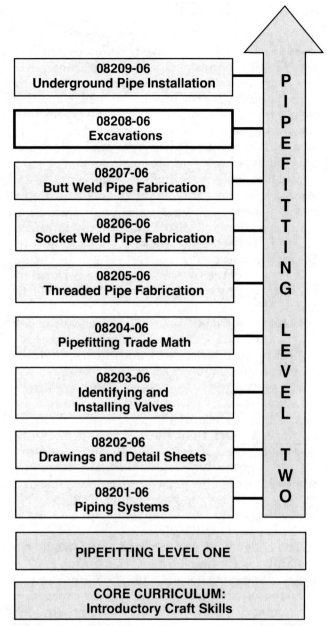

208CMAP.EPS

1.0.0 ◆ INTRODUCTION

Trenching and excavation work presents serious risks to all workers involved. On many job sites, the pipefitters must help shore the trench after it is dug, set the finish elevation and grade of the trench bottom, lay the pipe in the trench, and then help remove the shoring system. The greatest risk of working in a trench is the possibility of a cave-in. You must be aware of all current OSHA standards and follow all rules and regulations that apply to excavations to ensure the safety of yourself and your co-workers. This module explains different types of shoring materials and premanufactured shoring systems. It also explains how to assemble and disassemble support systems within the trench, how to determine and set elevations and grades within the trench, and how to perform the backfilling procedures required to properly fill the trench.

Safety is crucial during any excavation job. Excavations are done for a number of reasons, including to build cellars and during highway construction (*Figure 1*). During an excavation, earth is removed from the ground, creating a trench. A trench is a narrow excavation made below the surface of the ground in which the depth is greater than the width and the width does not exceed 15 feet. The soil that is removed from the ground is called spoil. When earth is removed from the ground, extreme pressures may be generated on the trench walls. If the walls are not properly secured by shoring, sloping, or shielding, they will collapse.

The collapse of unsupported trench walls can instantly crush and bury workers. This type of collapse happens because not enough material is available to support the walls of an excavation. One cubic yard of earth weighs approximately 3,000 pounds; that's the weight of a small car, and more than enough weight to seriously injure or kill a worker. In fact, each year in the United States, more than 100 people are killed and many more are seriously injured in cave-in accidents.

2.0.0 ◆ TRENCHING HAZARDS

Working in and around excavations is one of the most hazardous jobs you will ever do. Safety precautions must be exercised at all times to prevent injury to yourself and others. In addition to the hazards of collapse and falling equipment, there is the danger of striking underground utilities, such as power lines, water pipelines, and natural gas lines. Always contact the appropriate sources, such as Dig Safe, before you dig.

Some of the hazards you may encounter during an excavation include the following:

- Cave-ins due to trench failure
- Falls from workers coming too close to the trench edge
- Flooding from broken water or sewer mains
- Electrical shock from striking electrical cable in the trench or striking overhead lines
- Toxic liquid or gas leaks from nearby facilities or pipes
- Auto traffic if the excavation site is near a highway
- Falling dirt or rocks from an excavator bucket

2.1.0 Soil Hazards

The type of soil in and around a trench is also a factor that contributes to the collapse of trench walls. *Table 1* shows a sample list of 84 recorded trench failures broken down by the type of soil in which they occurred. Soil type is a major factor to consider in trenching operations. Only a company-assigned competent person has enough experience on the job, training, and education to determine if the soil in and around a trench is safe and stable. However, it is still your responsibility to know the basics about soil and its associated hazards.

WARNING!
Never enter a trench unless you have approval from your company-assigned competent person.

2.1.1 Properties of Soil

Soil is a mixture of organic materials, soil particles, air, and water. Soil particles, or grains, consist of chunks, pieces, fragments, and tiny bits of rock that are released by the weathering of parent

Table 1 Failures by Soil Type

Type of Soil	Number of Failures
Clay and/or mud	32
Sand	21
Wet dirt (probably silty clay)	10
Sand, gravel, and clay	8
Rock	7
Gravel	4
Sand and gravel	2

208T01.EPS

208F01.EPS

Figure 1 ◆ Excavation sites.

rocks. Weathering is a natural process of erosion that can be physical or chemical. Physical processes include freezing and thawing, gravity, and erosion by rivers and rainfall. Chemical processes include oxidation, hydration, and carbonation in which minerals are chemically broken down by the elements and removed via ground and rain water. Properties of a given soil include grain size, soil gradation, and grain shape.

2.1.2 Types of Soils

The soil found on most construction sites is a mixture of many mineral grains coming from several kinds of rocks. Average soils are usually a mixture of two or three materials such as sand and silt or silt and clay. The type of mixture determines the soil characteristics. For example, sand with small amounts of silt and clay may compact well and provide a very good excavation soil. In addition to the mineral grains, soil contains water, air, gases, chemicals, and other organic material.

2.1.3 Soil Behavior

Each of the various soil types, depending on the condition of the soil at the time of the excavation, will behave differently (*Table 2*). Sandy soil tends to collapse straight down. Wet clays and loams tend to slab off the side of the trench. These two conditions are shown in *Figure 2*.

Firm, dry clays and loams tend to crack. Wet sand and gravel tend to slide. These conditions are shown in *Figure 3*. You should be aware of the type of soil you are working in and know how it behaves. When you are working in or near a trench, stay alert to changes in the trench. Watch for developing cracks, moisture, or small movements in the trench material. Alert your supervisor and co-workers to any changes you notice in the trench walls. Changes in trench walls may be an early indication of a more severe condition.

2.1.4 Groundwater

Water that flows through the ground is called groundwater. It is normal to encounter groundwater in deep trenches and excavations, especially where the water table is high; that is, the groundwater is close to the surface. When groundwater begins to enter the trench or excavation, it creates several problems.

First, the soil of the trench walls becomes more likely to collapse. Second, the water makes it very difficult to keep digging. At some point, the water may fill the trench, so that workers cannot even enter.

Groundwater is likely to become contaminated with chemicals used on the dig site. These chemicals would otherwise not get so far down underground because they would be filtered out by the soil. On many construction jobs, monitoring wells are dug to check the quality of groundwater, especially where blasting or toxic chemicals may be involved. Many projects, especially pipelines, test the water in private wells within 200 to 400 feet of the line. Another reason for testing wells and springs is that pipelines and other long trenches, even after backfilling, can change the direction of flow, or block the flow of groundwater.

Before a large project is approved, an environmental survey is frequently required. The survey produces a map of known groundwater hazards such as buried chemical storage tanks or landfills containing possible pollutants. In many jurisdictions, plans to remedy such possible hazards must be supplied to the governing agencies before digging begins.

Finally, removing groundwater produces its own set of problems. If you use diversion wells, or well points, to lower the groundwater level, you may produce problems for other well owners in the area. More to the point, if you pump the water out of the trench, you have to use some kind of silt filtration to keep from clogging the storm drain systems or the nearby streams. These silt

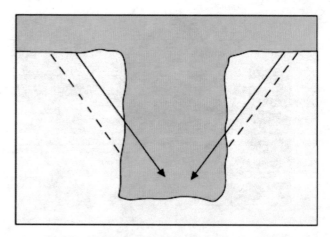

SANDY SOIL COLLAPSES STRAIGHT DOWN

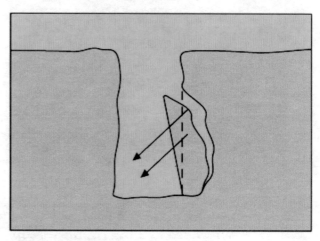

WET CLAY AND LOAMS SLAB OFF

208F02 EPS

Figure 2 ◆ Behavior of sandy soil and wet clay and loams.

fences, either fabric fences, straw bales, or straw rolls, will normally be specified in the drawings. It is common to surround catchment basins with coffer dams as temporary enclosures for the water that is pumped from the excavation, allowing it to be removed or to percolate into the ground in a normal manner. Whatever the required solution, you should be aware that it will be your company's responsibility to protect the storm drains and streams, and to protect the groundwater.

3.0.0 ◆ GUIDELINES FOR WORKING IN AND NEAR A TRENCH

When working in or around any excavation or trench, you are responsible for personal safety. You are also responsible for the safety of others in the trench. The following guidelines must be enforced to ensure everyone's safety:

• Never enter an excavation without the approval of the OSHA-approved competent person on site.

Table 2 Field Method for Identification of Soil Texture

Soil Texture	Visual Detection of Particle Size and General Appearance of Soil	Squeezed in Hand	Soil Ribboned Between Thumb and Finger
Sand	Soil has a granular appearance in which the individual grain sizes can be detected. It is free-flowing when in a dry condition.	When air dry: Will not form a cast and will fall apart when pressure is released. When moist: Forms a cast which will crumble when lightly touched.	Cannot be ribboned.
Sandy Loam	Essentially a granular soil with sufficient silt and clay to make it somewhat coherent. Sand characteristics dominate.	When air dry: Forms a cast which readily falls apart when lightly touched. When moist: Forms a cast which will bear careful handling without breaking.	Cannot be ribboned.
Loam	A uniform mixture of sand, silt, and clay. Grading of sand fraction quite uniform from coarse to fine. It is mellow, has a somewhat gritty feel, yet is fairly smooth and slightly plastic.	When air dry: Forms a cast which will bear careful handling without breaking. When moist: Forms a cast which can be handled freely without breaking.	Cannot be ribboned.
Silt Loam	Contains a moderate amount of the finer grades of sand and only a small amount of clay. Over half of the particles are silt. When dry it may appear quite cloddy, which can be readily broken and pulverized to a powder	When air dry: Forms a cast which can be freely handled. Pulverized, it has a soft, flourlike feel. When moist: Forms a cast which can be freely handled. When wet, soil runs together and puddles.	It will ribbon, but has a broken appearance, feels smooth, and may be slightly plastic.
Silt	Contain over 80% silt particles, with very little fine sand and clay. When dry, it may be cloddy, and readily pulverizes to powder with a soft flourlike feel.	When air dry: Forms a cast which can be handled without breaking. When moist: Forms a cast which can be freely handled. When wet, it readily puddles.	It has a tendency to ribbon with a broken appearance, feels smooth.
Clay Loam	Fine textured soil breaks into very hard lumps when dry. Contains more clay than silt loam. Resembles clay in a dry condition; identification is made on the physical behavior of moist soil.	When air dry: Forms a cast which can be freely handled without breaking. When moist: Forms a cast which can be freely handled without breaking. It can be worked into a dense mass.	Forms a thin ribbon which readily breaks, barely sustaining its own weight.
Clay	Fine textured soil breaks into very hard lumps when dry. Difficult to pulverize into a soft flourlike powder when dry. Identification is based on cohesive properties of the moist soil.	When air dry: Forms a cast which can be freely handled without breaking. When moist: Forms a cast which can be freely handled without breaking.	Forms long, thin, flexible ribbons. Can be worked into a dense, compact mass. Considerable plasticity.
Organic Soils	Identification based on the high organic content. Muck consists of thoroughly decomposed organic material with considerable amount of mineral soil finely divided with some fibrous remains. When considerable fibrous material is present it may be classified as peat. The plant remains or sometimes the woody structure can easily be recognized. Soil color ranges from brown to black. They occur in lowlands, swamps, or swales. They have high shrinkage upon drying,		

208T02.EPS

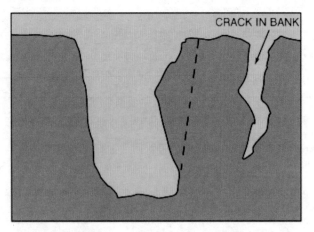

FIRM DRY CLAY AND LOAMS CRACK

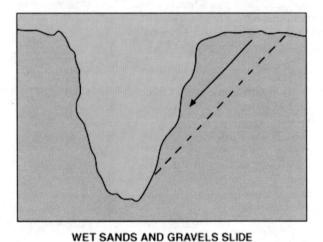

WET SANDS AND GRAVELS SLIDE

208F03.EPS

Figure 3 ◆ Behavior of dry clay and loams and wet sand and gravel.

- Inspect the excavation daily for changes in the excavation environment, such as rain, frost, or severe vibration from nearby heavy equipment.
- Never enter an excavation until the excavation has been inspected.
- Wear protective clothing and equipment such as hard hats, safety glasses, work boots, and gloves. Use respirator equipment if necessary.
- Wear traffic warning vests that are marked with or made of reflective or highly visible material if you are exposed to vehicle traffic.
- Get out of the trench immediately if water starts to accumulate in the trench.
- Do not walk under loads being handled by power shovels, derricks, or hoists.
- Stay clear of any vehicle that is being loaded.
- Be alert. Watch and listen for possible dangers.
- Do not work above or below a co-worker on a sloped or benched excavation wall.
- Barricade access to excavations to protect pedestrians and vehicles (*Figure 4*).

208F04.EPS

Figure 4 ◆ Barricaded trench.

- Check with your supervisor to see if workers entering the excavation need excavation entry permits.
- Ladders and ramps used as exits must be located every 25 feet in any trench that is over 4 feet deep.
- Make sure someone is on top to watch the walls when you enter a trench.
- Make sure you are never alone in a trench. Two people can cover each other's blind spots.
- Keep tools, equipment, and the excavated dirt at least 2 feet from the edge of the excavation.
- Make sure shoring, trench boxes, benching, or sloping is used for excavations and trenches over 5 feet deep.
- Stop work immediately if there is any potential for a cave-in. Note the cave-in point in the trench box in *Figure 5*. Make sure any problems are corrected before starting work again.

3.1.0 Ladders

There must be at least one method of entering and exiting all excavations over 4 feet deep. Ladders are generally used for this purpose (*Figure 6*). Ladders must be placed within 25 feet of each worker.

When ladders are used, there are a number of requirements that must be met.

- Ladder side rails must extend a minimum of 3 feet above the landing.
- Ladders must have nonconductive side rails if work will be performed near equipment or systems using electricity.
- Two or more ladders must be used where 25 or more workers are working in an excavation.
- All ladders must be inspected before each use for signs of damage or defects.

Figure 5 ◆ Trench box with cave-in point.

208F05.EPS

- Damaged ladders should be labeled *DO NOT USE* and removed from service until repaired.
- Use ladders only on stable or level surfaces.
- Secure ladders when they are used in any location where they can be displaced by excavation activities or traffic.
- While on a ladder, do not carry any object or load that could cause you to lose your balance.
- Exercise caution whenever using a trench ladder.

4.0.0 ◆ INDICATIONS OF AN UNSTABLE TRENCH

A number of stresses and weaknesses can occur in an open trench or excavation. For example, increases or decreases in moisture content can affect the stability of a trench or excavation. The following sections discuss some of the more frequently identified causes of trench failure. These conditions are illustrated in *Figure 7*.

Tension cracks usually form one-quarter to one-half of the way down from the top of a trench. Sliding or slipping may occur as a result of tension cracks. In addition to sliding, tension cracks can cause toppling. Toppling occurs when the trench's vertical face shears along the tension crack line and topples into the excavation. An unsupported excavation can create an unbalanced stress in the

Figure 6 ◆ Ladder in a trench.

208F06.EPS

soil, which in turn causes subsidence at the surface and bulging of the vertical face of the trench. If uncorrected, this condition can cause wall failure and trap workers in the trench, or greatly stress the protective system. Bottom heaving is caused by downward pressure created by the weight of adjoining soil. This pressure causes a bulge in the bottom of the cut. Heaving and squeezing can occur, even when shoring and shielding are properly installed.

 WARNING!

Protective systems are designed for even loads of earth. Heaving and squeezing can place uneven loads on the shielding system and may stress particular parts of the protective system.

Another indication of an unstable trench is boiling. Boiling occurs when water flows upward into the bottom of the cut. A high water table is one of the causes of boiling. Boiling can happen quickly and can occur even when shoring or trench boxes are used. If boiling starts, stop what you are doing and leave the trench immediately.

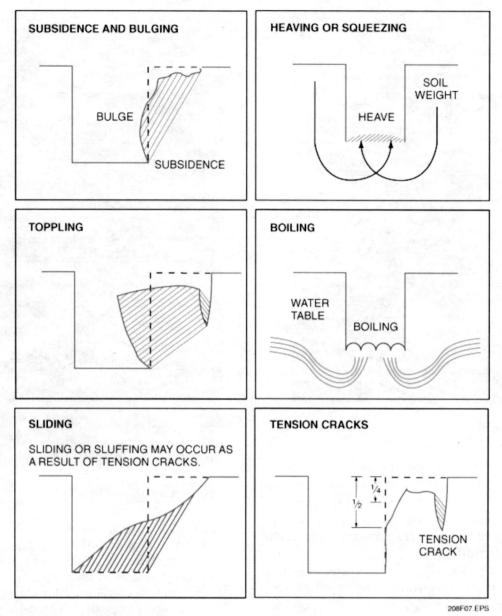

Figure 7 ◆ Indications of an unstable trench.

5.0.0 ◆ TRENCH FAILURE

The most common hazard during an excavation is trench failure or cave-in. Using common sense and following all applicable safety precautions will make the trench a much safer place to work.

NOTE

Excavations 5 feet or deeper require manufactured and engineered trench shielding and shoring devices. If shielding or shoring devices are not used, trench walls may be sloped to eliminate the risk of cave-ins. Each of these protective systems must be used in accordance with OSHA regulations.

To understand the seriousness of trench failure, consider what can happen when there is a shift in the earth that surrounds an unsupported trench. Workers could be buried when the following occurs:

• One or both edges of the trench cave in
• One or both walls slide in
• One or both walls shear away and collapse

Failure of unsupported trench walls is not the only cause of burial. Tons of dirt can be dumped on the workers if the spoil pile or excavated earth slides into the trench. Such slides occur when the pile is placed too close to the edge of the trench or when the ground beneath the pile gives way. There must be a minimum of 2 feet between the trench wall and the spoil pile. This area must also be kept free of any tools and materials.

The following conditions will likely lead to a trench cave-in. If you notice any of these conditions, immediately inform your supervisor. These conditions are listed in order of seriousness.

- Disturbed soil from previously excavated ground
- Trench intersections where large corners of earth can break away
- A narrow right-of-way causing heavy equipment to be too close to the edge of the trench
- Vibrations from construction equipment, nearby traffic, or trains
- Increased subsurface water that causes soil to become saturated, and therefore unstable
- Drying of exposed trench walls, which causes the natural moisture that binds together soil particles to be lost
- Inclined layers of soil dipping into the trench, causing layers of different types of soil to slide one upon the other and cause the trench walls to collapse

6.0.0 ◆ MAKING THE TRENCH SAFE

There are several ways to make the trench a safer place to work (*Figure 8*). Trench shoring, shielding, and sloping are different methods used to protect workers and equipment. It is important that you recognize the differences between them.

- *Shoring* – Shoring a trench supports the walls of the excavation and prevents their movement and collapse. Shoring does more than provide a safe environment for workers in a trench. Because it restrains the movement of trench walls, shoring also stops the shifting of adjacent soil formations containing buried utilities or on which sidewalks, streets, building foundations, or other structures are built.

- *Trench shields* – Trench shields, also called trench boxes, are placed in unshored excavations to protect personnel from excavation wall collapse. They provide no support to trench walls or surrounding soil, but for specific depths and soil conditions, will withstand the side weight of a collapsing trench wall.

- *Sloping* – Sloping an excavation means cutting the walls of the excavation back at an angle to its floor. This angle must be cut at least to the angle of repose for the type of soil being used. The angle of repose is the greatest angle above the horizontal plane at which a material will rest without sliding.

6.1.0 Shoring Systems

Shoring systems are metal, hydraulic, mechanical, or timber structures that provide a framework to support excavation walls. Shoring uses uprights, wales, and cross braces to support walls. *Figure 9* shows a shoring system in place.

When timber shoring is used, it must be built from the top down, and removed from the bottom up. Screw jacks or timbers used for horizontal bracing must be placed at a 90-degree angle to the walls, and secured firmly, so that the braces will not kick out or slip.

6.1.1 Hydraulic Shores and Spreaders

Hydraulic shores, shown in *Figure 10*, can be installed and removed quickly. Each shore consists of two vertical rails connected by a hydraulic cylinder. The shores are placed in the trench and the hydraulic cylinders are pumped up to push the vertical rails against the wall. Note that this system uses the hydraulic cylinders as cross bracing. This arrangement is commonly referred to as a skeleton shoring system. Hydraulic spreaders, shown in

208F08.EPS

Figure 8 ◆ Shoring methods.

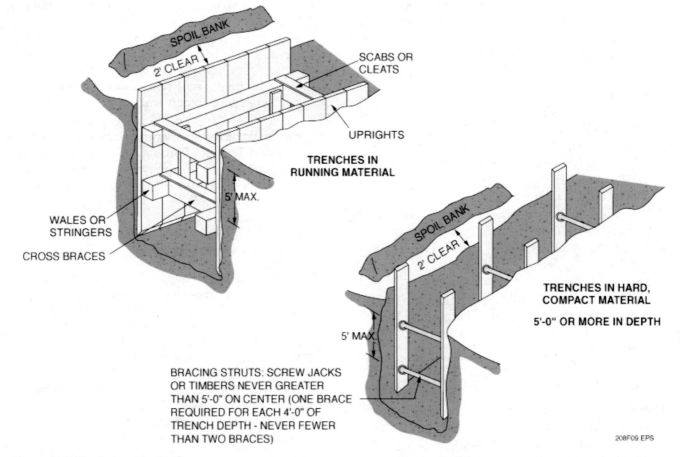

Figure 9 ◆ Shoring system in place.

Figure 11, can be added to a skeleton shoring system to provide additional cross bracing. Hydraulic shores come in 2-inch and 3-inch sizes; if the width of the trench is more than 8 feet, 2-inch hydraulic shores are required to have steel oversleeves to protect against seal failure or buckling.

6.1.2 Vertical Sheeting

When excavating near existing structures or performing long-term excavations, vertical sheeting may be used and supported with hydraulic wales, as shown in *Figure 12*. This method is referred to as tight sheet shoring. Other types of spreaders, including screw jacks, trench jacks, and hydraulic spreaders, are also available.

6.1.3 Interlocking Steel Sheeting

Interlocking steel sheeting (*Figure 13*) may be specified under certain conditions such as deep excavations and excavations near buildings or building foundations. Interlocking steel sheeting is commonly used on Department of Transportation (DOT) right-of-ways. It will prevent damage to sub-base pavement caused by vibration from vehicle traffic. Interlocking steel sheeting is required

when working in waterways. Steel sheeting consists of interlocking panels of steel reinforced with cross members. It is similar in design to tight sheeting. Steel sheeting is engineered for a particular application. It must be installed precisely in accordance with the engineer's specifications. Steel sheeting is commonly installed by driving it into the ground using a vibrating hydraulic hammer. It can also be driven with a drop hammer or backhoe bucket. In *Figure 13*, the sheeting at the top can be used with wales; the sheeting at the bottom must be completely braced to prevent buckling.

NOTE

The system just described identifies the common components used in shoring systems. You may encounter job-built systems, aluminum or wooden sheeting, or other approved methods.

6.1.4 Shoring Safety Rules

To avoid accidents and injury when shoring an excavation, special safety rules must be followed.

• Never enter an excavation before the shoring is in place.

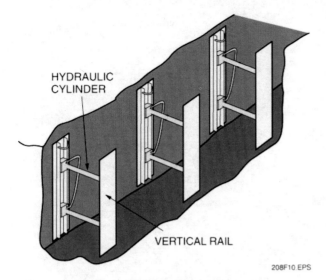

Figure 10 ◆ Hydraulic skeleton shoring system.

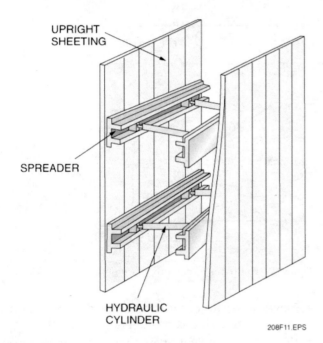

Figure 11 ◆ Hydraulic spreaders.

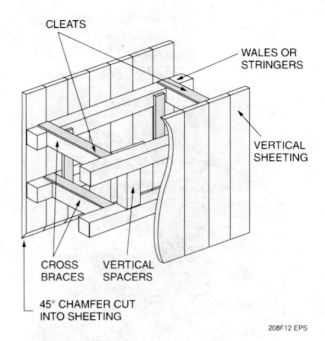

Figure 12 ◆ Tight sheet shoring.

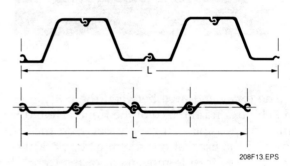

Figure 13 ◆ Interlocking steel sheeting.

- Do not install the shoring while you are inside the trench. All shoring must be installed from the top of the trench.
- The cross braces must be level across the trench.
- The cross braces should exert the same amount of pressure on each side of the trench.
- The vertical uprights must be drawn flat against the excavation wall.
- All materials used for shoring must be thoroughly inspected before use and must be in good condition.
- Shoring is removed by starting at the bottom of the excavation and going up.
- The vertical supports are pulled out of the trench from above.

- Every excavation must be backfilled immediately after the support system is removed.

6.2.0 Shielding Systems

A shielding system is a structure that is able to withstand the forces imposed on it by a cave-in, in order to protect employees within the structure. Shields can be permanent structures or can be designed to be portable and moved along as work progresses. The shielding system is also known as a trench box or trench shield. A trench box is shown in *Figure 14*. If the trench will not stand long enough to excavate for the shield, the shield can be placed high and pushed down as material is excavated.

NOTE

The trench in *Figure 14* is only partially protected against collapse. The area of the trench in the foreground has yet to be shored.

Figure 14 ◆ Trench boxes.

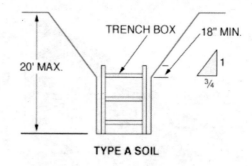

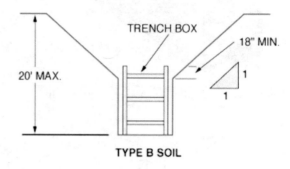

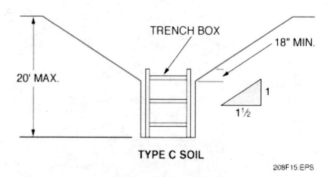

Figure 15 ◆ Proper trench-box placement.

The excavated area between the outside of the trench box and the face of the trench should be as small as possible. The space between the trench box and the excavation side must be backfilled to prevent side-to-side movement of the trench box. Remember that job site and soil conditions, as well as trench depths and widths, determine what type of trench protection system will be used. A single project can include several depth or width requirements and varying soil conditions. This may require that several different protective systems be used for the same site. A registered engineer must certify shields for the existing soil conditions. The certification must be available at the job site during assembly of the trench box. After assembly, certifications must be made available upon request.

6.2.1 Trench-Box Safety

If used correctly, trench boxes protect workers from the dangers of a cave-in. All safety guidelines for excavations also apply to trench boxes. Follow these safety guidelines when using a trench box:

- Be sure that the vertical walls of a trench box extend at least 18 inches above the lowest point where the excavation walls begin to slope. *Figure 15* shows proper trench box placement.
- Never enter the trench box during installation or removal.

- Never enter an excavation behind a trench box after the trench box has been moved.
- Backfill the excavation as soon as the trench box has been removed.
- If a trench box is to be used in a pit or a hole, all four sides of the trench box must be protected.
- An exit from the trench box and the excavation must be located within 25 feet of each worker.

6.3.0 Sloping Systems

A sloping system is a method of protecting workers from cave-ins. Sloping is accomplished by inclining the sides of an excavation. The angle of incline varies according to such factors as the soil type, environmental conditions of exposure, and nearby loads. There are three general classifications of soil types, and one type of rock. For each classification of soil type, OSHA defines maximum angles for the slope of the walls, as shown in *Table 3*. The designation and selection of the proper sloping system is far more complex than described in this section. Factors such as the depth of the trench, the amount of time the trench is to remain open, and other factors will affect the maximum allowable slope.

Step-back, also known as benching, is another method of sloping the excavation walls without the use of a support system. Step-back uses a series of steps that must rise on the approximate angle of repose for the type of soil being used. The same safety rules apply to step-back excavations that apply to sloped excavations.

6.4.0 Sloping Requirements for Different Types of Soils

Many excavations are started with a vertical cut. Although some soils will stand to considerable depths when cut vertically, most will not. When vertical slopes fall to a more stable angle, large amounts of material usually fall into the excavation. *Figure 16* shows methods of failure in excavation walls.

The slope of the excavation walls must be angled so that material will not fall into the excavation. The slope is figured by angle from the horizontal plane and by horizontal run to vertical rise. For example, a 45-degree slope would be a 1:1 slope, meaning that for each foot measured horizontally, the slope rises 1 foot vertically. The slope of the excavation walls varies according to the type and characteristics of the soil being used. Four basic classifications of soil have been identified. The classification is based on the stability of the soil and the maximum allowable slopes for excavations in each of the four types of soils. These four classifications, in decreasing order of stability, are stable rock, Type A, Type B, and Type C. *Figure 17* shows maximum allowable slopes for each soil type. Vertical shores without sheeting may only be used in Type A or Type B soils.

6.4.1 Stable Rock

Stable rock refers to natural solid mineral matter that can be excavated with vertical sides and remain intact while exposed. Stable rock includes solid rock, shale, or cemented sands and gravels. *Figure 18* shows an excavation cut in stable rock.

6.4.2 Type A Soil

Type A soil refers to solid soil with a compression strength of at least 1.5 tons per square foot. Cohesive soils are soils that do not crumble, are plastic when moist, and are hard to break up when dry. Examples of cohesive soils are clay, silty clay,

Table 3 Maximum Allowable Slopes

Soil or Rock Type	Maximum Allowable Slopes for Excavations Less Than 20' Deep
Stable rock	Vertical (90 degrees)
Type A	¾:1 (53 degrees)
Type B	1:1 (45 degrees)
Type C	1½:1 (34 degrees)

208T03.EPS

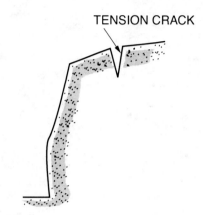

FORMATION OF TENSION CRACKS

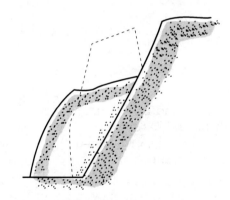

SLIDING OF SOIL INTO EXCAVATION

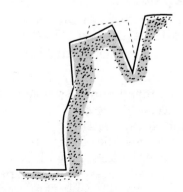

TOPPLING OF SOIL INTO EXCAVATION

208F16.EPS

Figure 16 ◆ Methods of failure in excavation walls.

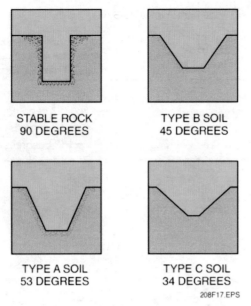

Figure 17 ◆ Maximum allowable slopes for soil types.

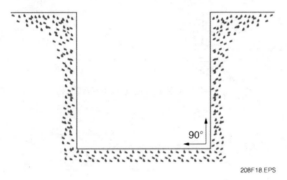

Figure 18 ◆ Excavation cut in stable rock.

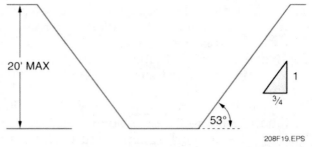

Figure 19 ◆ Simple slope excavation in Type A soil.

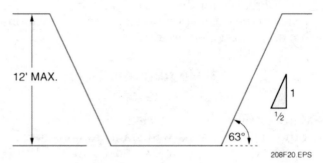

Figure 20 ◆ Simple slope, short-term excavation in Type A soil.

sandy clay, and clay loam. Cemented soils, such as caliche and hardpan, are also considered Type A. No soil can be considered Type A if any of the following conditions exist:

- The soil is fissured. Fissured means a soil material that has a tendency to break along definite planes of fracture with little resistance, or a material that has cracks, such as tension cracks, in an exposed surface.
- The soil can be affected by vibration from heavy traffic, pile driving, or other similar effects.
- The soil has been previously disturbed.

All sloped excavations that are 20 feet deep or less must have a maximum allowable slope of ¾ horizontal to 1 vertical. *Figure 19* shows a simple slope excavation in Type A soil.

An exception to this rule occurs when the excavation is 12 feet deep or less and will remain open 24 hours or less. These excavations in Type A soil can have a maximum allowable slope of ½ horizontal to 1 vertical.

Figure 20 shows a simple slope, short-term excavation in Type A soil.

6.4.3 Type B Soil

Type B soil refers to cohesive soils with a compression strength of greater than 0.5 ton per square foot but less than 1.5 tons per square foot. It also refers to granular soils, including angular gravel, which are similar to crushed rock, silt, sandy loam, unstable rock, and any unstable or fissured Type A soils. Type B soils also include previously disturbed soils, except those that would fall into the Type C classification. Excavations made in Type B soils have a maximum allowable slope of 1 horizontal to 1 vertical. Type B soils that are fairly firm, but which ravel slightly, can sometimes be held with vertical shoring and plywood. *Figure 21* shows simple slope excavations in Type B soil.

6.4.4 Type C Soil

Type C soil is the most unstable soil type. Type C soil refers to cohesive soil with a compression strength of 0.5 ton per square foot or less. Gravel, loamy soil, sand, any submerged soil or soil from which water is freely seeping, and unstable submerged rock are considered Type C soils. Excavations made in Type C soils have a maximum allowable slope of 1½ horizontal to 1 vertical. Type C soils cannot reliably be restrained with vertical shoring alone, but must be sloped and/or sheeted. *Figure 22* shows a simple slope excavation in Type C soil.

6.5.0 Combined Systems

Slide-rail systems can be considered a cross between trench boxes and steel sheeting. These

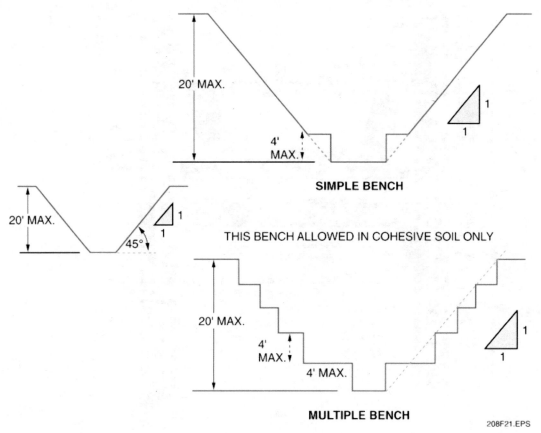

Figure 21 ◆ Simple slope excavations in Type B soil.

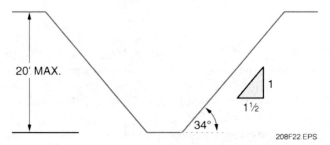

Figure 22 ◆ Simple slope excavation in Type C soil.

systems are designed to be used in shallow pits, tunnel pits, and trenches, or virtually anywhere that a trench box or sheeting systems can be used. The slide-rail system shown in *Figure 23* consists of the following components:

- *Slide rails* – The slide rails are the horizontal components of the system. They have a cavity to accept the lining plates and are fitted with cross braces. Slide rails are available in single-double-triple rail configurations. The triple-rail system is used at the greatest depth.
- *Lining plates* – Lining plates are used to provide the trench sidewall support. Lining plates are installed in the slide rails and are pushed into the ground as the excavation proceeds.

- *Cross braces* – Cross braces connect between the slide rails to provide lateral support to the system, usually at the end of the slide rail.

7.0.0 ◆ DETERMINING GRADE AND ELEVATION

The surveyor and trench digger set the initial grade of the trench, but before the pipe is laid, the finished grade must be determined and set in the trench. Never enter a trench until it has been properly shored, inspected, and approved by the competent person. Many job sites require you to obtain excavation entry permits from the competent person before entering the trench. Two methods commonly used to set the finished grade are the string-line method and the laser level method.

7.1.0 Setting Grade Using String Line

The quickest and easiest method of laying out a trench is to use a laser level. However, it is sometimes not convenient to use a laser level on a given job, so you will need to be able to do the job with a string line. The trench will only be as accurate as the string line is set. Nylon mason's line is best used for string lines. The surveyor

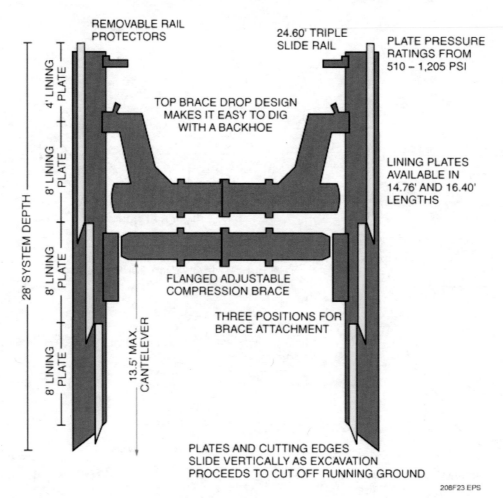

REMOVABLE RAIL PROTECTORS

24.60' TRIPLE SLIDE RAIL

PLATE PRESSURE RATINGS FROM 510 – 1,205 PSI

4' LINING PLATE

8' LINING PLATE

8' LINING PLATE

8' LINING PLATE

28' SYSTEM DEPTH

TOP BRACE DROP DESIGN MAKES IT EASY TO DIG WITH A BACKHOE

LINING PLATES AVAILABLE IN 14.76' AND 16.40' LENGTHS

13.5' MAX. CANTELEVER

FLANGED ADJUSTABLE COMPRESSION BRACE

THREE POSITIONS FOR BRACE ATTACHMENT

PLATES AND CUTTING EDGES SLIDE VERTICALLY AS EXCAVATION PROCEEDS TO CUT OFF RUNNING GROUND

208F23 EPS

Figure 23 ◆ Slide-rail system.

provides a bench mark elevation point. Having established the bench mark, the surveyor then sets other points, known as hubs, which can be used to set the finished grade in the trench. An information stake is located next to each hub, showing at what distance down from the hub the flow line of the pipe must be set. Information stakes always refer to the flow line, or invert elevation (the inside bottom), of the pipe. The surveyor should provide hubs all along the trench as references from which to run the grade line. *Figure 24* shows an information stake.

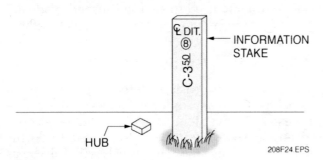

℄ DIT.
⑧
C-350

INFORMATION STAKE

HUB

208F24.EPS

Figure 24 ◆ Information stake.

String lines can be used to determine the depth of the trench and to set the finished grade in the trench after it has been dug. Pipefitters are responsible for setting string lines for sewer line projects before the trench is dug. They are also responsible for setting the finished grade in a trench before laying any type of underground pipe. Two types of string lines can be used to set finished grade. One type is a bottom grade line, which is a string line set in the bottom center of a trench, and the other is a top grade line, which is a string line set above the trench in the center.

7.1.1 Setting Grade Lines for Sewer Line Projects

When a trench is dug to tie into an existing sewer line, the surveyor usually does not provide a cut grade. If the trencher or backhoe operator wants a grade line to follow, you can provide this without the surveyor's assistance. After the service Y, tee, or point at which you will be tying into the sewer main is uncovered, you can set a grade line from that point back to the service location. Remember that the sewer line must be sloped down from the service location to the sewer main. The required slope of the line is given in the specifications.

Follow these steps to set a grade line for a sewer line project:

Step 1 Locate the service Y or tee on the sewer main and the service location to which the new line will be connected.

Step 2 Offset a grade pin at the service location and at the sewer main service tee.

NOTE

These grade pins will support the grade line string and need to be offset enough to allow a trencher or a backhoe to dig the trench. If a trencher is used, make sure that the grade pins are not on the side where the trencher will pile the removed dirt.

Step 3 Tie a string between the two grade pins.

Step 4 Level the string between the two grade pins using a string line level or a tubing water level.

Step 5 Measure the distance between the sewer main and the service location.

Step 6 Determine the minimum fall per foot required from the specifications.

Step 7 Multiply the minimum fall per foot by the length in feet of the run to determine the overall fall of the sewer line. If the minimum fall required is ¼ inch per foot, multiply ¼ inch by the length found in step 5.

Step 8 Move the string line at the service location grade pin up the amount of the overall fall. *Figure 25* shows an example of a sewer line grade line.

Step 9 Measure the distance between the sewer main grade pin string and the bottom of the sewer main service tee. This gives you the distance below the string line that the bottom of the pipe has to be at all points between the grade pins.

7.1.2 *Setting Bottom Grade Line*

In *Figure 26*, the information stake shows that the flow line of the pipe must be located 3.50 feet down from the hub. You must now determine the thickness of the pipe and if the specifications call for any bedding material before setting a string line in the trench. Suppose that the wall of the pipe is 2.40 inches thick and the specifications call for 6 inches of crushed rock to be under the pipe. Now one of you will measure at the hub; a second will hold a level across the trench, and a third will measure from the level down to the pin in the trench. Follow these steps to set a bottom grade line in a trench:

Step 1 Determine the initial measurement down from the hub at which the flow line of the pipe needs to be set. In this example, this will be 3.50 feet.

Step 2 Add the wall thickness of the pipe to this measurement.

3.50 feet + 2.4 inches (0.20 feet) = 3.70 feet

Step 3 Add the thickness of the bedding material as required by the specifications.

3.70 feet + 6 inches (0.5 feet) = 4.20 feet

The bottom of the trench must be 4.20 feet below the hub in order to add the specified backfill material and lay the pipe, keeping the flow line at the required elevation.

Step 4 Drive a grade pin into the bottom center of the trench directly across from each hub.

Step 5 Hold a straightedge level 1 foot above the hub extending over the trench (*Figure 26*).

Step 6 Measure down into the trench 4.20 feet from the straightedge and make a mark on the grade pin.

Step 7 Repeat steps 1 through 6 for each grade pin. Be sure to check each hub information stake before determining the measurement down to the flow line because each one will probably be different.

Step 8 Tie a string line between the grade pins at the line marks. This will give you a string line that is 1 foot above the required bottom elevation of the trench.

You can now grade the trench manually to achieve the finished grade. As you grade, continue to keep the bottom of the trench 1 foot below the grade line. After you grade the bottom of the trench to the correct elevation, you can begin to add the bedding material according to specifications. If the specifications call for 0.50 feet of bedding material, move your string line up at each grade pin 0.50 feet and fill in the bedding material so that it remains 1 foot below the string line for the length of the string line.

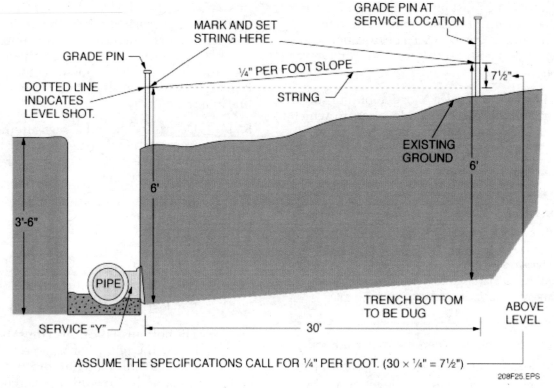

Figure 25 ◆ Sewer line grade line.

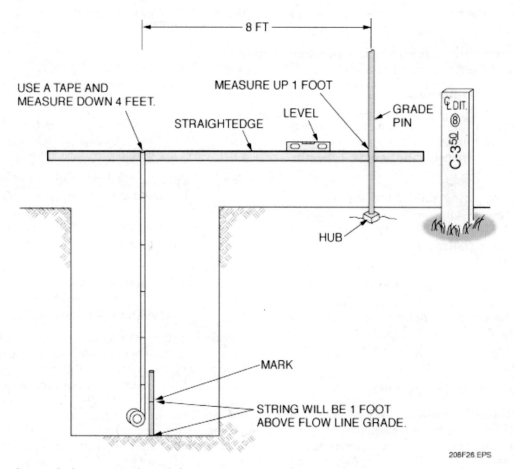

Figure 26 ◆ Setting the bottom trench grade line.

7.1.3 Setting Top Grade Line

Inspectors often prefer that a top grade line be set over the trench for performing manual finished grading and pipe laying. A top grade line is set in the same manner as a bottom grade line except that T-bars are used instead of grade pins. *Figure 27* shows a top grade line.

Follow these steps to set a top grade line:

> **NOTE**
>
> Use the same information stake in this exercise as was used in the previous section.

Step 1 Construct the T-bars using 2 × 4s.

Step 2 Place the T-bars over the trench at 20-foot intervals, and support them with 1 by 2-inch braces.

Step 3 Place dirt or sand bags over the ends of the legs of the T-bars to hold them in place.

Step 4 Determine the required elevation of the pipe from the information stake.

Step 5 Use a straightedge and a level to transfer the elevation from the hubs to the T-bars, and mark all the T-bars at the required elevation needed to use a grade rod, which is a small rod that is used in place of a rule.

Step 6 Drive nails just far enough into the T-bars to hold the string line at the marks.

Step 7 Wrap the string line around the first T-bar and tie it off on the nail.

Step 8 Pull the string line taut under the nail of the next bar, then wrap the string around the bar and back under the nail.

Step 9 Repeat Step 8 for all the T-bars.

Step 10 After the string is tight, go back and flip the string over the nails. Lap it over the bottom string to keep it from slipping (*Figure 28*).

> **NOTE**
>
> Remember that an undercut is needed for bedding material and the thickness of the pipe. You have to add this elevation to the grade rod length to perform finished grading.

7.2.0 Setting Grade Using Laser Level

Many grade setters are now using laser levels to mark and check grades and elevations. The accuracy and range of a laser level makes grade checking much easier. There are several types of laser levels. The best type for setting slopes and grades for pipeline in a trench is a laser level that can send out a level, declining, or inclined beam. With this type of level, each length of pipe can be set precisely on the slope desired.

All laser levels have an end receiver that intercepts the laser beam coming from the laser transmitter (*Figure 29*). This receiver is mounted on a

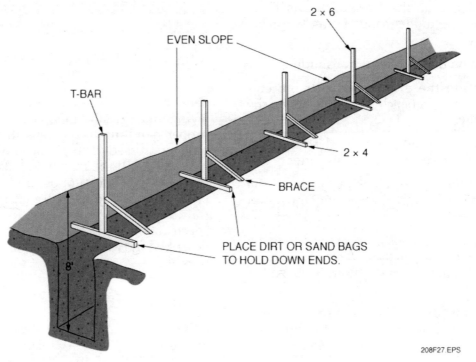

Figure 27 ◆ Top grade line.

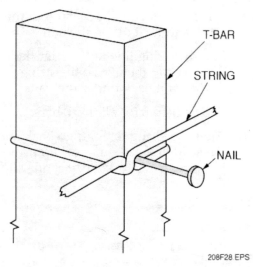

Figure 28 ◆ Tying string line on the T-bar.

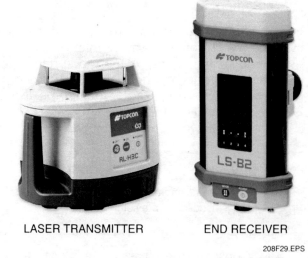

LASER TRANSMITTER   END RECEIVER

208F29.EPS

Figure 29 ◆ Laser level transmitter and end receiver.

grade rod to measure the elevation at that location. An arrow in the display window shows whether the receiver needs to be raised or lowered to intercept the laser beam exactly on center. When the end receiver is level with the laser transmitter, a horizontal line appears in the display window, or an audible signal is given.

Follow these steps to set the grade using a laser level:

Step 1 Position the laser level transmitter several feet from the work area, with the transmitter facing down the trench. Be sure that you set the transmitter where it cannot be disturbed by other workers or equipment in the area.

Step 2 Ensure that the transmitter is level according to the manufacturer's instructions.

Step 3 Turn on the laser level transmitter.

Step 4 Position the grade rod with the end receiver on a bench mark hub set by the surveyor.

Step 5 Turn the grade rod so that the end receiver faces the laser transmitter.

Step 6 Slide the end receiver up or down the grade rod until the receiver gives an on-grade signal.

Step 7 Measure the distance from the hub to the arrow pointer on the end receiver. This is the distance the laser beam is above the hub.

Step 8 Add this distance to the pipe invert elevation measurement from the hub that is found on the information stake at that hub.

Step 9 Measure this distance on a 1 × 2-inch rod, and draw a line indicating this distance on the rod.

Step 10 Clamp the end receiver to the 1 × 2-inch rod, with the arrow pointer level with the line just drawn.

Step 11 Set this rod into the trench to check for the proper depth to the flow line of the pipe.

NOTE

Remember that an undercut is needed to allow for bedding material and wall thickness of the pipe. You have to add this measurement to the grade rod length for finished elevation.

8.0.0 ◆ PERFORMING BACKFILLING PROCEDURES

Backfilling is one of the most important phases of laying underground pipe; careful attention to the proper handling of the backfill material cannot be overstressed. The purpose of the backfill is not only to fill the trench, but also to protect the pipe and provide support underneath the pipe in the trench. The backfill material and compaction requirements are given in the engineering specifications. The backfilling procedure can be described in two phases: initial backfill and final backfill.

8.1.0 Initial Backfill

After the pipe is laid on the bedding material as required by the engineering specifications, backfill the trench to a height of about 12 inches above the top of the pipe. This is considered the initial backfill. This procedure must be carefully performed to avoid disturbing the pipe alignment and to protect the pipe from damage during final backfilling. The initial backfill should be sliced under the sides or haunches of the pipe to fill the voids in this area. Slicing should be done when the backfill material is no higher than about one-fourth of the pipe diameter. *Figure 30* shows proper slicing of the initial backfill.

The type of material used for the initial backfill is stated in the engineering specifications and should always be free of large lumps or stones. Fittings should be positioned and supported by tamping the soil around and under the haunches of the fitting. Care must be taken to prevent rocks or lumps from damaging the pipe or disturbing the alignment of the newly installed line. Care must also be used not to disturb any pipe coatings.

8.2.0 Final Backfill

Final backfilling is normally done in layers known as lifts or fills. The engineering specifications give the compaction requirements, called the Proctor test, for the final backfill material by specify-

ing the maximum lift thicknesses. Each lift is spread in the trench and is compacted before the next lift is added. Compacting equipment forces smaller soil particles to move into spaces between the larger particles, thus decreasing the voids or spaces and increasing soil density. Soils are compacted to make the fill so dense that failure by water saturation, imposed loads, or other causes is not likely. During the final backfilling procedure, the following factors will affect the degree of compaction:

- Moisture content
- Type of material to be compacted
- Material gradation
- Lift thickness
- Number of passes by the compaction equipment
- Type of compaction equipment

In some cases, the native material cannot be compacted to satisfy the engineering specifications. In these cases, backfill material must be brought in from another area. Tests on compaction, density, and moisture are made during the backfilling operation to determine if the backfill is acceptable.

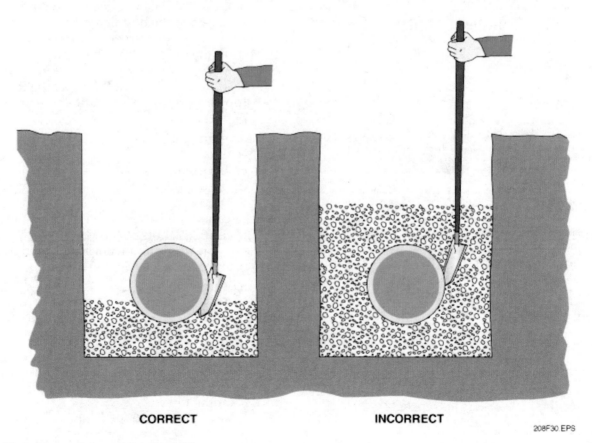

CORRECT   INCORRECT

208F30.EPS

Figure 30 ◆ Proper slicing of initial backfill.

1. A trench is a narrow excavation in which the depth is greater than the width and the width is no greater than _____.
   a. 10 feet
   b. 15 feet
   c. 25 feet
   d. 30 feet

2. One cubic yard of earth weighs approximately _____ pounds.
   a. 1,000
   b. 2,000
   c. 3,000
   d. 4,000

3. Flooding is considered a trenching hazard.
   a. True
   b. False

4. Dry sandy soil tends to _____.
   a. slab off
   b. crack
   c. collapse
   d. stay in place

5. Tools and equipment must be kept at least _____ from the edge of a trench.
   a. 1 foot
   b. 2 feet
   c. 3 feet
   d. 4 feet

6. In any trench that is over 4 feet deep, ladders and ramps used as exits in a trench must be located every _____.
   a. 10 feet
   b. 15 feet
   c. 20 feet
   d. 25 feet

7. Shoring, sloping, or other acceptable protections should be used to guard the walls of all excavations more than _____ deep.
   a. 1 foot
   b. 3 feet
   c. 4 feet
   d. 5 feet

8. Which of the following is *not* a form of protection for workers in trenches?
   a. Shoring
   b. Sloping
   c. Benching
   d. Vertical walls

9. The most common hazard during an excavation is _____.
   a. a cave-in
   b. an electrical strike
   c. a fall
   d. flooding

10. Always install shoring from _____.
    a. the top of the trench
    b. the bottom of the trench
    c. inside the shoring
    d. the side of the trench slope

11. The sides of a simple trench in type C soil should be sloped at a _____ angle from the horizontal.
    a. ¾ to 1
    b. 1 to 1
    c. 1½ to 1
    d. 2 to 1

12. An information stake next to the surveyor's bench mark tells the distance _____.
    a. down to the flow line of the pipe
    b. down to the bottom of the trench
    c. down to the center line of the pipe
    d. to the other side of the trench

13. To determine the grading of a pipeline from top to bottom, the pipefitter ties a string between two _____.
    a. grade pins
    b. T-bars
    c. information stakes
    d. grade rods

14. Because the information stake shows the invert elevation of the pipeline, you need not consider the thickness of the pipe wall or the bedding to dig the trench.
    a. True
    b. False

15. Slicing of backfill should be started when the backfill is about _____ of the pipe.
    a. halfway to the top
    b. one-third of the way to the top
    c. at the top
    d. one-quarter of the way up from the bottom

Summary

Working in and around excavations is the most dangerous type of work you will encounter. You must be aware of your surroundings and be alert at all times while in the trench. Even though the competent person on the job site is the person responsible and liable for the design and working conditions of the excavation, everyone involved must take responsibility to ensure personal safety.

Shoring materials and premanufactured support systems must meet or exceed all OSHA requirements and be thoroughly inspected before they are used. If you are ever unsure about a situation in a trench, get out and find the competent person immediately to report your concerns.

When calculating and setting any grade line, be sure to check every figure. Avoid any distraction while figuring or marking grades. If you are not quick with numbers, make it a practice to use a pencil and paper to do calculations or use a calculator.

Notes

Trade Terms Introduced in This Module

Angle of repose: The greatest angle above the horizontal place at which a material will adjust without sliding.

Bedding: The surface or foundation on which a pipe rests in a trench.

Bench mark: A point of known elevation from which the surveyors establish all of the grades.

Benching: A method of protecting workers from cave-ins by excavating the sides of an excavation to form one or a series of horizontal levels or steps, usually with vertical or near-vertical surfaces between levels.

Carbonation: A chemical process in which carbon accumulates and causes a breakdown of surrounding minerals, such as dirt.

Catchment basin: A bowl area constructed to capture and retain runoff water.

Coffer dam: A temporary dam constructed to divert or contain water.

Compaction: The action of forcing small soil particles between larger ones until most of the air space or voids and moisture have been squeezed out. The result is a hard, solid, rock-like mass.

Competent person: A person who is capable of identifying existing and predictable hazards in the area or working conditions that are unsanitary, hazardous, or dangerous to employees, and who has the authority to take prompt corrective measures to fix the problem.

Compression strength: The ability of soil to hold a heavy weight.

Cross braces: The horizontal members of a shoring system installed perpendicular to the sides of the excavation, the ends of which bear against either uprights or wales.

Disturbed soil: Soil that has been previously backfilled or removed from the earth and replaced by any means.

Excavation: Any man-made cut, cavity, trench, or depression in the earth's surface, formed by earth removal.

Finished grade: Any surface that has been cut or built to the elevation indicated for that point.

Grade: The surface level required by the plans or specifications.

Grade pin: Steel rod driven into the ground at each surveyor's hub. A string is attached between them at the grade indicated on the information stakes.

Grade rod: A small length of round or rectangular wood or metal used in place of a rule for checking grades.

Hub: A point-of-origin stake that identifies a point on the ground. The top of the hub establishes the point from which soil elevations and distances are computed.

Hydration: A chemical process in which water or liquids accumulate and cause a breakdown of surrounding minerals, such as dirt.

Information stake: A marker that shows elevations and grades that must be established and the distances to them.

Lift: A layer of fill that can be either loose or compacted.

Monitoring well: A well dug to test the effect of construction on groundwater.

Oxidation: A chemical process in which oxygen accumulates and causes a breakdown of surrounding minerals, such as dirt.

Protective system: A method of protecting employees from cave-ins, from material that could fall or roll from an excavation face or into an excavation, or from the collapse of adjacent structures. Protective systems include support systems, sloping and benching systems, and shielding systems.

Raveling: The process of small particles breaking away from excavation walls.

Screw jack: A screw- or hydraulic-type jack that is used as cross bracing in a trench.

Sheeting: Plywood or sheet metal that is placed against and braced between the walls of an excavation.

Shield: A structure that is able to withstand the forces imposed on it by a cave-in and thereby protect employees within the structure. Shields can be permanent structures or can be designed to be portable and moved along as work progresses. Additionally, shields can be either pre-manufactured or job-built in accordance with *29 CFR 1926.652 (c)(3)* or *(c)(4)*.

Shore: Timber or other material used as a temporary prop for excavations. It may be sloping, vertical, or horizontal.

Shoring: A structure such as a metal hydraulic, mechanical, or timber shoring system that supports the sides of an excavation and is designed to prevent cave-ins.

Skeleton: A condition that occurs when individual timber uprights or individual hydraulic shores are not placed in contact with the adjacent member.

Sloping: A method of protecting employees working in excavations from cave-ins by cutting the excavation walls to the angle of repose of the soil being excavated.

Spoils: Excavated soil that is removed from the hole and piled next to the excavation. The distance between the spoils and the excavation is regulated by OSHA.

String line: A nylon line usually strung tightly between supports to indicate both direction and elevation.

Subsidence: A depression in the earth that is caused by unbalanced stresses in the soil surrounding an excavation.

Support system: A structure, such as an underpinning, bracing, or shoring, that provides support to an adjacent structure, underground installation, or the walls of an excavation.

T-bars: T-shaped wood frames used in place of steel pins to support a string line over trenches.

Tight sheeting: The use of specially edged timber planks, such as tongue-and-groove planks, that are at least 3" thick. Steel sheet piling or similar construction that resists the lateral pressure of water and prevents loss of backfill is called tight sheeting.

Trench: A narrow excavation made below the surface of the ground. In general, a trench is no wider than it is deep and should never be more than 15 feet wide.

Trench box: A premanufactured steel box that provides a safe means of shoring by allowing workers to lay pipe within the box while it is inside the excavation.

Uprights: The vertical members of a trench shoring system placed in contact with the earth and usually positioned so that individual members do not contact each other. Uprights placed so that individual members are closely spaced, in contact with, or interconnected to each other, are often called sheeting.

Void: A space in a soil mass not occupied by solid material.

Wales: Horizontal members of a shoring system placed parallel to the excavation face whose sides bear against the vertical members of the shoring system or the earth.

Resources & Acknowledgments

Additional Resources

This module is intended to be a thorough resource for task training. The following reference works are suggested for further study. These are optional materials for continued education rather than for task training.

Excavators Handbook Advanced Techniques for Operators, 1999. Reinar Christian. Addison, IL: The Aberdeen Group, A division of Hanley-Wood, Inc.

Basic Equipment Operator, 1994 Edition. John T. Morris (preparer), NAVEDTRA 14081, Naval Education and Training Professional Development and Technology Center.

Figure Credits

Trench Shoring Services, Inc., 208F01, 208F04, 208F05, 208F06, 208F08, 208F14

Topcon Positioning Systems, Inc., 208F29

Cianbro Coporation, Module opener

NCCER CURRICULA — USER UPDATE

NCCER makes every effort to keep its textbooks up-to-date and free of technical errors. We appreciate your help in this process. If you find an error, a typographical mistake, or an inaccuracy in NCCER's curricula, please fill out this form (or a photocopy), or complete the online form at **www.nccer.org/olf**. Be sure to include the exact module ID number, page number, a detailed description, and your recommended correction. Your input will be brought to the attention of the Authoring Team. Thank you for your assistance.

Instructors – If you have an idea for improving this textbook, or have found that additional materials were necessary to teach this module effectively, please let us know so that we may present your suggestions to the Authoring Team.

NCCER Product Development and Revision
13614 Progress Blvd., Alachua, FL 32615

Email: curriculum@nccer.org
Online: www.nccer.org/olf

❏ Trainee Guide   ❏ Lesson Plans   ❏ Exam   ❏ PowerPoints   Other _____

Craft / Level: _____   Copyright Date: _____

Module ID Number / Title: _____

Section Number(s): _____

Description: _____

_____

_____

_____

_____

Recommended Correction: _____

_____

_____

_____

_____

Your Name: _____

Address: _____

_____

Email: _____   Phone: _____

08209-06

Underground Pipe Installation

08209-06
Underground Pipe Installation

Topics to be presented in this module include:

Overview

Most municipal piping systems that deliver water and gas and oil, and the systems that carry storm drain water and sewerage, are underground. In this module, you will learn about the connection systems used underground, and the way they are assembled. You will learn how to work with different pipe materials, and the uses of each kind. You will also learn a little about some new technologies for laying pipe underground, called trenchless pipelaying.

Objectives

When you have completed this module, you will be able to do the following:

1. Identify and explain the types of underground piping materials.
2. Identify the size classifications of underground pipe.
3. Identify and explain the use of underground pipe fittings.
4. Explain the joining methods for underground pipe.
5. Explain the storage and handling requirements of underground pipe.
6. Identify and explain underground pipe installation guidelines.
7. Join CPVC and PVC.
8. Join ductile iron.

Trade Terms

Acrylonitrile-butadiene-styrene (ABS)
Bell-and-spigot pipe
Bituminous coating
Boring
Branch
Cellular core wall
Chemically inert
Chlorinated polyvinyl chloride (CPVC)
Compression collar
Cross-linked polyethylene (PEX)
Culvert
Elastomeric
Elliptical
Exfiltration
Fusion fitting
Gasket
High-density polyethylene (HDPE)
Hydronic
Infiltration
Inside diameter (ID)
Interference fit

Invert elevation
Lead
Mastic
Neoprene
Outside diameter (OD)
Polybutylene (PB)
Polyethylene (PE)
Polyvinyl chloride (PVC)
Portland cement
Pounds per square inch (psi)
Pressure rating
Resin
Ring-tight gasket fitting
Size dimension ratio (SDR)
Solid wall
Solvent
Solvent weld
Tap
Thermoplastic pipe
Thermosetting pipe
Thrust block
Transition fitting

Required Trainee Materials

1. Pencil and paper
2. Appropriate personal protective equipment

Prerequisites

Before you begin this module, it is recommended that you successfully complete *Core Curriculum*; *Pipefitting Level One*; and *Pipefitting Level Two*, Modules 08201-06 through 08208-06.

This course map shows all of the modules in the second level of the *Pipefitting* curriculum. The suggested training order begins at the bottom and proceeds up. Skill levels increase as you advance on the course map. The local Training Program Sponsor may adjust the training order.

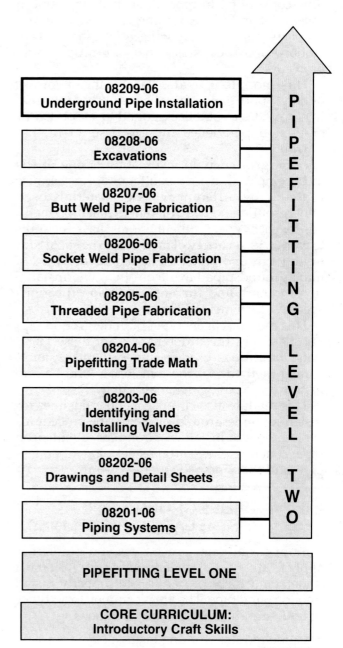

08209-06
Underground Pipe Installation

08208-06
Excavations

08207-06
Butt Weld Pipe Fabrication

08206-06
Socket Weld Pipe Fabrication

08205-06
Threaded Pipe Fabrication

08204-06
Pipefitting Trade Math

08203-06
Identifying and Installing Valves

08202-06
Drawings and Detail Sheets

08201-06
Piping Systems

PIPEFITTING LEVEL TWO

PIPEFITTING LEVEL ONE

CORE CURRICULUM: Introductory Craft Skills

209CMAP.EPS

1.0.0 ◆ INTRODUCTION

Underground piping is truly a job that you want to do only once for the planned lifetime of the project. By thoroughly understanding the basics in this module, you will be able to safely lay underground piping that will last.

Underground piping is an important part of the pipefitter's job. Installing underground piping properly is a complex process requiring attention to detail and specific knowledge of the installation process.

Bad piping installations are expensive and time-consuming to repair. Bad installations can cause system failure, leading to loss of life and property damage. Following established installation procedures is essential to limit potential problems.

This module provides basic information on underground piping installations. This information provides a solid background to support your on-the-job experience and is a base for further learning.

Underground pipe installation is one of the most complex duties a pipefitter performs because of the many different types of materials used underground and the different installation methods for each type. A pipefitter must have the skills to work with a variety of materials to be successful in underground pipe installation. Many types of underground pipe have more than one installation method, and the engineer's specifications specify which method to use.

This module covers types of underground piping materials, the different sizes of these materials, the fittings used in underground systems, joining methods for these materials, and how to handle these piping materials safely.

This module also covers general guidelines for installing underground pipe. Specific procedures explain how to install cast iron pipe, ductile iron pipe, concrete pipe, carbon steel pipe, fiberglass pipe, and thermoplastic pipe.

2.0.0 ◆ UNDERGROUND PIPE INSTALLATION GUIDELINES

Certain installation procedures and guidelines apply to all types of underground pipe. You must understand these procedures to properly install underground pipe. The areas common to all types of underground pipe include the following:

- Guidelines for working in and near a trench
- Preparing the trench
- Preparing the bedding material
- Laying the pipe
- Installing thrust blocks

2.1.0 Guidelines for Working in and near a Trench

When working in or around any excavation or trench, you are responsible for personal safety. You are also responsible for the safety of others in the work trench. The following guidelines must be enforced to ensure everyone's safety:

- Never enter an excavation without the approval of the OSHA-approved competent person on site.
- Inspect the excavation daily for changes in the excavation environment, such as rain, frost, or severe vibration from nearby heavy equipment.
- Never enter an excavation until the excavation has been inspected.
- Wear protective clothing and equipment, such as hard hats, safety glasses, work boots, and gloves. Use respirator equipment if necessary.
- If you are exposed to vehicle traffic, wear traffic warning vests that are marked with or made of reflective or highly visible material.
- Get out of the trench immediately if water starts to accumulate in the trench.
- Do not walk under loads being handled by power shovels, derricks, or hoists.
- Stay clear of any vehicle that is being loaded.
- Be alert. Watch and listen for possible dangers.
- Do not work above or below a co-worker on a sloped or benched excavation wall.
- Barricade access to excavations to protect pedestrians and vehicles (*Figure 1*).
- Check with your supervisor to see if workers entering the excavation need excavation entry permits.

209F01.EPS

Figure 1 ◆ Barricaded trench.

- Ladders and ramps used as exits must be located every 25 feet in any trench that is over 4 feet deep.
- Make sure someone is on top to watch the walls when you enter a trench.
- Make sure you are never alone in a trench. Two people can cover each other's blind spots.
- Keep tools, equipment, and the excavated dirt at least 2 feet from the edge of the excavation.
- Make sure shoring, trench boxes (*Figure 2*), benching, or sloping are used for excavations and trenches over 5 feet deep.
- Stop work immediately if there is any potential for a cave-in. Make sure any problems are corrected before starting work again.

2.1.1 Indications of an Unstable Trench

A number of stresses and weaknesses can occur in an open trench or excavation. For example, increases or decreases in moisture content can affect the stability of a trench or excavation. The following sections discuss some of the more frequently identified causes of trench failure. These conditions are illustrated in *Figure 3*.

Tension cracks usually form one-quarter to one-half of the way down from the top of a trench. Sliding or slipping may occur as a result of tension cracks. In addition to sliding, tension cracks can cause toppling. Toppling occurs when the trench's vertical face shears along the tension crack line and topples into the excavation. An unsupported excavation can create an unbalanced stress in the soil, which in turn causes subsidence at the surface and bulging of the vertical face of the trench. If uncorrected, this condition can cause wall failure and trap workers in the trench, or greatly stress the protective system. Bottom heaving is caused by downward pressure created by the weight of adjoining soil. This pressure causes a bulge in the bottom of the cut. Heaving and squeezing can occur even when shoring and shielding are properly installed.

CAUTION

Protective systems are designed for even loads of earth. Heaving and squeezing can place uneven loads on the shielding system and may stress particular parts of the protective system.

Another indication of an unstable trench is boiling. Boiling is when water flows upward into the bottom of the cut. A high water table is one of the causes of boiling. Boiling can happen quickly and can occur even when shoring or trench boxes are used. If boiling starts, stop what you are doing and leave the trench immediately.

www.SHORING.com

209F02.EPS

Figure 2 ◆ Trench box.

2.2.0 Preparing a Trench

The trench must be designed, excavated, and shored according to OSHA standards and must be inspected and approved by the on-site competent person. In most cases, the engineer or surveyor establishes the line and grade of the pipe to be laid and installs information stakes that the equipment operator uses to dig the trench at the specified location and depth.

The trench width and depth determines the load that the pipe has to support once backfilling is complete. The trench width at the top of the pipe cannot exceed the dimensions specified on the design drawings because the increased load on the pipe may cause a structural failure. Even a small increase in the trench width causes a large increase in loading. For example, if a 2-foot trench width is increased by 6 inches, the load on the pipe increases by 50 percent. To meet the specified trench width at the top of the pipe, different trench configurations can be used. A narrow step trench or subtrench can be excavated near the trench bottom after a wider trench is used above the top of the pipe. A vee or modified vee trench can also be used. *Figure 4* shows load increases and trench configurations.

For additional details on excavations, see the *Pipefitting Level Two* module *Excavations*.

2.3.0 Preparing Bedding Material

The engineer's standards for providing uniform support of the pipe on a firm foundation must be followed. The most effective method of providing

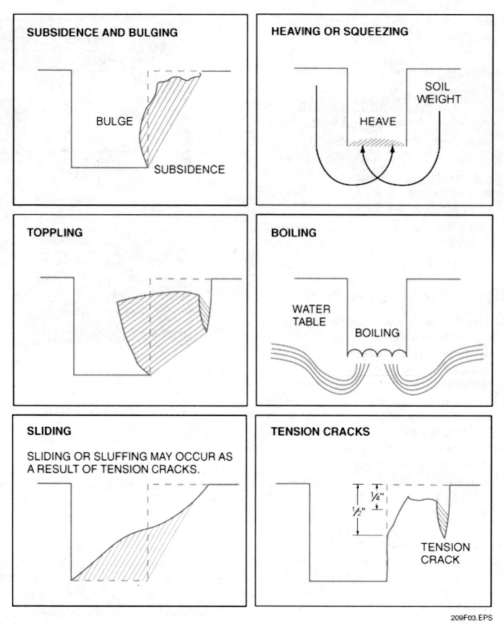

Figure 3 ◆ Indications of an unstable trench.

a firm foundation is to over-excavate the trench and add the proper amount of bedding material in order to bring the invert elevation of the pipe to the proper grade. Holes must be dug at each bell or coupling so that the load is supported by the pipe barrel instead of the pipe bell. The holes must be no larger than necessary to relieve pressure from the bell or coupling. *Figure 5* shows bell holes in the bedding material.

The engineering specifications dictate what type of bedding support is needed in the trench. The pipe strength and the bedding support work together to withstand the trench load. Engineers use a load factor based on a designated type of bedding support to determine the loadbearing capacity or pipeline strength. The loadbearing capacity consists of the pipe itself increased by

the support of the particular bedding foundation. The engineer has given special consideration to the selection of a load factor to meet the calculated trench load. This is why it is so important that the trench width be maintained and bedding material be placed according to specifications.

2.4.0 Laying Pipe

To ensure an efficient, high-quality job, follow these guidelines when laying pipe in a trench:

- Handle pipe and fittings carefully to protect them from damage due to impact, shocks, and free fall.
- Examine each pipe or fitting carefully before installation.

- Handle pipe so that premolded installing surfaces or attached couplings do not support the weight of the pipe.
- Do not damage the installing surfaces or couplings by dragging them, allowing them to come in contact with hard materials, or by using pipe hooks to transport the pipe.
- Clean joint surfaces immediately before installing them.
- Use joint lubricants and installing methods recommended by the manufacturer.
- Lay all pipe straight between changes in alignment and at uniform grade between changes in grade.
- Excavate bell holes for each pipe joint as shown in *Figure 5*.
- Ensure that the pipe forms a smooth, straight line when joined in the trench.
- Start laying pipe at the lowest point and work upstream, with the bell end of the pipe laid upstream when possible.

- Shovel slice bedding material under the pipe haunches after each pipe has been aligned, graded, and brought to final grade. *Figure 6* shows the haunch area.

2.5.0 Installing Thrust Blocks

When a pipeline is in service, whether it runs underground or aboveground, it is subject to movement because of the pressure of the fluid flowing through it. This pressure is strong enough to cause the joints to separate and leak. Thrust forces in pressurized systems are created when the pipeline changes direction, stops, or changes in size.

Thrust blocks, designed by engineers, help restrain the forces in an underground piping system. They are usually made of concrete that has a certain compression strength specified by the design engineer. The size and volume of the concrete determine how the thrust block is to be formed because the concrete has to transmit forces from the fitting to the undisturbed soil. *Figure 7* shows typical placement of thrust blocks at fittings. The arrows show the direction of force.

2.6.0 Underground Pipe Sizes

Pipe and fittings are also identified by size and strength. Pipe size is measured by the diameter, or

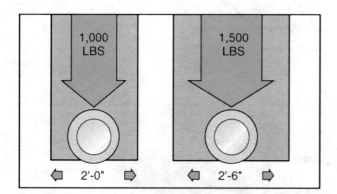

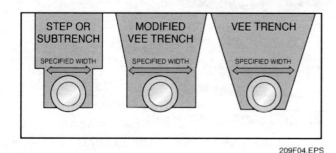

209F04.EPS

Figure 4 ◆ Load increases and trench configurations.

HAUNCH AREA

209F06.EPS

Figure 6 ◆ Haunch area.

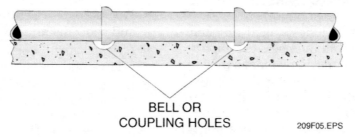

BELL OR
COUPLING HOLES

209F05.EPS

Figure 5 ◆ Bell holes in bedding material.

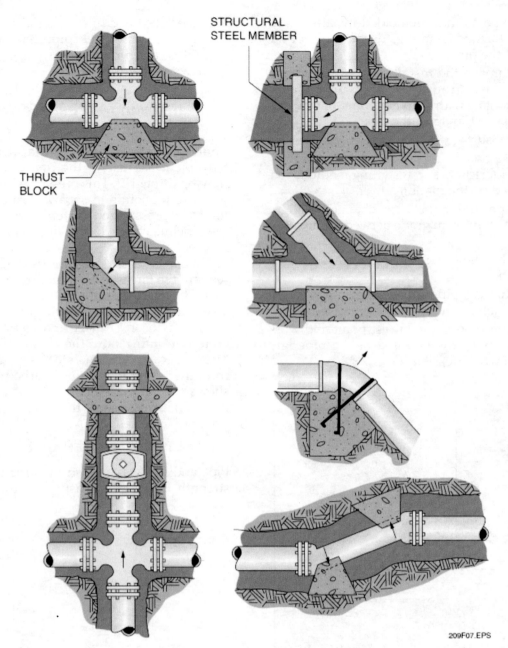

STRUCTURAL STEEL MEMBER

THRUST BLOCK

Figure 7 ◆ Thrust block placement.

the distance across the end of the pipe. The inside diameter (ID) of a pipe is the fluid-carrying area of the pipe. The outside diameter (OD) is the measurement of the actual width of the pipe. The difference between the inside and outside diameter is the total wall thickness. *Figure 8* shows the inside and outside diameter of pipe.

Total wall thickness includes a wall on each side. *Figure 9* shows the wall thickness of a pipe.

Another way to denote the diameter of a pipe is to use the nominal pipe size, which is the approximate ID of a pipe in Schedule 40 pipe. This nominal size is used by manufacturers. The actual pipe size is usually several thousandths of an inch larger than the nominal size.

2.7.0 Trenchless Pipelaying

Trenchless pipelaying is the general name for methods of burying pipelines that do not involve digging trenches. There are three systems of trenchless pipelaying:

- *Pipe jacking* – Used for small to medium pipes and short distances
- *Pipe ramming* – Used most often when the pipe is large and the distances are less than 150 feet
- *Horizontal directional drilling (HDD)* – A favored method of placing and replacing pipe in cities, especially, which can place piping as large as 36 inches in diameter, and deliver pipelines as much as a mile long (See *Figure 10*.)

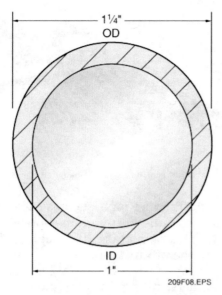

Figure 8 ◆ Inside and outside diameter of a pipe.

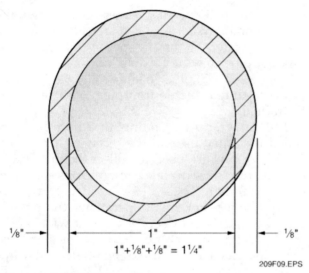

Figure 9 ◆ Wall thickness of pipe.

Pipe jacking, as the name implies, uses a hydraulic jack to force a capped pipe through the soil. The contractor digs a pit, long enough to hold the jack and a joint of pipe. The cap on the pipe keeps the dirt out, and the jack forces the pipe through the required distance. This is strictly a method for steel pipe, as the strength of the pipe is needed to sustain the forces involved.

Pipe ramming uses a hydraulic ram to hammer pipe through the soil. The pipe is usually left open, allowing soil to be pushed into the pipe as the pipe is pushed through. The technique involves repeated blows, and is much like the technique used for setting pilings. Again, the operation starts from a pit, and is then attached, a joint at a time, to the pipe as it feeds through. The method is useable in any sort of soil except solid rock, but is best used in relatively unconsolidated soils. Once again, the strength of steel pipe

Figure 10 ◆ Drill pipe.

is required for ramming pipe, and it must be in a straight line.

HDD is becoming more frequent than the other methods, and is replacing trenching for many city applications. HDD is faster, and frequently more economical than trenching; there are not as many drilling contractors as trenchers, but this is a very fast and successful process. Plastic, steel, and ductile iron pipe can all be buried with drilling equipment.

2.7.1 Horizontal Directional Drilling

HDD jobs can vary from installing pipes that are as small as 2 inches up to those with diameters as large as 48 inches. The size of the drill rigs used will also vary. Rigs can be small, medium, or large (*Figure 11*). Small rigs are used for installing utility cable and smaller pipes in crowded city areas. Medium-sized rigs are used to install municipal pipelines. Large rigs are used to install large diameter pipes.

During a typical boring operation, sections of pipe are joined together at each end and pushed through the ground. The drill head (*Figure 12*), which is placed on the front of the drill pipe, rotates and pushes through the ground to make the bore hole. The drill is usually started at an angle from 8 to 20 degrees from the horizontal, and generally makes a small hole called a pilot hole. Once the drill head reaches the end of the bore hole, a back reamer (*Figure 13*) is attached and the drill pipe is pulled back through the bore hole. See *Figure 14* for a picture of the pilot drill-

SMALL RIG

MEDIUM RIG

LARGE RIG

209F11.EPS

Figure 11 ◆ Drill rigs.

209F12.EPS

Figure 12 ◆ Drill heads.

209F13.EPS

Figure 13 ◆ Back reamer.

ing system. Product pipe, such as cable or utility lines, may be pulled through the bore at the same time. The back reamer is usually at least 1.5 times the size of the product pipe. The reaming process may be repeated several times with progressively larger reamers, until the product pipe can be freely pulled through. *Figure 15* shows the back-reaming process. Special lubricants, called muds, are used to help the drill and the reamers go through; the most common materials used are either bentonite clay and water, or certain special polymers. The product pipe may be flexible high-density polyethylene (HDPE) pipe, in large rolls; welded steel; or even jointed ductile iron pipe, with the joints restrained.

The entry and exit points of a bore hole are called the entry pit and the receiving pit (*Figure 16*). These pits are used to collect drilling fluids. Drilling fluids are drained from the pits at the end of the operation. These pits vary in size, depending on the job. For small operations, the pits are typically 1 to 3 feet deep, 1 to 2 feet wide, and 2 to 6 feet long. For larger operations, the entry and receiving pits may require trenching and shoring.

A technique that is similar to HDD is pipe bursting. This method is used when replacing PVC or clay pipe. A special pneumatically or hydraulically powered bursting head is sent through the existing pipe. As it returns, it expands and bursts the old pipe while pulling the replacement pipe through.

3.0.0 ◆ STORING AND HANDLING UNDERGROUND PIPE

When a shipment of pipe arrives at the job site, it must be unloaded, inspected, and restacked. This section describes the procedures for handling underground pipe.

3.1.0 Unloading and Transporting Pipe

Slings and pipe hooks can be used for unloading and handling cast iron, ductile iron, carbon steel, and galvanized pipe. The sling may have a sliding hook or may be made of nylon. Nylon web slings

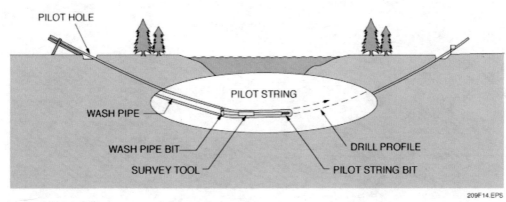

Figure 14 ♦ Pilot bore.

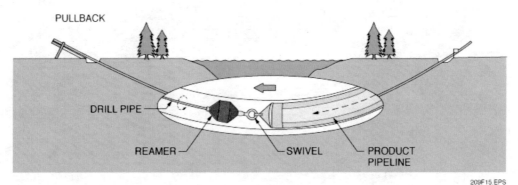

Figure 15 ♦ Back-reaming process.

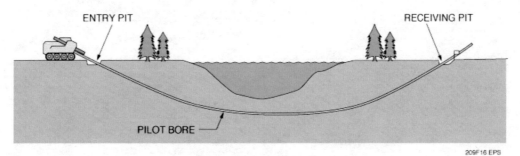

Figure 16 ♦ Entry and receiving pits.

are recommended for unloading ductile iron pipe that has been specially coated. Pipe hooks are usually made of thick, round stock and are shaped to fit both the bell-and-spigot ends without damaging the pipe. The hooks must be padded when used to handle cement-lined pipe. *Figure 17* shows slings and pipe hooks. Follow these safety guidelines when handling pipe:

• Do not handle the ends of the pipe. Personal injury or damage to the pipe ends could occur.
• Beware of pinch points between the pipes.
• Be aware of the weight of the load to be lifted.
• Match rigging to the weight and size of the load being lifted.
• Never get underneath a load that is suspended.
• Handle all loads with a tag line.
• When lifting pipe, keep it low to the ground.

3.2.0 Handling Concrete Pipe

There are a number of ways to handle concrete pipe, depending on its size and the availability of rigging equipment. Slings and chokers can be used, and slings with sliding hooks are very useful. Be sure that the wire rope is of sufficient size and strength to handle the size of concrete pipe you are moving. Larger sizes of concrete pipe are cast with a hole in the barrel located at the exact center of gravity. A rigging device is inserted into the hole and secured by a bearing plate or something similar. Some of these rigging devices are commercially available, while others can be job-fabricated. All job-fabricated lifting devices must be engineered and certified for weight capacity.

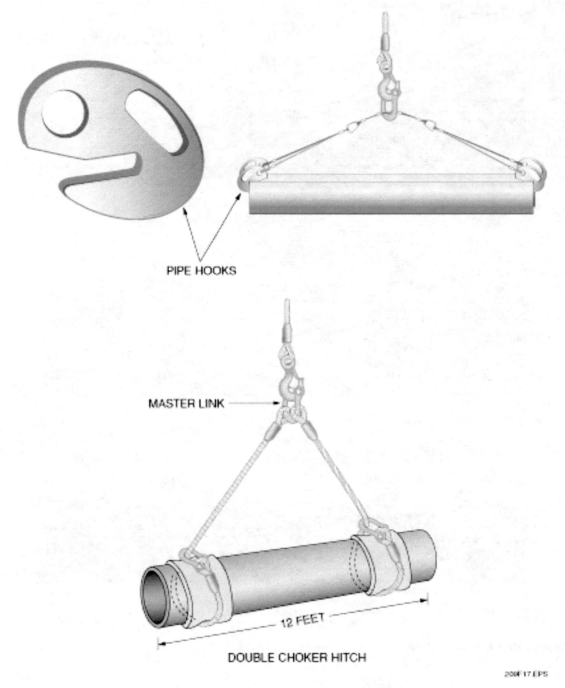

PIPE HOOKS

MASTER LINK

12 FEET

DOUBLE CHOKER HITCH

209F17.EPS

Figure 17 ◆ Slings and pipe hooks.

3.3.0 Commercial Lifting Devices

One type of commercial lifting device is known as a hairpin. A hairpin is a large steel hook that slides into the pipe from one end to lift the pipe. The lower part of the hook extends well into the pipe and provides support over a large portion of the pipe to prevent the pipe ends from breaking and to help keep the pipe level during lifting. Often a softener, such as rubber from an old tire, is attached to the hairpin to prevent the steel on the hairpin from coming in contact with the concrete pipe. *Figure 18* shows a hairpin lifting device.

3.4.0 Job-Fabricated Lifting Devices

Job-fabricated lifting devices can also be used to lift concrete pipe if they are designed and approved by a registered professional engineer. A choker with eye on the end is inserted into the lifting hole. A hard steel bar, a bullpoint from a jackhammer, or, in some cases, a long hardwood 4-inch by 4-inch timber is placed through the eye of the choker to bear against the pipe. Be sure these devices are strong enough to support the weight of the pipe. Do not use ordinary pipe hooks to lift concrete pipe because they will break the ends of the pipe.

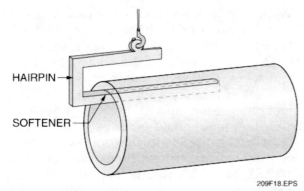

HAIRPIN

SOFTENER

209F18.EPS

Figure 18 ◆ Hairpin lifting device.

3.5.0 Handling Fiberglass and Plastic Pipe

Fiberglass and plastic pipe are lightweight and come in many different lengths. They should be unloaded and transported by hand, unless they arrive on the job site secured to a pallet. If the pipe comes on a pallet, it can be lifted and transported with a forklift. Do not throw or drop fiberglass or plastic pipe and fittings.

3.6.0 Inspecting Pipe

When a shipment of pipe arrives on the job site, it should be inspected immediately. Check the shipment for the following:

- Be sure the correct size, quantity, and type of pipe has been shipped.
- Check the contents of the load against the shipping papers.
- Check each pipe for signs of abuse, including abrasions, dents, gouges, chips, and other damage.
- Check the bell ends for damage.
- Check the condition of the cement lining if the pipe is lined.
- Check the concrete pipe barrels for voids, called honeycombs.

If a length of pipe is found to be unacceptable, it should be flagged and isolated. Never place a rejected pipe in a trench. Removing a length of pipe is always harder than installing it.

3.7.0 Stacking Pipe

If at all possible, avoid stacking pipe because of the safety hazards involved. If space is not available to lay pipe on the ground, follow these guidelines to stack pipe:

- Store all pipe on level ground.
- Stack all pipe on timbers or elevated concrete supports to keep the bottom layer off the ground.

- Protect the ends of all pipe with plastic covers or end protectors.
- Stack the same sizes and types of pipe together.
- Alternate layers of pipe by placing the bell end of a pipe on the spigot end of a pipe in successive layers.
- Always place timbers between layers.
- Chock the sides of each layer on each end to prevent movement.
- Do not stack dissimilar metal pipes together. This prevents cross-contamination of metals.
- When stacking pipe, use a designed rack that is load-rated to withstand the load of the pipe.

Concrete pipe should be stored or stacked as close to the installation area as possible. Set the pipe on blocks placed under the pipe barrel to avoid damaging the ends. If storing only one layer of pipe, alternate the bell-and-spigot ends. When stacking concrete pipe in multiple layers, place all the bells at the same end in one layer and at opposite ends in the next layer. Bells should project beyond spigots. Be sure to block the bottom layer securely to guard against movement.

4.0.0 ◆ TYPES OF UNDERGROUND PIPING MATERIALS

Most underground piping is used for utility systems, including sanitary drainage and vent piping, storm water drainage piping, fire and cooling water, and pump-out systems. Underground utility systems are classified as either gravity-flow systems or pressurized systems. Gravity-flow systems consist of piping that is not pressurized and is sloped to allow flow. These systems usually include storm and sanitary drain services. The more common underground, pressurized piping systems include fire water systems, cooling water systems, closed process drains, and pump-out systems. A wide variety of materials can be used for underground piping systems:

- Cast iron
- Ductile iron
- Iron alloy
- Carbon steel
- Concrete
- Fiberglass
- Thermoplastic

5.0.0 ◆ CAST IRON SOIL PIPE

Cast iron soil pipe and fittings are manufactured from grey cast iron, a strong and corrosion-resistant material. Cast iron pipe is centrifugally cast. In this process, molten cast iron is poured into a spinning pipe mold, and the centrifugal force causes the molten cast iron to form on the sides of the mold, where it solidifies to the pipe shape. This method of casting ensures a straight length of pipe with a smooth interior surface. Cast iron pipe fittings are cast in permanent metal molds to produce fittings with a uniform wall thickness. After casting, the pipe and fittings are coated with coal tar pitch to prevent rust during storage and use and to improve their appearance.

Cast iron soil pipe and fittings have the advantage of being made of a strong material. They do not leak or absorb water; they are easily cut and joined; and they are economical and soil-resistant. Cast iron soil pipes are also one of the quietest piping systems available, because they do not transmit the sound of water draining through them as do some of the other thinner-walled materials. The major disadvantage of cast iron pipe is that it is a very heavy material with low tensile strength, and it is easily broken if mishandled. Cast iron soil pipe is available in three basic types: single-hub soil pipe, double-hub soil pipe, and hubless soil pipe. *Figure 19* shows cast iron soil pipe.

5.1.0 Single- and Double-Hub Cast Iron Pipe Sizes

Single- and double-hub cast iron soil pipe is made with inside diameters that range from 2 to 15 inches. Single-hub soil pipe is also known as bell-and-spigot pipe or hub-and-spigot pipe. The spigot end is the plain end without a hub. Single-hub cast iron soil pipe is made in lengths of 5 and 10 feet. Double-hub pipe is made in lengths of 5 feet or 30 inches. The double-hub pipe is an economical means of providing short lengths of bell-and-spigot pipe with a minimal amount of waste. The double-hub pipe should never be installed into a system in its entire length with both hubs attached except when correcting a previously incorrectly installed flow system. The lengths of bell-and-spigot pipe refer to the laying length, which is the total end-to-end length minus the length of the bell. Bell-and-spigot pipe is currently available in two different wall thicknesses or weights: service weight and extra-heavy weight. Service weight cast iron pipe is marked SV, and extra-heavy weight cast iron pipe is marked XH. Extra-heavy and service weight pipe and fittings are not interchangeable because the outside diameter of extra-heavy pipe is larger. *Table 1* shows the dimensional details of service weight bell-and-spigot cast iron soil pipe. *Table 2* shows the dimensional details of extra-heavy weight bell-and-spigot cast iron soil pipe.

> **NOTE**
> Hub configurations from different manufacturers vary.

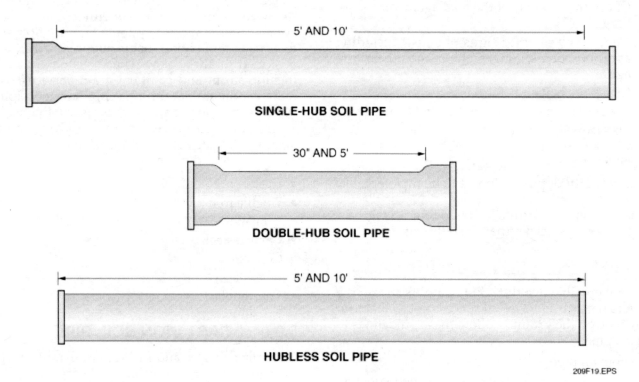

Figure 19 ◆ Cast iron soil pipe.

Table 1 Dimensional Details of Service Weight Bell-and-Spigot Cast Iron Soil Pipe

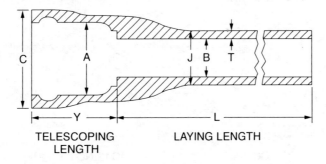

SV (SERVICE WEIGHT)
CAST IRON SOIL PIPE

TELESCOPING LENGTH   LAYING LENGTH

SIZE	TELESCOPING LENGTH Y	HUB ID A	HUB OD C	BARREL OD J	BARREL ID B	NOMINAL THICKNESS T
2"	2.50	2.94	3.62	2.30	1.96	0.17
3"	2.75	3.94	4.68	3.30	2.96	0.17
4"	3.00	4.94	5.68	4.30	3.94	0.18
5"	3.00	5.94	6.68	5.30	4.94	0.18
6"	3.00	6.94	7.68	6.30	5.94	0.18
8"	3.50	9.25	10.13	8.38	7.94	0.23
10"	3.50	11.38	12.44	10.50	9.94	0.28
12"	4.25	13.50	14.56	12.50	11.94	0.28
15"	4.25	16.75	17.91	15.62	15.00	0.31

5-FT AND 10-FT SINGLE AND DOUBLE HUB

SIZE	LAYING LENGTH (L) 5-FT SINGLE HUB	LAYING LENGTH (L) 5-FT DOUBLE HUB	LAYING LENGTH (L) 10-FT SINGLE HUB	SHIPPING WEIGHT 5-FT SINGLE HUB	SHIPPING WEIGHT 5-FT DOUBLE HUB	SHIPPING WEIGHT 10-FT SINGLE HUB
2"	5'-0"	4'-9½"	10'-0"	20	21	38
3"	5'-0"	4'-9¼"	10'-0"	30	31	56
4"	5'-0"	4'-9"	10'-0"	25	54	75
5"	5'-0"	4'-9"	10'-0"	52	54	98
6"	5'-0"	4'-9"	10'-0"	65	68	124
8"	5'-0"	4'-8½"	10'-0"	100	105	185
10"	5'-0"	4'-8½"	10'-0"	145	150	270
12"	5'-0"	4'-7¾"	10'-0"	190	200	355
15"	5'-0"	4'-7¾"	10'-0"	255	270	475

209T01.EPS

5.2.0 Hubless Cast Iron Pipe Sizes

Hubless cast iron pipe is cast in lengths of 5 feet and 10 feet with inside diameters ranging from 1½ inches to 8 inches. A standard length of cast iron pipe is cast with plain spigots on each end. Some pipes have a small bead and a positioning lug at each end to ensure proper positioning of the gasket and clamp, which are used to form the joints. Hubless pipe is used in sanitary and storm drainage systems above and below ground. It is not suitable for many industrial and laboratory chemical processes.

5.3.0 Cast Iron Soil Pipe Fittings

Underground piping systems consist of pipes, pumps, valves, and other parts, including fittings. Pipe fittings come in various sizes, materials, strengths, and designs to match the various piping systems. Pipe fittings can be either plain or banded. Pipe fittings are attached to a pipe and may be used to change the direction of fluid flow, to connect a branch line to a main line, to close off the end of a line, or to join two pipes of the same or different sizes.

Table 2 Dimensional Details of Extra-Heavy Weight Bell-and-Spigot Cast Iron Soil Pipe

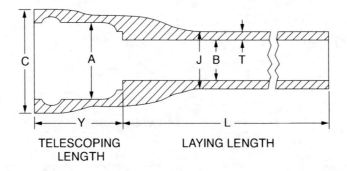

XH (EXTRA-HEAVY)
CAST IRON SOIL PIPE

TELESCOPING LENGTH

LAYING LENGTH

SIZE	TELESCOPING LENGTH Y	HUB ID A	HUB OD C	BARREL OD J	BARREL ID B	NOMINAL THICKNESS T	
2"	2.50	3.06	3.80	2.38	2.00	0.19	
3"	2.75	4.19	5.05	3.50	3.00	0.25	
4"	3.00	5.19	6.05	4.50	4.00	0.25	+.06 −.03
5"	3.00	6.19	7.05	5.50	5.00	0.25	
6"	3.00	7.19	8.05	6.60	6.00	0.25	
8"	3.50	9.50	10.68	8.62	8.00	0.30	
10"	3.50	11.62	12.92	10.75	10.00	0.37	
12"	4.25	13.75	15.05	12.75	12.00	0.37	+.09 −.06
15"	4.25	17.00	18.42	15.88	15.00	0.44	

5-FT AND 10-FT SINGLE AND DOUBLE HUB

SIZE	LAYING LENGTH (L) 5-FT SINGLE HUB	5-FT DOUBLE HUB	10-FT SINGLE HUB
2"	5'-0"	4'-9½"	10'-0"
3"	5'-0"	4'-9¼"	10'-0"
4"	5'-0"	4'-9"	10'-0"
5"	5'-0"	4'-9"	10'-0"
6"	5'-0"	4'-9"	10'-0"
8"	5'-0"	4'-8½"	10'-0"
10"	5'-0"	4'-8½"	10'-0"
12"	5'-0"	4'-7¾"	10'-0"
15"	5'-0"	4'-7¾"	10'-0"

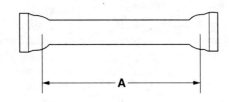

DOUBLE HUB 30" LENGTHS		
ASA 02	SIZE	A
0082	2"	25
0084	3"	25
0086	4"	25

209T02.EPS

There are various cast iron pipe fittings manufactured for use with bell-and-spigot and hubless cast iron pipe. These fittings are also currently manufactured in service weight and extra-heavy weight sizes. It is critical to match the weights of the fittings to the pipe. Bell-and-spigot cast iron soil pipe fittings have a bell or hub cast at one end of the fitting and a spigot on the other end. *Figure 20* shows bell-and-spigot cast iron soil pipe fittings.

Hubless cast iron fittings have spigot ends that match and are joined to the hubless pipe by a no-hub clamp. *Figure 21* shows hubless cast iron soil pipe fittings.

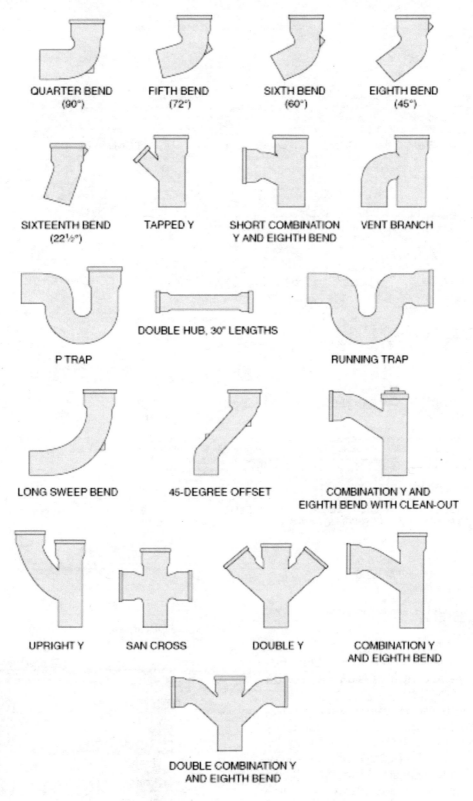

QUARTER BEND
(90°)

FIFTH BEND
(72°)

SIXTH BEND
(60°)

EIGHTH BEND
(45°)

SIXTEENTH BEND
(22½°)

TAPPED Y

SHORT COMBINATION
Y AND EIGHTH BEND

VENT BRANCH

P TRAP

DOUBLE HUB, 30" LENGTHS

RUNNING TRAP

LONG SWEEP BEND

45-DEGREE OFFSET

COMBINATION Y AND
EIGHTH BEND WITH CLEAN-OUT

UPRIGHT Y

SAN CROSS

DOUBLE Y

COMBINATION Y
AND EIGHTH BEND

DOUBLE COMBINATION Y
AND EIGHTH BEND

209F20.EPS

Figure 20 ◆ Bell-and-spigot cast iron soil pipe fittings.

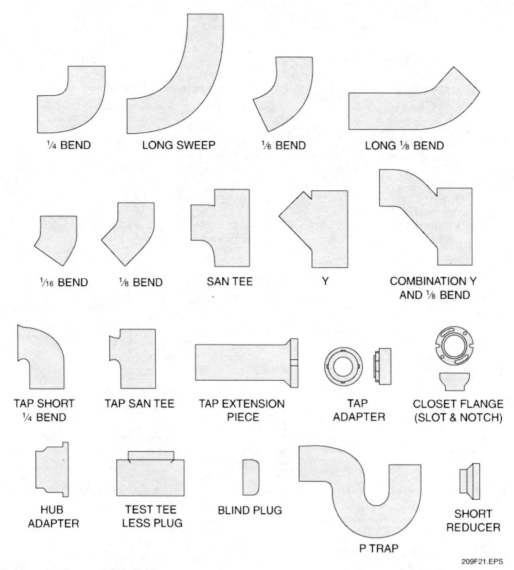

Figure 21 ◆ Hubless cast iron soil pipe fittings.

5.4.0 Joining Bell-and-Spigot Cast Iron Pipe

The most common method used to join bell-and-spigot cast iron pipe and fittings is a compression joint.

Compression joints use a rubber gasket, making this a low-cost method of joining bell-and-spigot pipe. The compression gasket is inserted into the bell end of the pipe and lubricated, and the spigot end of the mating pipe is inserted into the bell end with the gasket. This method requires pipe without a bead on the spigot end. The joint is sealed by displacement and compression of the rubber gasket. *Figure 22* shows a compression joint.

5.5.0 Joining Hubless Cast Iron Pipe

Hubless cast iron pipe is joined with a mechanical joint, known as a no-hub joint, composed of a

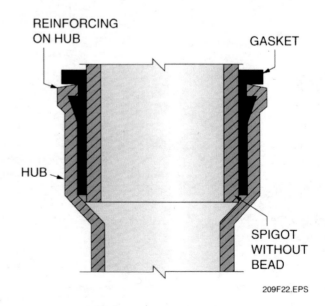

Figure 22 ◆ Compression joint.

lubricated neoprene sleeve gasket and a stainless steel band with screw clamps. The ends of the pipe are inserted into the neoprene gasket and butted together. They are held in place by a stainless steel band and screw clamps, which are tightened to 60 inch-pounds of torque. The hubless cast iron pipe with this type of joint is used mainly for temporary repairs in underground systems. *Figure 23* shows joining hubless cast iron pipe.

5.6.0 Cutting Cast Iron Pipe

At some point in the installation of cast iron soil pipe, cut pieces are needed. When a short piece of a cast iron pipe with a hub is needed, it should be cut from a double-hub pipe so that the piece of pipe left after cutting is not wasted. Cast iron pipe can be cut using a soil pipe cutter.

5.6.1 Cutting Cast Iron Pipe Using Soil Pipe Cutter

The easiest way to cut cast iron pipe is with a soil pipe cutter. The cutter consists of a ratchet handle with a length of chain permanently attached to one side of the cutter. Each link of the chain contains

a cutting wheel. The chain is wrapped around the pipe and locked into the side of the cutter. As the ratchet handle is pumped, the cutter tightens the chain around the pipe. As the chain tightens, the cutting wheels act as small chisels that penetrate the pipe until it is cut. *Figure 24* shows a soil pipe cutter and a hydraulically powered pipe cutter.

Follow these steps to cut cast iron pipe, using a soil pipe cutter:

Step 1 Measure and mark the pipe to be cut. The pipe should be cut while it lies on the ground.

Step 2 Lift the pipe, and place the chain underneath the cut mark.

Step 3 Place the cutter on top of the cut mark, and pull the chain around the pipe.

Step 4 Attach the end of the chain into the open side of the cutter.

Step 5 Turn the chain tension knob clockwise to tighten the chain on the cut mark. Pull the chain as tightly as possible.

Step 6 Pull out the lock knob, and turn it so that the arrow points toward the handle.

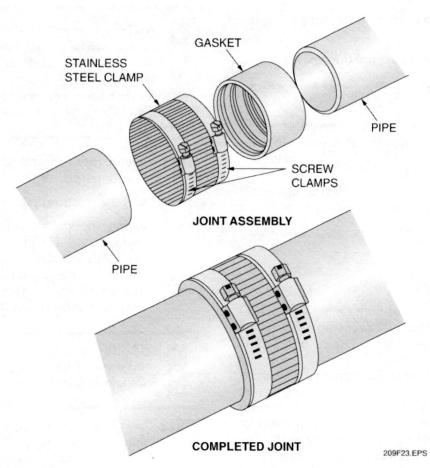

Figure 23 ◆ Joining hubless cast iron pipe.

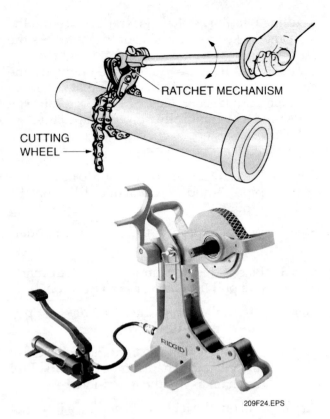

CUTTING WHEEL

RATCHET MECHANISM

209F24.EPS

Figure 24 ◆ Pipe cutters.

Step 7   Push down the handle to start cutting the pipe.

Step 8   Pump the handle until the pipe is broken.

NOTE
The ratchet allows you to lift the handle without affecting the chain tension.

Step 9   Inspect the soil pipe cutter for any obvious damage.

Step 10 Store the cutter.

5.6.2 *Cutting Cast Iron Pipe Using a Demolition Saw*

Cast iron pipe can be cut with an abrasive wheel saw or a rotary wheel cutter. The abrasive wheel saw is used for cutting pipe out of the trench. The rotary wheel cutter is used in or out of the trench on pipe up to 30 inches in diameter. Cut ends and rough edges must be ground smooth before installing the pipe. When the pipe is used in compression joints, the end should be slightly beveled. *Figure 25* shows cast iron pipe-cutting tools. The demolition saw is pulled backward across the pipe while cutting, so that it is less likely to catch and pull away from you.

5.7.0 Installing Cast Iron Pipe

The method used to install bell-and-spigot cast iron pipe and fittings is a compression joint. Hubless cast iron pipe is joined using a mechanical joint known as a no-hub joint. Always install cast iron pipe according to the pipe manufacturer's recommendations.

Compression joints use a rubber gasket, which makes this a low-cost method of installing bell-and-spigot pipe. The compression gasket is inserted into the bell end of the pipe, and the spigot end of the mating pipe is inserted into the bell end with the gasket. The joint is sealed by displacing and compressing the rubber gasket.

The gaskets are manufactured in two weights: service weight, for use with service-weight soil pipe and fittings, and extra heavy, for use with extra-heavy soil pipe and fittings. Like the two types of pipe, the gaskets are not interchangeable.

5.7.1 *Installing Cast Iron Pipe with Chain Hoists*

When installing larger sizes of cast or ductile iron pipe, the best tool can be a pair of chain hoists. The procedure for using the hoists is as follows:

NOTE
For this exercise you need two 1-ton chain hoists, each with 25 feet of chain and two bell choker slings for 4- to 24-inch diameter pipe or two 2½-ton chain hoists, each with 25 feet of chain and two bell choker slings for 30- to 54-inch sizes.

Step 1   Clean the hub and spigot, and ensure that the inside of the pipe is clean.

Step 2   Cut the cast iron pipe to length.

Step 3   Grind the sharp, cut end of the pipe, using a portable grinder, until the edge is slightly beveled. If you do not remove the sharp edge from the pipe, it may jam against the gasket. This can harm the gasket and make it difficult to install the pipe.

WARNING!
Always wear gloves when grinding the end of the pipe to prevent injuring your fingers and hands.

Step 4   Hold the gasket upright, with your thumbs at the bottom, and fold the bottom of the gasket up through the top as if you were going to turn it inside out.

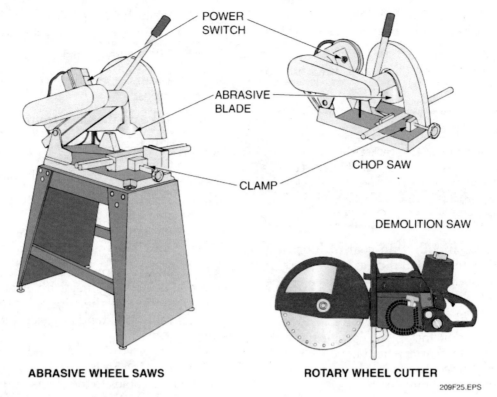

POWER
SWITCH

ABRASIVE
BLADE

CHOP SAW

CLAMP

DEMOLITION SAW

ABRASIVE WHEEL SAWS

ROTARY WHEEL CUTTER

209F25.EPS

Figure 25 ◆ Cast iron pipe-cutting tools.

Step 5 Place the gasket in the hub, making sure that the gasket ring is in the groove of the hub.

Step 6 Release the gasket so that it unfolds into the pipe hub.

Step 7 Brush a smooth, thin coat of lubricant completely around the gasket. Be sure to lubricate the entire inner surface of the gasket. Pay special attention to the inner seal.

Step 8 Lubricate the spigot end of the mating pipe.

Step 9 Position the spigot in the first seal of the gasket.

Step 10 Wrap the bell choker slings around the existing pipe just behind the bell so that the free ends are located on the horizontal center line of the pipe on opposite sides of the bell, with the loose ends projecting in front of the bell face.

Step 11 Double-wrap a choker around the mating pipe approximately 6 feet from the spigot end so that the free ends are located on the horizontal center line of the pipe on opposite sides of the mating pipe.

Step 12 Connect the hooks of the chain hoists on the horizontal center line of the pipe on opposite sides of the spigot.

Step 13 Attach the hook of each chain into the eye of each choker sling.

Step 14 Pull evenly with both chain hoists, keeping the pipe in alignment, to install the joint.

5.7.2 Other Methods of Installing Compression Joints

Follow these steps to install a compression joint without a chain hoist:

Step 1 Prepare the pipe ends for installation as explained previously. File any sharp edges. If you do not remove the sharp edge from the pipe, it may jam against the gasket. This can harm the gasket and make it difficult to install the pipe.

 WARNING!
Always wear gloves when filing the end of the pipe to prevent injuring your fingers and hands.

Step 2 Hold the gasket upright, with your thumbs at the bottom, and fold the bottom of the gasket up through the top as if you were going to turn it inside out.

Step 3 Place the gasket in the hub, making sure that the gasket ring is in the groove of the hub.

Step 4 Release the gasket so that it unfolds into the pipe hub.

NOTE

Another way to install the gasket into the hub is to place the gasket in position to be inserted into the hub and strike it with the heel of your hand or with a board.

Step 5 Brush a smooth, thin coat of lubricant completely around the gasket. Be sure to lubricate the entire inner surface of the gasket. Pay special attention to the inner seal.

Step 6 Push or pull the spigot end all the way into the hub of the mating pipe until it has bottomed out. This is known as pushing or pulling the pipe home.

There are several methods used to push or pull the pipe home. The method chosen depends on the size of the pipe, the trench conditions, and the availability of special tools. The following sections describe some of the tools and methods used for mating pipe.

Lever tool – The lever tool is used to install cast iron and ductile iron pipe ranging in size from 3 to 8 inches. The tool consists of a lever and a chain that is double wrapped around the bell of one pipe and the spigot of the second pipe. *Figure 26* shows a lever tool.

Follow these steps to pull pipe home using a lever tool:

209F26.EPS

Figure 26 ◆ Lever tool.

Step 1 Place the lever on the bell end of the existing pipe just behind the hub.

Step 2 Tilt the lever back about 45 degrees.

Step 3 Wrap the chain twice around the mating pipes.

Step 4 Take up the slack from the spigot chain.

Step 5 Tilt the lever toward the mating pipe, and hook the chain wrapped around the spigot into the slot at the base of the lever.

Step 6 Pull the lever toward the bell pipe to pull the spigot pipe home. If one pull is not enough to pull the pipe home, readjust the chain in the lever and pull again.

Pipe can also be pulled using a battery-powered unit that is installed inside an existing large pipe. Once installed in the pipe, a hydraulic ram in the vertical arm extends to secure the unit to the existing pipe.

Then, the crosspiece is placed across the outside end of the pipe to be pulled. The crosspiece is attached via cables to the clevis-end of a hydraulic piston on the pulling unit.

The control cable extends outside the pipe that will be pulled. When actuated, the hydraulic piston contracts, pulling the cable, crosspiece, and the pipe toward the existing pipe, thus joining the two pipes.

Deadblow hammer – A deadblow hammer is a good tool for installing fittings onto lengths of pipe. *Figure 27* shows installing a fitting using a deadblow hammer.

Bar – For pipe 8 inches and smaller laid in dry trenches with a firm, clay bottom, a bar or shovel can be used to push the pipe home. This method is not effective in sandy soils, wet soils, or rock.

Follow these steps to push pipe home using a bar:

Step 1 Center the spigot end of the mating pipe into the lubricated gasket.

Step 2 Drive a large pry bar 6 to 8 inches into the ground underneath the hub end of the mating pipe. *Figure 28* shows the bar method.

Step 3 Place a block of wood between the bar and the end of the pipe.

CAUTION

If you push against the pipe without the block of wood, the pipe will break.

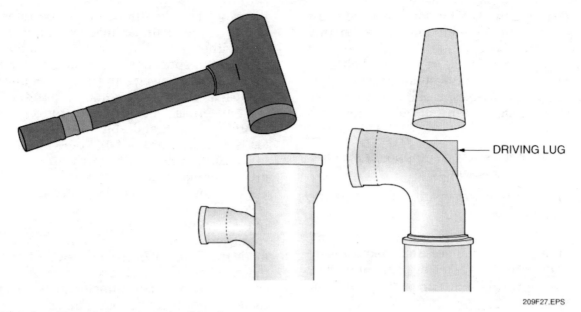

DRIVING LUG

209F27.EPS

Figure 27 ◆ Installing a fitting using a deadblow hammer.

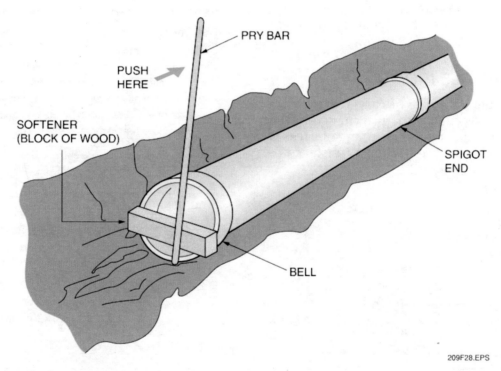

PRY BAR

PUSH
HERE

SOFTENER
(BLOCK OF WOOD)

SPIGOT
END

BELL

209F28.EPS

Figure 28 ◆ Bar method.

Step 4 Use the bar as a lever, and push against the bell end of the mating pipe to push the pipe home. Have another person stand over the spigot end to determine when the pipe is pushed home.

WARNING!

Maintain your balance as you push the pipe. The bar can slip or come out of the ground, causing you to fall forward.

Step 5 Make all adjustments for line and grade after the pipe is pushed home.

5.8.0 Installing Hubless Cast Iron Pipe

Hubless cast iron pipe is joined with a mechanical joint, known as a no-hub joint, that is composed of a neoprene sleeve gasket and a stainless steel band with screw clamps. The ends of the pipe are inserted into the neoprene gasket and butted together. The ends are held in place by the stainless steel band and screw clamps, which are tightened to 60 inch-pounds of torque. The hubless cast iron pipe with this type of joint is used mainly for commercial aboveground service systems. It can also be used to make repairs to bell-and-spigot systems if job specifications permit. *Figure 29* shows installing hubless cast iron pipe.

Follow these steps to install hubless cast iron pipe:

Step 1 Loosen the screw clamps on the stainless steel clamp assembly.

Step 2 Ensure that the insides of the mating pipes are clean.

Step 3 Insert the spigot ends of the pipes or fittings into the stainless steel clamp assembly with the gasket in place until they butt against the separator ring inside the gasket.

Step 4 Tighten the screw clamps on the stainless steel clamp assembly, using a preset T-handle torque wrench, to 60 inch-pounds of torque.

6.0.0 ◆ DUCTILE IRON PIPE

Ductile iron pipe is made of cast iron that has undergone a molecular structure change in the molten process. This molecular change increases the strength, ductility, and impact resistance of the pipe. The ductility of the pipe refers to its ability to bend under loads without breaking and without losing its strength. Ductile iron is used to transport raw and treated water, natural gas, domestic sewage, industrial chemicals and wastes, and many other liquids, gases, and even solid materials. Ductile iron is stronger, more corrosion-resistant, and able to withstand greater loads than standard cast iron pipe. These qualities make ductile iron particularly suited to underground applications.

Unless otherwise specified, all ductile iron pipe is coated on the outside with a bituminous coating. This type of coating gives ductile iron pipe its characteristic flat-black appearance. The coating is made from asphalt or coal tars. According to manufacturing standards, the coating must be smooth, continuous, and 1 millimeter thick throughout the length of the pipe. Bituminous coating is not affected by temperature extremes.

NOTE

In very corrosive applications, such as cases where the groundwater tends to be acidic, a sacrificial anode is used. This is a metal that is galvanically corrupted by the ductile iron pipe, so that it is corroded by the galvanic current that would otherwise destroy the pipe.

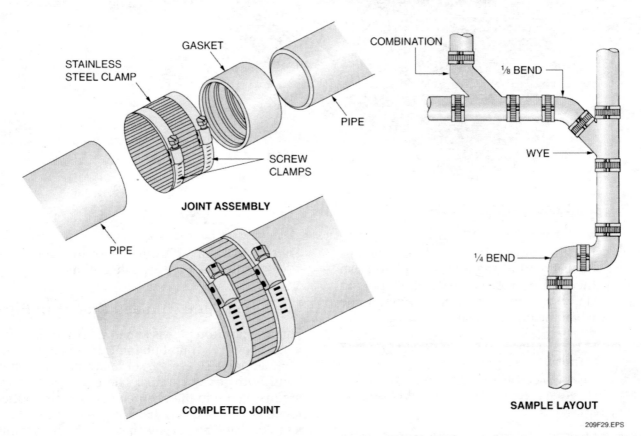

Figure 29 ◆ Installing hubless cast iron pipe.

Ductile iron pipe used for water service is usually lined with cement. For other applications, pipe may be unlined or lined with a bituminous coating. To meet the needs of particular applications, linings may be special ordered from the manufacturer. The increased weight of ductile iron pipe with inside linings requires special handling and care.

> **NOTE**
> In many localities, especially where the groundwater tends to be acidic or caustic, iron pipe is wrapped in polyethylene tubes or bags. The tube is slid on as each pipe is laid, and the bell end is overlapped doubly, in the order in which the pipes were laid.

6.1.0 Ductile Iron Pipe Sizes

Ductile iron pipe is available in nominal sizes ranging from 3 inches to 54 inches and comes in 20-foot lengths. Wall thickness is described by standard class numbers that range from 50 to 56 (53 and 56 are the most common). The larger class numbers designate greater wall thickness. There is a considerable difference in nominal sizes of ductile iron pipe and the actual outside diameters. *Table 3* shows ductile iron pipe sizes.

6.2.0 Ductile Iron Pipe Fittings

The fittings used for ductile iron pipe are similar in appearance, design, and function to those used with other types of pipe. Fittings are available in the same class thicknesses, pressure ratings, linings, and joint designs as standard ductile iron pipe. The standard class numbers of the fittings must match those of the pipe being used. *Figure 30* shows ductile iron pipe fittings.

6.3.0 Joining Ductile Iron Pipe

The three most common types of ductile iron pipe connections are the compression joint, the mechanical joint, and the flange joint. This section describes each of these types of connections.

6.3.1 Compression Joints

Compression joints, which are used with cast iron soil pipe, are used with bell-and-spigot pipe only. A special gasket is inserted into the bell of the pipe, and lubricated. The spigot end of the mating pipe is slightly tapered to ensure that it is centered inside the bell. The spigot end is also greased with a special lubricant to make it easier to assemble into the bell. The joint is sealed by displacement and compression of the rubber gasket.

Table 3 Ductile Iron Pipe Sizes

SIZE INCHES	OUTSIDE DIAMETER INCHES	STANDARD CLASSES – WALL THICKNESSES IN INCHES						
		50	51	52	53	54	55	56
3	3.96	–	0.25	0.28	0.31	0.34	0.37	0.40
4	4.80	–	0.26	0.29	0.32	0.35	0.38	0.41
6	6.90	0.25	0.28	0.31	0.34	0.37	0.40	0.43
8	9.05	0.27	0.30	0.33	0.36	0.39	0.42	0.45
10	11.10	0.29	0.32	0.35	0.38	0.41	0.44	0.47
12	13.20	0.31	0.34	0.37	0.40	0.43	0.46	0.49
14	15.30	0.33	0.36	0.39	0.42	0.45	0.48	0.51
16	17.40	0.34	0.37	0.40	0.43	0.46	0.49	0.52
18	19.50	0.35	0.38	0.41	0.44	0.47	0.50	0.53
20	21.60	0.36	0.39	0.42	0.45	0.48	0.51	0.54
24	25.80	0.38	0.41	0.44	0.47	0.50	0.53	0.56
30	32.00	0.39	0.43	0.47	0.51	0.55	0.59	0.63
36	38.30	0.43	0.48	0.53	0.58	0.63	0.68	0.73
42	44.50	0.47	0.53	0.59	0.65	0.71	0.77	0.83
48	50.80	0.51	0.58	0.65	0.72	0.79	0.86	0.93
54	57.10	0.57	0.65	0.73	0.81	0.89	0.97	1.05

209T03.EPS

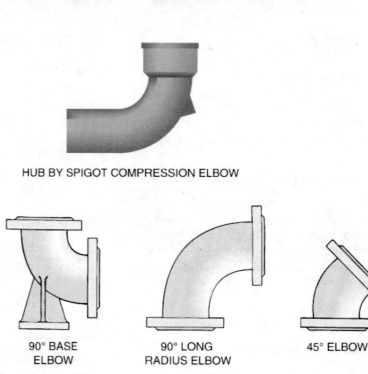

HUB BY SPIGOT COMPRESSION ELBOW

TRUE "Y"   90° BASE   90° LONG   45° ELBOW
                  ELBOW   RADIUS ELBOW

TEE   45° Y LATERAL   CROSS

209F30.EPS

Figure 30 ◆ Ductile iron pipe fittings.

6.3.2 Mechanical Joints

Ductile iron piping that is joined with mechanical joints is cast with a flange on the bell end of the pipe. The principal parts of a mechanical joint are a rubber gasket, a gland, or follower ring that is used to compress the gasket, and T-bolts and nuts. *Figure 31* shows a mechanical joint.

6.3.3 Flanged Joints

Ductile iron pipe with flanged joints are standard for in-plant piping, such as is found in pump rooms and sewage treatment plants. However, it is seldom used for underground piping, except for valves and fittings in large meter settings. Flanged joints are also common when a run of underground pipe comes out of the ground. Because of the rigidity of the joint, flanged joints are not used in areas where they may be subject to excessive vibration. *Figure 32* shows a ductile iron flange joint.

6.4.0 Cutting Ductile Iron Pipe

Ductile iron pipe can be cut with an abrasive wheel saw or a rotary wheel cutter. The abrasive wheel saw is used for cutting pipe out of the trench. The rotary wheel cutter is used in or out of the trench on pipe up to 30 inches in diameter. Cut ends and rough edges must be ground smooth before installing the pipe. When the pipe is used in compression joints, the end should be slightly beveled. *Figure 33* shows ductile iron pipe-cutting tools.

Mueller jig and corporation thread drills and taps are used for small branch connections. National Pipe Threads cannot be used with ductile iron pipe because they will pull out.

6.5.0 Installing Ductile Iron Pipe

The three most common methods used to install ductile iron pipe are compression joints, flanged joints, and mechanical joints.

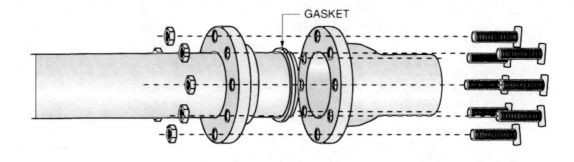

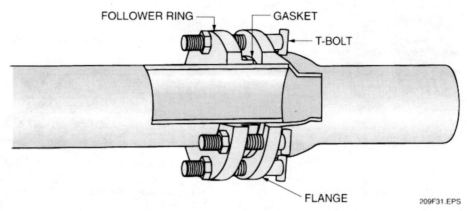

Figure 31 ◆ Mechanical joint.

6.5.1 Installing Compression Joints

The compression joint, which is also used with cast iron soil pipe, is used with bell-and-spigot ductile iron pipe. The joint has a special gasket that is inserted into the bell of the pipe. The spigot end of the mating pipe is slightly tapered to ensure that it is centered inside the bell. The spigot end is also greased with a special lubricant to make it easier to install into the bell. The joint is sealed by displacing and compressing the rubber gasket. The compression joint for ductile iron pipe is installed the same way as compression joints for cast iron pipe. The only differences are the tools used to pull larger ductile iron pipe home. Ductile iron pipe is available in much larger sizes than cast iron pipe, so special methods are used to pull the pipe home. One of these methods includes using chain hoists. Follow these steps to install a compression joint, using chain hoists (use correct lube or lanolin-rich soap to lube the gasket and pipe end):

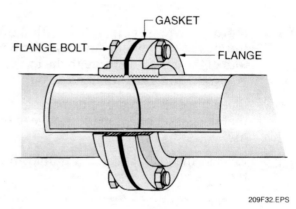

Figure 32 ◆ Ductile iron flange joint.

Step 1 Clean the hub and spigot, and ensure that the inside of the pipe is clean.

Step 2 Cut the ductile iron pipe to length.

Step 3 Grind the sharp, cut end of the pipe, using a portable grinder, until the edge is slightly beveled. If you do not remove the sharp edge from the pipe, it may jam against the gasket. This can harm the gasket and make it difficult to install the pipe.

 NOTE

For this exercise you need two 1-ton chain hoists, each with 25 feet of chain and two bell choker slings for 4- to 24-inch diameter pipe or two 2½-ton chain hoists, each with 25 feet of chain and two bell choker slings for 30- to 54-inch sizes.

 WARNING!

Always wear gloves when grinding the end of the pipe to prevent injuring your fingers and hands.

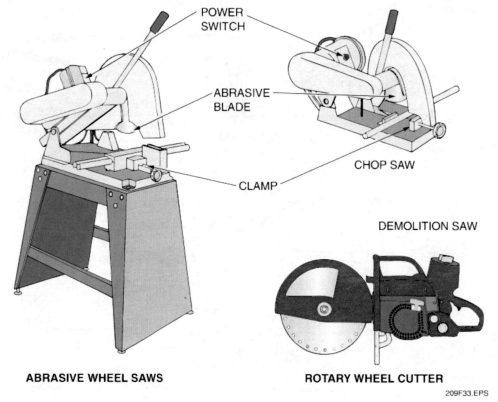

POWER SWITCH

ABRASIVE BLADE

CHOP SAW

CLAMP

DEMOLITION SAW

ABRASIVE WHEEL SAWS

ROTARY WHEEL CUTTER

209F33.EPS

Figure 33 ◆ Ductile iron pipe-cutting tools.

Step 4 Hold the gasket upright, with your thumbs at the bottom, and fold the bottom of the gasket up through the top as if you were going to turn it inside out.

Step 5 Place the gasket in the hub, making sure that the gasket ring is in the groove of the hub.

Step 6 Release the gasket so that it unfolds into the pipe hub.

NOTE

Another way to install the gasket into the hub is to place the gasket in position to be inserted into the hub and strike it with the heel of your hand or with a board.

Step 7 Brush a smooth, thin coat of lubricant completely around the gasket. Be sure to lubricate the entire inner surface of the gasket. Pay special attention to the inner seal.

Step 8 Lubricate the spigot end of the mating pipe.

Step 9 Position the spigot in the first seal of the gasket.

Step 10 Wrap the bell choker slings around the existing pipe just behind the bell so that the free ends are located on the horizontal center line of the pipe on opposite sides of the bell, with the loose ends projecting in front of the bell face.

Step 11 Double-wrap a choker around the mating pipe approximately 6 feet from the spigot end so that the free ends are located on the horizontal center line of the pipe on opposite sides of the mating pipe.

Step 12 Connect the hooks of the chain hoists on the horizontal center line of the pipe on opposite sides of the spigot.

Step 13 Attach the hook of each chain into the eye of each choker sling.

Step 14 Pull evenly with both chain hoists, keeping the pipe in alignment, to install the joint.

A backhoe can also be used to install compression joints on larger pipe. The spigot end of the mating pipe should be carefully guided into the bell end of the existing pipe. The bucket of the backhoe can then be used to push the bell end of the mating pipe until it is fully seated. A timber header should always be placed between the bucket of the backhoe and the pipe to prevent damaging the pipe.

WARNING!

Be sure to keep body parts from between the end of the pipe and the backhoe bucket, and beware of other pinch points.

6.5.2 Installing Flanged Joints

Ductile iron pipe with flanged joints is standard for in-plant piping, such as is found in pump rooms and sewage-treatment plants. However, it is seldom used for underground piping, except for valves and fittings in large, metered systems. Flanged joints are also common when a run of underground pipe comes out of the ground. Because of the rigidity of the joint, flanged joints are not used in areas where they may be subject to excessive vibration. Ductile iron screwed flanges are not interchangeable with carbon steel or other alloy flanges because of the difference in the outside diameters of pipes made from these types of materials. *Figure 34* shows a ductile iron flanged joint.

Gaskets must be used between flanges. Gaskets for ductile iron pipe systems can be red rubber or black neoprene rubber. Dovetail gaskets are commonly used with large-bore ductile iron pipe. These gaskets come in sections that are pieced together to form the gasket.

Flange bolts must be tightened carefully to avoid warping the flange. If the bolts are not tightened correctly, the joint may crack or leak. Pressure must be applied evenly and in the correct amount to make a good seal. A torque wrench is used to apply equal pressure at all points around the flange. With the aid of a torque wrench, a pipefitter can properly adjust and tighten a flanged joint. Use the following guidelines when tightening flange bolts:

- Make sure the threads of the flange nuts and bolts are cleaned and lightly oiled before torquing.

- If the bolts of a flange have already been tightened, loosen and retighten them. Loosen each bolt one full turn, using an open-end or socket wrench.

- Make sure that the torque wrench is correctly adjusted. Most companies calibrate precision tools at regular intervals.

- Always place the torque wrench on the nut, not on the bolt. Use an open-end or socket wrench to hold the bolt while torquing.

- Always use a smooth, even motion when using the torque wrench. A hurried, uneven motion does not give a correct measurement.

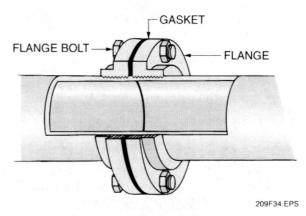

Figure 34 ◆ Ductile iron flanged joint.

Each piping system may use different torque settings, depending on the sizes of the flanges and bolts, the pressure or temperature of the system, and the type of gasket used. Consult your supervisor or the piping drawings for the proper torque setting. Follow these steps to install a flanged joint:

Step 1 Clean the flanges, bolts, and gaskets thoroughly.

Step 2 Ensure that the insides of the mating pipes are clean.

Step 3 Align the mating flanges. Be sure to support the pipe, and use a drift pin to line up the holes.

WARNING!

Never use your fingers to align the bolt holes.

Step 4 Place the gasket between the full face flanges, and hold it in place.

Step 5 Insert the bottom bolts into the flanges.

Step 6 Insert all the bolts into the flanges.

Step 7 Place the nuts on the bolts, and tighten them finger-tight. To properly torque a flanged joint, you must use a crossover method of tightening the bolts. *Figure 35* shows the proper tightening sequences for torquing different flanges.

CAUTION

Do not begin torquing bolts to the full pressure the first time around the flange. This would cause unequal pressure on the flanges, which will cause leaks.

Step 8 Set the torque wrench at 25 percent of the required torque.

Step 9 Tighten the top center bolt to this torque.

Step 10 Tighten the bolt directly opposite this bolt, the bottom center bolt, to this torque.

Step 11 Tighten the side bolt to this torque.

Step 12 Tighten the bolt directly opposite of the bolt tightened in step 11 to this torque.

Step 13 Continue this process until all the bolts are tightened to this torque.

Step 14 Reset the torque wrench to 50 percent of the required torque.

Step 15 Tighten all the bolts to this torque following the crossover method.

Step 16 Reset the torque wrench to 75 percent of the required torque.

Step 17 Tighten all the bolts to this torque following the crossover method.

Step 18 Reset the torque wrench to 100 percent of the required torque.

Step 19 Tighten all bolts to this torque, starting at the top center bolt and moving clockwise around the flange.

Step 20 Repeat Step 19 for final tightening of the bolts.

6.5.3 Installing Mechanical Joints

Ductile iron piping that is joined with mechanical joints is cast with a flange on the bell end of the pipe. The principal parts of a mechanical joint are a rubber gasket, a gland, or follower ring, that is used to compress the gasket, and T-bolts and nuts. *Figure 36* shows a mechanical joint.

Follow these steps to install a mechanical joint:

Step 1 Clean the plain end and socket of the mating pipes and the gasket, using a rag. This helps the gasket seat better and allows the parts of the joint to slide together more easily.

Step 2 Ensure that the inside of the pipe is clean.

Step 3 Place the follower ring on the plain end, with the lip pointing toward the plain end of the pipe.

Step 4 Place the gasket on the plain end so that the narrow edge of the gasket points toward the end of the pipe, and lubricate the gasket. There are specific brand lubricants, but many pipefitters use liquid soap and water.

Step 5 Push the gasket back from the end of the pipe so that the end can be properly inserted into the socket.

Step 6 Push the end of the mating pipe into the socket of the existing pipe.

Step 7 Position the gasket into the flange of the socket to seat the gasket properly. If the gasket does not seat properly, the pipe will leak. Make sure that the gasket is evenly seated around the flange. Excessive resistance in installation may indicate gasket has rolled.

Step 8 Push the follower ring up to the socket, and align the holes.

Step 9 Insert the T-bolts into the bolt holes from the socket side and tighten the nuts onto them finger-tight.

Step 10 Tighten the nuts on the T-bolts following the proper tightening sequence.

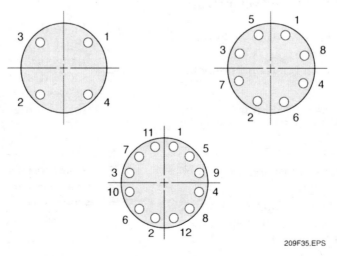

209F35.EPS

Figure 35 ◆ Proper tightening sequences.

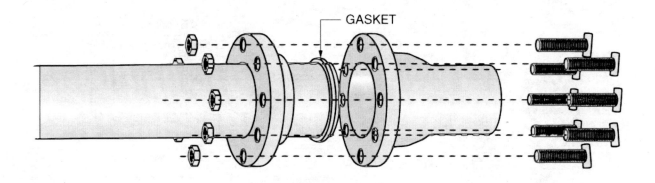

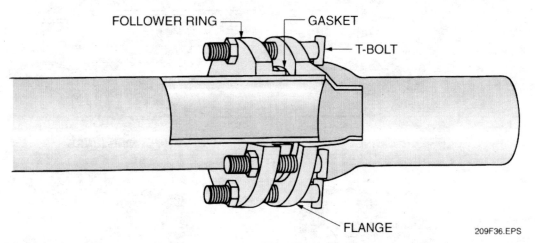

Figure 36 ◆ Mechanical joint.

209F36.EPS

CAUTION

Each bolt should be tightened a little at a time, following the same crossover sequence that was used in tightening flange bolts. This prevents warping or damaging the flange and prevents leaks. Tighten the bolts only until they are snug. Do not overtighten the bolts.

Step 11 Tighten the bolts to the specified torque, using a calibrated torque wrench and following the proper tightening sequence. The required torque for an application depends on the size of the flange and the grade of bolts used. Contact the design engineer if you are not sure of the torque requirements. *Table 4* lists the recommended torque ranges for mechanical joints.

CAUTION

Do not overtighten the bolts because this could damage the gasket, the flange, or the bolts.

Table 4 Recommended Torque Ranges for Mechanical Joints

Bolt Size in Inches	Torque in Foot-Pounds
5/8	45 to 60
3/4	75 to 90
1	100 to 120
1 1/4	120 to 150

209T04.EPS

Ductile iron pipe with mechanical or compression joints may have to be restrained to withstand the forces within the pipe. Restraint methods include the use of rodding, thrust blocks, or retaining rings with set screws. Check your job specifications to determine the type of restraint method required. *Figure 37* shows restraint methods.

7.0.0 ◆ CONCRETE PIPE

Concrete pipe is used underground for sanitary and storm drainage lines. It is manufactured in two types: standard non-reinforced concrete sewer pipe and standard reinforced concrete sewer pipe. Each type is available in several different grades. If the pipe is to be reinforced, a steel cage is placed

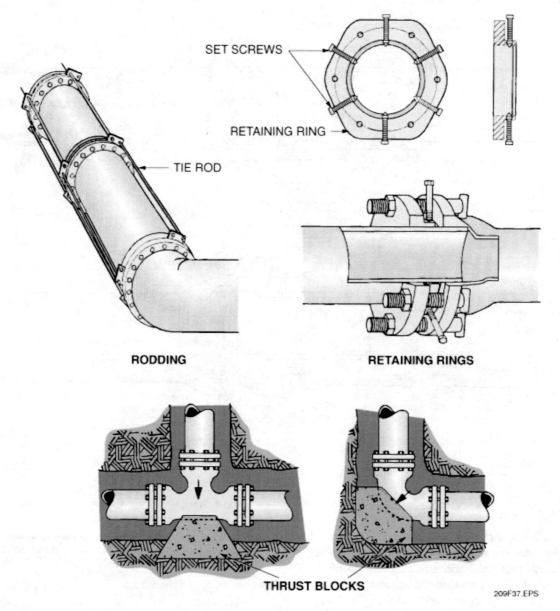

SET SCREWS

RETAINING RING

TIE ROD

RODDING

RETAINING RINGS

THRUST BLOCKS

209F37.EPS

Figure 37 ◆ Restraint methods.

inside the mold. The cage may be either a single circle or two concentric circles, depending on the desired strength of the finished pipe. Concrete pipe is available in four different shapes: circular, horizontal elliptical, vertical elliptical, and arch. The shape chosen for a particular system is a design consideration based on many factors.

Circular pipe is the most frequently used type of concrete pipe. For circular pipe, the span and rise are equal. The span of the pipe is the greatest internal dimension of the pipe measured horizontally. The rise of the pipe is the greatest internal dimension of the pipe measured vertically.

Elliptical pipe is oval in cross section. In heavy-loading situations, elliptical pipe can be installed with the longer cross-section vertical. In less heavily loaded situations, the pipe can be laid with the longer section horizontal.

Arch pipe is useful in minimum-cover situations or when vertical clearance problems are encountered. Because of its shape, arch pipe offers greater capacity for the same depth of flow than other shapes of pipe.

Two types of joints are used on concrete pipe: bell-and-spigot and tongue-and-groove. Elliptical pipe and arch pipe have tongue-and-groove ends, and circular pipe can have either type of end. Concrete pipe is joined by using rubber gaskets, cement or mortar sealants, mastic sealants, or external bonds. The joint must be able to provide the following:

• Resistance to infiltration of ground water or backfill material

• Resistance to exfiltration of sewage or storm water

- Flexibility to accommodate lateral deflection or longitudinal movement without leaking
- Resistance to shear stresses between pipes without leaking
- Hydraulic continuity and a smooth flow line
- Ease of installation

7.1.0 Concrete Pipe Sizes

Non-reinforced concrete pipe is manufactured in sizes ranging from 4 to 24 inches in diameter and in lengths of 2, 2½, 3, and 4 feet. The reinforced concrete pipe is used for the larger sewer lines and is available in sizes ranging from 12 to 180 inches in diameter. Diameters given for concrete pipe refer to inside diameters. This is standard throughout the industry for concrete pipe. *Tables 5* through *9* list the minimum required dimensions and wall thicknesses for concrete pipe, based on standards written by ASTM International. Since there may be variations among manufacturers, always consult the literature received from the specific manufacturer of the concrete pipe used

on the project. The wall thicknesses given in the following tables are the minimum wall thickness required by ASTM International.

There are three different classes of non-reinforced concrete: culvert, storm drain, and sewer pipe with bell-and-spigot joints. They are designated Class 1, Class 2, and Class 3. The larger class numbers indicate thicker walls. *Table 5* lists the dimensions of non-reinforced concrete pipe with bell-and-spigot joints.

> **NOTE**
> These tables are based on concrete weighing 150 pounds per cubic foot and vary with heavier or lighter weight concrete.

Reinforced concrete culvert, storm drain, and sewer pipe with bell-and-spigot joints are manufactured in two different wall thicknesses. These are known as Wall A and Wall B, with Wall B being thicker. *Table 6* lists the dimensions of reinforced concrete pipe with bell-and-spigot joints.

Table 5 Non-Reinforced Concrete Pipe with Bell-and-Spigot Joints

| Internal Diameter (inches) | Class 1 | | | Class 2 | | Class 3 | |
	Minimum Wall Thickness (inches)	Average Weight (pounds per foot)	Average Weight (pounds per foot)	Minimum Wall Thickness (inches)	Minimum Wall Thickness (inches)	Average Weight (pounds per foot)
4	⅝	9½	¾	13	⅞	15
6	⅝	17	¾	20	1	24
8	¾	27	⅞	31	1⅛	36
10	⅞	37	1	42	1¼	50
12	1	50	1⅜	68	1¾	90
15	1¼	78	1⅝	100	1⅞	120
18	1½	105	2	155	2¼	165
21	1¾	159	2¼	205	2¾	260
24	2⅛	203	3	315	3¾	350

209T05.EPS

Table 6 Reinforced Concrete Pipe with Bell-and-Spigot Joints

| Internal Diameter (inches) | Wall A | | Wall B | |
	Minimum Wall Thickness (inches)	Average Weight (pounds per foot)	Minimum Wall Thickness (inches)	Average Weight (pounds per foot)
12	1¾	90	2	106
15	1⅞	120	2¼	148
18	2	155	2½	200
21	2¼	205	2¾	260
24	2½	265	3	325
27	2⅝	310	3¼	388
30	2¾	363	3½	459

209T06.EPS

Reinforced concrete culvert, storm drain, and sewer pipe with tongue and groove joints are available in three wall thicknesses, with Wall C being the thickest. *Table 7* lists the dimensions of reinforced concrete pipe with tongue and groove joints.

Reinforced concrete elliptical pipe sizes are given as if they were for round pipe. This is the standard for the industry. The minor axis is the smallest dimension of the pipe, regardless of the pipe orientation in the trench. Likewise, the major axis is the largest dimension. *Table 8* lists the dimensions of reinforced concrete elliptical pipe.

Reinforced concrete arch culvert, storm drain, and sewer pipe is also sized as if it were round pipe. Dimensions for span, rise, and wall thickness are the minimum requirements of ASTM International. *Table 9* lists the dimensions of reinforced concrete arch pipe.

Table 7 Reinforced Concrete Pipe with Tongue and Groove Joints

Internal Diameter (inches)	Minimum Wall Thickness (inches)	Average Weight (pounds per foot)	Minimum Wall Thickness (inches)	Average Weight (pounds per foot)	Minimum Wall Thickness (inches)	Average Weight (pounds per foot)
12	1¾	79	2	93	-	-
15	1⅞	103	2¼	127	-	-
18	2	131	2½	168	-	-
21	2¼	171	2¾	214	-	-
24	2½	217	3	264	3¾	366
27	2⅝	255	3¼	322	4	420
30	2¾	295	3½	384	4¼	476
33	2⅞	336	3¾	451	4½	552
36	3	383	4	524	4¾	654
42	3½	520	4½	686	5¼	811
48	4	683	5	867	5¾	1,011
54	4½	864	5½	1,068	6¼	1,208
60	5	1,064	6	1,295	6¾	1,473
66	5½	1,287	6½	1,542	7¼	1,735
72	6	1,532	7	1,811	7¾	2,015
78	6½	1,797	7½	2,100	8¼	2,410
84	7	2,085	8	2,409	8¾	2,660
90	7½	2,395	8½	2,740	9¼	3,020
96	8	2,710	9	3,090	9¾	3,355
102	8½	3,078	9½	3,480	10¼	3,760
108	9	3,446	10	3,865	10¾	4,160

Large Sizes of Pipe with Tongue and Groove Joints			
Internal Diameter (inches)	Internal Diameter (inches)	Wall Thickness (inches)	Average Weight (pounds per foot)
114	9½	9½	3,840
120	10	10	4,263
126	10½	10½	4,690
132	11	11	5,148
138	11½	11½	5,627
144	12	12	6,126
150	12½	12½	6,647
156	13	13	7,190
162	13½	13½	7,754
168	14	14	8,339
174	14½	14½	8,945
180	15	15	9,572

209T07.EPS

Table 8 Reinforced Concrete Elliptical Pipe

Equivalent Round Size (inches)	Minor Axis (inches)	Major Axis (inches)	Minimum Wall Thickness (inches)	Waterway Area (square feet)	Approximate Weight (pounds per foot)
18	14	23	2¾	1.8	195
24	19	30	3¼	3.3	300
27	22	34	3½	4.1	365
30	24	38	3¾	5.1	430
33	27	42	3¾	6.3	475
36	29	45	4½	7.4	625
39	32	49	4¾	8.8	720
42	34	53	5	10.2	815
48	38	60	5½	12.9	1,000
54	43	68	6	16.6	1,235
60	48	76	6½	20.5	1,475
66	53	83	7	24.8	1,745
72	58	91	7½	29.5	2,040
78	63	98	8	34.6	2,350
84	68	106	8½	40.1	2,680
90	72	113	9	46.1	3,050
96	77	121	9½	52.4	3,420
102	82	128	9¾	59.2	3,725
108	87	136	10	66.4	4,050
114	92	143	10½	74.0	4,470
120	97	151	11	82.0	4,930
132	106	166	12	99.2	5,900
144	116	180	13	118.6	7,000

209T08.EPS

Table 9 Reinforced Concrete Arch Pipe

Equivalent Round Size (inches)	Minimum Rise (inches)	Minimum Span (inches)	Minimum Wall Thickness (inches)	Waterway Area (square feet)	Approximate Weight (pounds per foot)
15	11	18	2¼	1.10	127
18	13½	22	2½	1.65	170
21	15½	26	2¾	2.20	225
24	18	28½	3	2.80	320
30	22½	36¼	3½	4.40	450
36	26⅝	43¾	4	6.40	595
42	31⁵⁄₁₆	51⅛	4½	8.80	740
48	36	58½	5	11.40	880
54	40	65	5½	14.30	1,090
60	45	73	6	17.70	1,320
72	54	88	7	25.60	1,840
84	62	102	8	34.60	2,520
90	72	115	8½	44.50	2,750
96	77¼	122	9	51.70	3,110
108	87⅛	138	10	66.00	3,850

209T09.EPS

7.2.0 Joining Concrete Pipe

Two types of joints are used on concrete pipe: bell-and-spigot and tongue and groove. Elliptical pipe and arch pipe have tongue and groove ends, and circular pipe can have either type of end. Concrete pipe is joined by four different methods: rubber gaskets, mastic sealants, cement or mortar sealants, or external bands.

7.2.1 Rubber Gaskets

The rubber gaskets used with concrete pipe require lubrication. Special lubricants are available from the manufacturers of concrete pipe. One type of gasket is a flat gasket that is cast into place during manufacturing. Another type of rubber gasket is an O-ring gasket that fits into a groove cast into the tongue or spigot end. Roll-on gaskets are placed around the spigot end of the pipe and rolled into place as the spigot is inserted into the bell of the mating pipe.

7.2.2 Mastic Sealants

Mastic sealants, made from coal tar and mineral fillers, are usually applied cold. These sealants require that a primer be applied to both mating surfaces and allowed to dry for a specified amount of time. After the primer has dried, the sealant is applied to one of the mating surfaces. Enough sealant should be used so that there is excess sealant squeezed out of the joint when the pipes are joined.

7.2.3 Cement or Mortar Sealants

The difference between cement and mortar lies in the mix. Cement is made by adding 1 to 1½ gallons of water to a sack of portland cement. Mortar is made by mixing one part portland cement to two parts sand and adding 6½ to 7 gallons of water per sack of cement. These sealants are applied to the lower half of the bell end and the upper half of the spigot end. After the joint is made, any spaces between the pipe should be filled with the sealant, and in the case of large pipe, the sealant on the inside of the joint should be finished smooth.

7.2.4 External Bands

Some specifications may call for bands of portland cement to be placed around the outside of the pipe joints. The joint should be backfilled as soon as possible after the cement has set. Specifications may also call for bands of rubber to be placed around the joint. These bands are held in place by the weight of the backfill.

7.3.0 Cutting Concrete Pipe

Concrete pipe can be cut using a gas-powered cut-off (demolition) saw using a masonry wheel.

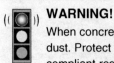

WARNING!

When concrete is cut, it produces hazardous dust. Protect yourself by using an OSHA-compliant respirator.

7.4.0 Installing Concrete Pipe Using Rubber Gaskets

Installing concrete pipe with rubber gaskets is the easiest, quickest, and cleanest way to install the pipe. The gaskets used with concrete pipe require lubrication. Special lubricants are available from the manufacturers of concrete pipe. One type of rubber gasket is a flat gasket that is cast into place during manufacturing. Another type of rubber gasket is an O-ring gasket that fits into a groove cast into the tongue or spigot end. Roll-on gaskets are placed around the spigot end of the pipe and rolled into place and lubricated. Then the spigot is inserted into the bell of the mating pipe. *Figure 38* shows concrete pipe joined with rubber gaskets.

7.4.1 Installing Concrete Pipe Using Cement or Mortar Sealants

Laying pipe that requires a cement or mortar joint is much more difficult than laying pipe installed with rubber gaskets. The difference between cement and mortar is how they are mixed. Cement is made by adding 1 to 1½ gallons of water to a sack of portland cement. Mortar is made by mixing one part portland cement to two parts sand and adding 6½ to 7 gallons of water per sack of cement. These sealants are applied to the lower half of the bell end and the upper half of the spigot end. After the joint is made, any spaces between the pipe should be filled with the sealant, and the sealant on the inside of the joint should be finished smooth.

Laying concrete pipe in a high-production operation requires several people. Two people must work at the top of the trench, one to hook the pipe and one to mix the mortar or cement. Two people are required in the trench to lay the pipe, and one other is needed to apply the mortar. When laying pipe that is larger than 48 inches in diameter, two people are needed to apply the mortar, one on each side of the pipe. An additional two workers work ahead of the pipe-laying operation, with one person in the trench to grade the bedding material and another person on top to check the grade and

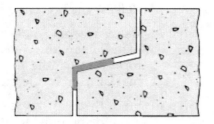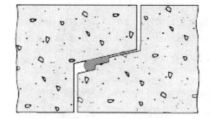

TYPICAL CROSS SECTIONS OF BASIC COMPRESSION-TYPE RUBBER GASKET JOINTS

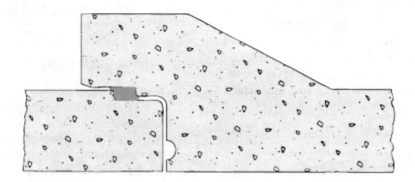

TYPICAL CROSS SECTION OF OPPOSING SHOULDER-TYPE JOINT WITH O-RING GASKET

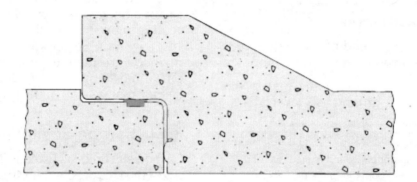

TYPICAL CROSS SECTION OF SPIGOT GROOVE-TYPE JOINT WITH O-RING GASKET

209F38.EPS

Figure 38 ◆ Concrete pipe joined with rubber gaskets.

supply bedding material to the trench. Follow these steps to install concrete pipe, using cement or mortar sealant:

NOTE

For this exercise assume that you are adding to an existing pipeline that is already in the trench with the grooved or bell end facing upstream.

Step 1   Mix the mortar to the desired consistency.

NOTE

The water content of the mortar is very important to provide the creamy consistency needed. One way to check the consistency of the mortar is to hold a shovel of mortar 3 feet above the mortar box and let the mortar slide off the shovel. If the mortar disappears far below the surface, it is too wet, and if it clumps in a mound on top of the surface, it is too dry. If it submerges only partly below the surface, it is properly mixed.

Step 2 Dig a small impression for the bell in front of the existing pipe in the trench. The impression should be the width of the part of the bell that is touching the ground.

Step 3 Fill the impression with mortar.

Step 4 Splash water into the groove of the bell of the existing pipe, using a short-handled concrete brush.

Step 5 Smear mortar inside the groove of the bottom half of the bell.

Step 6 Move the buckets and tools out of the way so that the next piece of pipe can be laid. The next piece of pipe should be hooked up and ready to be lowered into the trench at this time. Steps 7 and 8 are performed while the pipe is still out of the trench.

Step 7 Splash water onto the spigot end of the pipe.

Step 8 Apply mortar to the top half of the spigot end of the pipe.

Step 9 Lower the pipe into position in front of the existing pipe until it is touching the trench bottom lightly. Keep just enough weight on the hoist line so that the pipe can be moved by hand.

Step 10 Guide the spigot end of the mating pipe into the bell end of the existing pipe.

Step 11 Push the pipe home using a steel bar driven into the ground as a lever.

CAUTION

Always use a timber between the steel bar and the pipe to prevent damage to the pipe end.

NOTE

The pipes must be in line with each other, or they will not fit together easily. When the pipes have been joined correctly, there will always be a space between the pipes ranging from ⅜ inch for 12-inch pipe to 2 inches for 60-inch pipe. In a properly aligned joint, mortar squeezes from between the two pipes and fills this space as they are shoved together.

Step 12 Check the flow line of the pipe using a level, a string line level, or a laser level. You may have to lift the pipe slightly and add or remove bedding material to correct the flow line of the pipe.

Step 13 Release the hoist line from the pipe.

Step 14 Use a sweeping motion around the inside of the joint with a concrete brush attached to a long handle to remove the excess mortar from inside the pipe.

Step 15 Smooth away the excess mortar from around the outside of the joint, using a short-handled concrete brush.

Step 16 Grout the lifting hole with mortar, and smooth it on the inside and outside.

Step 17 Repeat the entire procedure for each piece of pipe.

8.0.0 ◆ CARBON STEEL PIPE

Carbon steel pipe is widely used throughout the industry and in various applications. It is durable, machinable, and less expensive than most other types of pipe. Carbon steel pipe is preferred by many design engineers because it is easy to install. In underground applications where carbon steel pipe is exposed to external corrosion, the pipe must be coated to ensure long life. One method of coating carbon steel pipe, referred to as tar and feathers, is coating the pipe with tar and felt paper before installation. However, this coating can be easily damaged in fabrication. Better methods of coating pipe that are done by the pipe manufacturer include Scotchkote™ and Plicoflex. These methods are more expensive, but they make the pipe almost impossible to damage and enable it to last for many years in the ground.

8.1.0 Carbon Steel Pipe Sizes

Carbon steel pipe is listed in inches by its nominal size. For sizes of Schedule 40 pipe up to and including 12 inches, nominal size is an approximation of inside diameter. From 14 inches on, nominal size reflects the outside diameter of the pipe. There are times when the nominal size of a pipe and its actual inside or outside diameter differ greatly, but nominal size is used to describe pipe.

Wall thickness of carbon steel pipe can be described in two ways. The first is by schedule. As the schedule numbers increase, the wall thickness gets larger; therefore, the pipe is stronger and can withstand more pressure. Schedule numbers for carbon steel pipe are 40, 80, and 160. The schedule numbers refer to the wall thickness and the inside diameters only. The outside diameters are the same for each of the schedule numbers. A 2-inch pipe has an outside diameter of 2.375 inches in Schedule 40, 80, and 160 pipe. Since the Schedule 160 pipe has a thicker wall, the inside diameter is smaller. *Figure 39* shows pipe schedule difference.

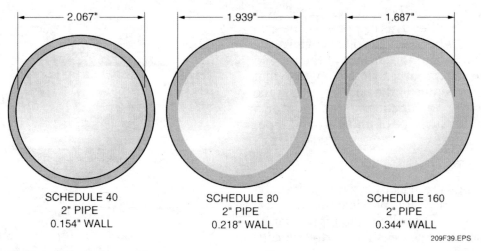

Figure 39 ◆ Pipe schedule difference.

It is important to remember that the schedule number only refers to the wall thickness of a pipe of a given nominal size. A ¾ inch Schedule 40 pipe will not have the same thickness as a 1-inch Schedule 40 pipe.

The next way to describe pipe wall thickness is by manufacturer's weight. From smallest to largest, there are three classifications in common use today. These are as follows:

- *STD* – Standard wall
- *XS* – Extra-strong wall
- *XXS* – Double extra-strong wall

STD wall pipe is approximately the same wall dimension as Schedule 40 pipe of the same nominal size up to 10-inch nominal size. XS is approximately the same wall thickness as Schedule 80 pipe of the same nominal size through 8-inch pipe, and then thinner walled. XXS pipe is heavier than Schedule 160 pipe through 6-inch pipe, then thinner walled than Schedule 160 in the larger sizes.

Table 10 lists commercial pipe sizes and wall thicknesses.

8.2.0 Joining Carbon Steel Pipe

Carbon steel underground pipe can be joined by butt welding, socket welding, flanges, screwed fittings, and dresser couplings. In some applications, carbon steel pipe is lined with concrete, starting at 12-inch pipe. The pipe is cut and joined in a manner similar to ordinary carbon steel pipe, except the brittleness of the concrete requires special care in handling.

8.2.1 Butt Welding

Butt welding forms the strongest, most permanent types of joints. Butt welding refers to welding two straight ends of pipe together end to end. Butt welding is normally used on pipe that is 2 inches in diameter or larger.

8.2.2 Socket Welding

Socket welding involves inserting the plain end of a length of pipe into the socket of a fitting and fillet welding joint. This joint is not as strong as the butt welded joint and is not used in high-pressure systems. However, this fitting is easier to assemble than the butt weld and can be used in many small diameter systems.

8.2.3 Flanges

A flange is a rim that can be welded or screwed onto the end of a fitting or pipe. The flanges of the mating pipes and fittings are joined together with bolts. Most flanges require gaskets between them to prevent leaking.

8.2.4 Threaded Joints

Threaded joints are another method of joining carbon steel pipe. A pipe joint compound or Teflon® tape should be used to lubricate the threads to prevent leaking. The fittings used in underground systems are usually cast iron recessed fittings that provide a smooth waterway inside when the pipe is threaded into them. *Figure 40* shows a threaded joint in a recessed cast iron fitting.

8.2.5 Dresser Couplings

A dresser coupling consists of one steel middle ring, two steel follower rings, two gaskets, and a set of steel trackhead bolts that secure the follower rings together. The middle ring has a conical flare on each side to receive the two gaskets, and the follower rings confine the outer rings of the gaskets.

Tightening the trackhead bolts draws the follower rings toward each other. This compresses the gaskets into the spaces between the follower rings, the middle ring flares, and the surface of the pipe to form a flexible, leakproof seal. If it is possible that the pipe will move out of the coupling, the pipe must be properly anchored. Dresser couplings may be used with any type of underground pipe.

8.3.0 Installing Dresser Couplings

Dresser couplings are usually used below grade only for repair work. If it is possible that the pipe will move out of the coupling, the pipe must be properly anchored. *Figure 41* shows a dresser coupling.

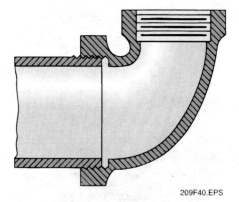

209F40.EPS

Figure 40 ◆ Threaded joint in recessed cast iron fitting.

Table 10 Commercial Pipe Sizes and Wall Thicknesses

Nominal Pipe Size	Outside Diam.	Nominal Wall Thickness												
		Sched. 10	Sched. 20	Sched. 30	STD	Sched. 40	Sched. 60	XS	Sched. 80	Sched. 100	Sched. 120	Sched. 140	Sched. 160	XXS
⅛	0.405	-	-	-	0.068	0.068	-	0.095	0.095	-	-	-	-	-
¼	0.540	-	-	-	0.088	0.088	-	0.119	0.119	-	-	-	-	-
⅜	0.675	-	-	-	0.091	0.091	-	0.126	0.126	-	-	-	-	-
½	0.840	-	-	-	0.109	0.109	-	0.147	0.147	-	-	-	0.188	0.294
¾	1.050	-	-	-	0.113	0.113	-	0.154	0.154	-	-	-	0.219	0.308
1	1.315	-	-	-	0.133	0.133	-	0.179	0.179	-	-	-	0.250	0.358
1¼	1.660	-	-	-	0.140	0.140	-	0.191	0.191	-	-	-	0.250	0.382
1½	1.900	-	-	-	0.145	0.145	-	0.200	0.200	-	-	-	0.281	0.400
2	2.375	-	-	-	0.154	0.154	-	0.218	0.218	-	-	-	0.344	0.436
2½	2.875	-	-	-	0.203	0.203	-	0.276	0.276	-	-	-	0.375	0.552
3	3.500	-	-	-	0.216	0.216	-	0.300	0.300	-	-	-	0.438	0.600
3½	4.000	-	-	-	0.226	0.226	-	0.318	0.318	-	-	-	-	-
4	4.500	-	-	-	0.237	0.237	-	0.337	0.337	-	0.438	-	0.531	0.674
5	5.563	-	-	-	0.258	0.258	-	0.375	0.375	-	0.500	-	0.625	0.750
6	6.625	-	-	-	0.280	0.280	-	0.432	0.432	-	0.562	-	0.719	0.864
8	8.625	-	0.250	0.277	0.322	0.322	0.406	0.500	0.500	0.594	0.719	0.812	0.906	0.875
10	10.750	-	0.250	0.307	0.365	0.365	0.500	0.500	0.594	0.719	0.844	1.000	1.125	1.000
12	12.750	-	0.250	0.330	0.375	0.406	0.562	0.500	0.688	0.844	1.000	1.125	1.312	1.000
14 OD	14.000	0.250	0.312	0.375	0.375	0.438	0.594	0.500	0.750	0.938	1.094	1.250	1.406	-
16 OD	16.000	0.250	0.312	0.375	0.375	0.500	0.656	0.500	0.844	1.031	1.219	1.438	1.594	-
18 OD	18.000	0.250	0.312	0.438	0.375	0.562	0.750	0.500	0.938	1.156	1.375	1.562	1.781	-
20 OD	20.000	0.250	0.375	0.500	0.375	0.594	0.812	0.500	1.031	1.281	1.500	1.750	1.969	-
22 OD	22.000	0.250	0.375	0.500	0.375	-	0.875	0.500	1.125	1.375	1.625	1.875	2.125	-
24 OD	24.000	0.250	0.375	0.562	0.375	0.688	0.969	0.500	1.218	1.531	1.812	2.062	2.344	-
26 OD	26.000	0.312	0.500	-	0.375	-	-	0.500	-	-	-	-	-	-
28 OD	28.000	0.312	0.500	0.625	0.375	-	-	0.500	-	-	-	-	-	-
30 OD	30.000	0.312	0.500	0.625	0.375	-	-	0.500	-	-	-	-	-	-
32 OD	32.000	0.312	0.500	0.625	0.375	0.688	-	0.500	-	-	-	-	-	-
34 OD	34.000	0.312	0.500	0.625	0.375	0.688	-	0.500	-	-	-	-	-	-
36 OD	36.000	0.312	0.500	0.625	0.375	0.750	-	0.500	-	-	-	-	-	-
42 OD	42.000	-	-	-	0.375	-	-	0.500	-	-	-	-	-	-

209T10.EPS

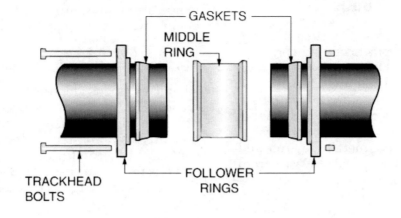

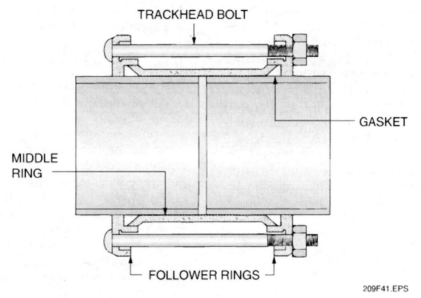

209F41.EPS

Figure 41 ◆ Dresser coupling.

Follow these steps to install a dresser coupling:

Step 1 Ensure that the ends of the mating pipes are cut square.

Step 2 Clean the ends of the mating pipes and the dresser coupling components, and ensure that the insides of the mating pipes are clean.

Step 3 Slide a follower ring onto each of the mating pipe ends, with the follower ring lip facing the end of the pipe.

Step 4 Lubricate the gaskets, using a lubricant recommended by the coupling manufacturer.

Step 5 Slide a gasket onto each of the mating pipe ends, with the smaller end facing the end of the pipe.

Step 6 Slide the ends of the mating pipes into the middle ring until the pipes meet in the middle of the ring.

Step 7 Slide the two follower rings and gaskets against the middle ring.

Step 8 Insert the trackhead bolts through the two follower rings.

Step 9 Place the nuts on the trackhead bolts and tighten them finger-tight.

Step 10 Tighten the nuts to the torque recommended by the manufacturer using the crossover method.

 CAUTION

Do not tighten the bolts to full torque the first time around the coupling. Tighten them in increments of at least 25 percent of full torque each time around the coupling to reach full torque after four tightenings.

Step 11 Coat the dresser coupling and bolts with a waterproof mastic-type coating.

9.0.0 ◆ IRON ALLOY PIPE

Iron alloy pipe is an acid-resistant form of cast iron pipe that is resistant to practically all corrosive wastes. The pipe is available in two types, known by their trade names Duriron and Durichlor 51. Both types contain approximately 15 percent silicon with small quantities of carbon and magnesium. The acid resistance is contained throughout the entire thickness of the metal. Duriron and Durichlor 51 are used for disposal of wastes from chemical and allied industries.

Of the two types of iron alloy pipes, Duriron is most widely used. It is almost completely unaffected by the corrosive wastes from most processes, with the exception of hydrofluoric acid or fluorides which attack Duriron pipe. Duriron pipe is of the bell-and-spigot type and is similar in appearance to cast iron pipe. No protective coating is used or required inside or outside the Duriron pipe. Allowance for expansion of Duriron pipe is required when using it to convey high-temperature solutions. Duriron bell-and-spigot pipe and fittings are intended for use in gravity-fed applications. They are not designed for use under pressure. Duriron also is available with special mechanical joint fittings. Duriron pipe is too hard to cut with a blade, but can be cut with a chain cutter.

9.1.0 Duriron Hub and Plain End Pipe Sizes

Duriron pipe has different pipe sizes, depending on whether it is joined as a bell-and-spigot type of pipe or with the special type of mechanical joint used for Duriron. *Table 11* shows the sizes for hub and plain end pipe.

9.2.0 Duriron Mechanical Joint Pipe Sizes

The Duriron mechanical joint is a mechanical joint that is similar to ductile iron in the way it joins the pipes. *Table 12* shows the sizes for those pipes.

10.0.0 ◆ FIBERGLASS PIPE

Fiberglass pipe used for underground installations is considered a flexible pipe. It has the ability to deflect under a load without causing structural damage to the pipe. Fiberglass pipe is made up of layers of a fiberglass and resin mixture. The resin provides corrosion resistance, and the fiberglass provides strength. Different types of fiberglass pipe are manufactured, using many different resins and glass-to-resin ratios to make them suitable for different applications. The type of fiberglass pipe used depends on the temperature, pressure, and type of substance to be conveyed by the pipe.

10.1.0 Installing Fiberglass Pipe

Fiberglass pipe is identified by different methods according to the manufacturer. When installing fiberglass pipe, the manufacturer's recommendations for installing pipe sections and curing the joints must be carefully followed. The following sections review the basic guidelines for installing three types of fiberglass pipe joints:

- Bell-and-spigot
- Redi-Thread
- Threaded-and-bonded

10.1.1 Types of Adhesive

Each type of fiberglass pipe requires a certain type of adhesive. You must follow the manufacturer's recommendations when selecting an adhesive to join pipes. Before selecting an adhesive, you must consider the type, quantity, service temperature, and working life of the adhesive. *Table 13* shows general adhesive kit information.

Many adhesive kits have two parts that must be mixed in order to harden. When an epoxy resin is mixed with a hardener, also known as a catalyst, the mixture produces heat and makes a hard, strong glue. When mixing any adhesive, be sure to follow the step-by-step instructions supplied with the adhesive kits.

Most adhesive mixing charts use 75°F as the standard temperature at which to mix the adhesive. The adhesive hardens faster if the temperature is above 75°F and hardens more slowly if the temperature is below 75°F. Do not heat any adhesive above 80°F. Most adhesive charts also list the number of fittings that a kit can bond. This number depends on the size of pipe being joined. Most kits can bond only one 12-inch pipe or can bond several smaller pipe joints. Always pay close attention to the working life of the adhesive. This is the time that it takes the adhesive to harden from the time that it is mixed. One large bond can easily be made in 10 minutes, but the adhesive may harden before several small joints can be made. Always have the pipe ready to install before starting to mix the adhesive.

Blanket heaters, which are similar to electric heating pads, are used to keep the temperature of the adhesive within the proper range for curing.

10.1.2 Installing Bell-and-Spigot Joints

When installing bell-and-spigot fiberglass pipe, several tools can be used to cut the pipe to the desired length. An electric saw with an abrasive

Table 11 Duriron Hub and Plain End Pipe Sizes

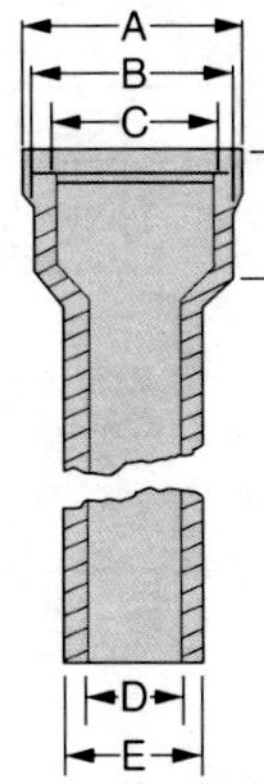

Hub and Plain End Pipe Sizes	Size in (mm)	A in (mm)	B in (mm)	C in (mm)	D in (mm)	E in (mm)	F in (mm)	Wt. lbs. (kg)	Working Length (m)	Overall Length ft/in (m)
	2 (50)	4⁹⁄₁₆ (116)	4³⁄₁₆ (106)	3⁵⁄₁₆ (84)	2¹⁄₃₂ (52)	2¹¹⁄₁₆ (68)	2⁵⁄₈ (67)	62 (28.1)	7 (2.1)	7'-2⁵⁄₈" (2200)
	3 (80)	5⁵⁄₁₆ (135)	5³⁄₁₆ (132)	4⁵⁄₁₆ (110)	3⅛ (79)	3¹⁷⁄₃₂ (96)	2⁵⁄₈ (67)	90 (40.8)	7 (2.1)	7'-2⁵⁄₈" (2200)
	4 (100)	6⅜ (162)	6³⁄₁₆ (157)	5⁵⁄₁₆ (135)	4⅛ (105)	4¹⁷⁄₃₂ (121)	2⁵⁄₈ (67)	114 (51.7)	7 (2.1)	7'-2⁵⁄₈" (2200)
	6 (150)	8¹⁷⁄₃₂ (217)	8¹¹⁄₃₂ (212)	7⁵⁄₁₆ (186)	5¹⁵⁄₁₆ (151)	6¹¹⁄₁₆ (170)	3 (76)	169 (76.7)	7 (2.1)	7'-2⁵⁄₈" (2200)
	8 (200)	11¼ (286)	10¾ (273)	9⅝ (244)	8¼ (210)	9 (229)	3 (76)	234 (106.1)	7 (2.1)	7'-2⁵⁄₈" (2200)
	10 (250)	14¼ (362)	13¾ (349)	12¼ (311)	10 (254)	11¼ (286)	3⅞ (98)	340 (154)	5 (1.5)	5'-3⁵⁄₈" (1616)
	12 (300)	16¾ (425)	16 (406)	14½ (368)	12 (305)	13¼ (337)	4 (102)	470 (213)	5 (1.5)	5'-4" (1626)
	15 (380)	20¼ (514)	19¾ (502)	17¾ (451)	15 (381)	16¾ (425)	4⅛ (451)	800 (363)	5 (1.5)	5'-4⅛" (1629)

Provided by Flowserve Corporation, a global leader in the fluid motion and control business. More information on Flowserve and its products can be found at www.flowserve.com.

209T11.EPS

wheel is the best tool for cutting fiberglass pipe. If this saw is not available, a hacksaw with fine teeth can be used. A blade with at least 32 teeth per inch should be used. When the fiberglass has been cut, use a mandrel to shape the end to the appropriate taper, and clean the surface with a sander or sandpaper. Any overcut or roughness should be repaired with a surfacer, a compound rather like thick paint. Follow the instructions of the manufacturer of the surfacer to get a smooth surface. If the bell-and-spigot fiberglass pipe has been stored outdoors for more than 3 days and the ends have been exposed to the weather, they must be resurfaced. About 1 inch should be cut off the spigot end, and the end should be retapered. The bell should be sanded until it looks like new pipe. *Figure 42* shows a fiberglass bell-and-spigot joint.

Follow these steps to install bell-and-spigot fiberglass pipe:

Step 1 Ensure that the ends and the insides of the mating pipes are clean and dry.

Step 2 Sand the bonding surfaces using a ¼-inch drill motor and flapper sander. All bonding surfaces must be clean and dry and must be sanded within 2 hours of assembly. Sanded surfaces should show a dull, fresh finish and not a polished look. The end of the spigot must also be sanded.

Table 12 Duriron Mechanical Joint Pipe Sizes

Type "MJ" Pipe

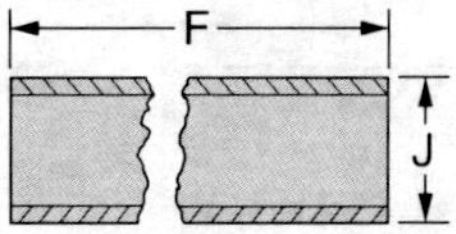

Size in (mm)	Part No.	J in (mm)	F in (mm)
1½ (40)	AS30196A	2³⁄₁₆ (56)	84 (2134)
2 (50)	AS30196B	2⅝ (67)	84 (2134)
3 (80)	AS30196C	3¾ (95)	84 (2134)
4 (100)	AS30196D	4¾ (121)	84 (2134)

Provided by Flowserve Corporation, a global leader in the fluid motion and control business. More information on Flowserve and its products can be found at www.flowserve.com.

209T12.EPS

Step 3 Clean the bell and spigot with the cleaner provided in the adhesive kit.

Step 4 Allow the cleaner to evaporate.

Table 13 General Adhesive Kit Information

Description	Quantity (Mixed Adhesive)	Working Life	Max. Service Temp.	Use With	Appropriate Number of Bonds Per Kit										
					1"	1½"	2"	3"	4"	6"	8"	10"	12"	14"	16"
General-purpose standard kit	6.8 oz.	25 min.		Redi-Thread® Bell-and-Spigot TAB joint			2	18	10	6	3	2	1		
General-purpose large kit	9.95 oz.	25 min.	150°F							8	4	3	2		
Twin-pack small job	2.50 oz. each	25 min.					1	10	6	2					
Bag kit for repairs	1.6 oz.	25 min.						6	4	2	1				
General purpose	6.8 oz.	20 min.	210°F	Silver thread® Bell-and-Spigot TAB joint			2	18	10	6	3	2	1	1	1
General-purpose standard kit	6.35 oz.	15 min.		Redi-Thread® Green thread® Silver thread® Chemline® Bell-and-Spigot TAB joint	45	27	2	15	8	5	3	2	1		
General-purpose large kit	9.22 oz.	15 min.	225°F							8	4	3	2	1	1
Twin-pack small job	2.66 oz. each	15 min.			40	23	1	12	7	4					
Bag kit for repairs	1.3 oz.	15 min.			13	7	3	2	1						
POLY THREAD standard kit	6.2 oz.	20 min.	200°F	POLY THREAD®			2	18	10	6	3	2	1		
POLY THREAD twin pack	3.1 oz. each	20 min.	200°F	POLY THREAD®			2	18	10	6	3	2	1		
POLY THREAD large kit	9.5 oz.	20 min.	200°F	POLY THREAD®						5	4	3	2	1	

209T13.EPS

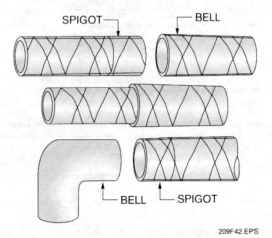

Figure 42 ◆ Fiberglass bell-and-spigot joint.

209F42.EPS

CAUTION

Use special care when handling parts to avoid contamination. Handle the parts with new gloves that are used only for this purpose or with clean, dry cotton cloths.

Step 5   Mix the adhesive according to the manufacturer's instructions.

WARNING!

When mixing the adhesive wear a proper OSHA-compliant respirator, safety goggles and chemical-resistant rubber gloves that are compatible with the adhesive being mixed. Adhesive in the eyes can cause permanent damage.

Step 6   Apply an even, thin coat of adhesive to the inside of the bell and the outside of the spigot, using a small paintbrush.

CAUTION

Too much adhesive in the bell restricts flow in the pipe.

Step 7   Align the pipes, and insert the spigot into the bell.

Step 8   You may need to push the spigot into the bell, and turn the spigot one-half to one full turn. However, follow the manufacturer's instructions when at this stage of installation. Make sure that the spigot is

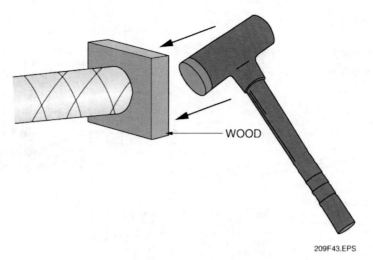

WOOD

209F43.EPS

Figure 43 ◆ Driving pipes together.

fully seated in the bell. On 4-inch pipe and larger sizes, a come-along or a hammer and strong piece of wood (*Figure 43*) may be needed to force the spigot into the bell.

Step 9 Allow the joint to harden. This normally takes 4 to 6 hours at 75°F. In circumstances where the joint is at a much lower temperature, use a heating blanket to keep the joint warm.

10.1.3 Installing Redi-Thread Joints

The Redi-Thread method is a screwed joint that can be quickly joined, removed, and reused. A collar is screwed onto the threads of the installing pipe, and a gasket seals the joint. The advantages of this type of joint include ease of installation, absence of adhesives to mix and cure, and the ability for the finished joint to deflect off center up to 2 degrees. Because this joint can be reused, it is often used for temporary pipe runs. *Figure 44* shows a Redi-Thread joint.

Follow these steps to install a Redi-Thread joint:

Step 1 Make sure that the rubber gasket is inside the female end of the pipe.

Step 2 Place the male and female ends together.

Step 3 Lubricate the threads on the locking sleeve using lanolin soap or a lithium-type grease.

Step 4 Slide the threaded locking sleeve into the female threads and tighten the collar.

10.1.4 Installing Threaded-and-Bonded Joints

The threaded-and-bonded joint is made with screwed threads and an adhesive. This type of joint requires bell-and-spigot-type pipe that has male threads on the spigot end and female threads inside the bell end. To make this joint, the bell and spigot threads are coated with glue and the threads are screwed together. This makes a very strong joint. *Figure 45* shows a threaded-and-bonded joint.

Follow these steps to install a threaded-and-bonded joint:

Step 1 Measure the length of the threads in the bell.

Step 2 Mark the threaded spigot end of the mating pipe ¼ inch longer than the length of the threads in the bell.

Step 3 Clean the joint parts using the primer included in the adhesive kit.

Step 4 Allow the primer to evaporate.

CAUTION

Use special care when handling parts to avoid contamination. Handle the parts with new gloves that are used only for this purpose or with clean, dry cotton cloths.

Step 5 Mix the adhesive according to the manufacturer's instructions.

WARNING!

When mixing the adhesive, wear safety goggles and chemical-resistant rubber gloves that are compatible with the adhesive being mixed. Adhesive in the eyes can cause permanent damage.

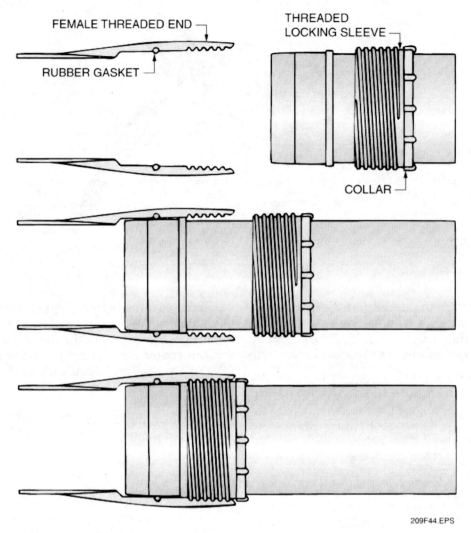

FEMALE THREADED END

RUBBER GASKET

THREADED LOCKING SLEEVE

COLLAR

209F44.EPS

Figure 44 ◆ Redi-Thread joint.

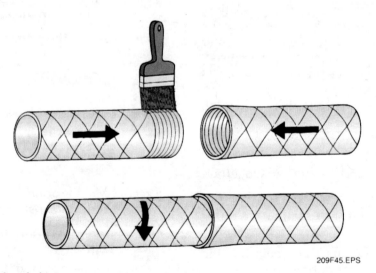

209F45.EPS

Figure 45 ◆ Threaded-and-bonded joint.

Step 6 Apply the adhesive to the threads inside the bell end of the pipe.

Step 7 Apply the adhesive to the male threads on the spigot end mating pipe.

Step 8 Screw the pipes together, using a strap wrench, until the mark on the threaded spigot is inside the bell.

Step 9 Check the joint for any movement between the bell and spigot. If movement occurs, tighten the spigot until no movement occurs, but do not overtighten the joint.

11.0.0 ◆ THERMOPLASTIC PIPE

Thermoplastic pipe is used in many systems in which heat and pressure are not a problem. In the past few years, the use of plastic pipe has increased significantly. Thermoplastic pipe is made by an extrusion process in which the heat-softened plastic is forced through a circular-shaped die to form the round pipe shape. The advantages of plastic pipe are that it is inexpensive, lightweight, easily joined, corrosion-resistant, and has smooth internal walls that do not restrict liquid flow. The disadvantages of plastic pipe are that it has a low resistance to heat and pressure, it is not as strong as many metal pipes, it requires an added number of hangers and supports because it is so flexible, and it expands and contracts at a very high rate when heated and cooled. The most widely used thermoplastics for underground service include polyvinyl chloride (PVC), chlorinated polyvinyl chloride (CPVC), HDPE, and acrylonitrile-butadiene-styrene (ABS). Various types of thermoplastic pipe can be glued, fusion-welded, threaded, bell-and-spigot (also called push-on) or no-hub.

Plastic pipe has both advantages and disadvantages that you need to know. In addition, you need to understand the different properties of plastic pipe and how the industry measures plastic piping. The following sections discuss these characteristics and properties, as well as the standard labeling practice for plastic piping and the most common plastic pipe and fitting manufacturers. You will also learn about the specific types of plastic pipe that you will use in piping applications.

11.1.0 Advantages and Disadvantages of Using Plastic Pipe

Most plastic pipe and fittings are chemically inert, meaning they do not react with other substances or materials, so they can withstand most chemicals that are used in the home, office, or factory. Strong and durable, plastic pipe and fittings also require little maintenance because they are gener-

ally resistant to corrosion and do not pit or scale. In addition, they are easier to handle than metals because of their relative light weight, which can make them more cost-effective to install.

The use of plastic pipe helps to eliminate one of the most common hazards associated with installation—fire. During the installation process, no arc-welding or soldering is involved. Therefore, it is less likely that fires will occur. In the past, when pipe was joined using molten **lead**, accidents involving spilled lead and tipped-over heating furnaces were a constant concern.

Plastic does have some drawbacks. Temperature changes cause it to expand and contract more than metal. Plastic pipe can also be flammable and give off fumes when it burns. In addition, fumes from solvent chemicals used to join plastic pipes are harmful; you must be familiar with the types of safety hazards each variety of pipe and chemical solvent presents. It is your responsibility to know how to protect yourself on the job site. Always consult the manufacturer's specifications regarding the properties of the pipe you are using.

All plastic pipe is affected by ultraviolet (UV) radiation to some degree. That means that you should store plastic piping under cover during a construction project to protect it from the sun. In general, short-term sun exposure will not harm the piping, but long-term exposure will degrade the plastic material. Always consult the manufacturer's instructions regarding the proper way to select and protect the particular piping you are using.

11.2.0 Properties of Plastic Pipe

Plastics contain polymers, long chains of molecules that manufacturers can mold, cast, or force through a die into desired shapes. Manufacturers can heat and shape resins in the form of pellets, powders, and solutions in various combinations to give the finished product the desired properties.

Each type of plastic pipe has its own unique properties. These properties determine where and how you can install it, how you can join it to other piping, and how you should support it. There are two general categories of plastic: thermoplastic and thermosetting. Thermoplastic pipe can be repeatedly softened by heating and hardened by cooling. When softened, thermoplastics can be molded into desired shapes. Thermosetting pipe changes chemically when heated, so that once hardened by heat or chemicals, it is hardened permanently.

In addition, plastic pipe can be either rigid or flexible. Rigid pipe is straight and maintains its shape. Flexible pipe bends, so although it requires fewer fittings and joints, it requires more support. Flexible pipe is often referred to as tubing. Manufacturers sell flexible pipe in coils.

The construction of plastic piping can vary as well. Pipes can be solid wall or cellular core wall construction. Solid wall does not contain trapped air. A pipe with cellular core construction, also called foam core construction, has walls that contain trapped air, so it is lighter weight. Cellular core wall piping consists of three layers of plastic: solid inner and outer layers and a foam middle layer. Because it consists of less material, it is also less expensive than solid wall pipe. Solid wall and cellular core wall pipe can be used interchangeably for drain, waste, and vent (DWV) applications. Several varieties of thermoplastic are used to repair pipe that is already in the ground, by forming an insertion sleeve, usually of polyester felt impregnated with thermosetting plastic, and using hot water, steam or air pressure to cure the plastic material in place against the existing pipe wall.

11.3.0 Plastic Pipe Sizing

Plastic pipe is manufactured with various wall thicknesses, commonly called schedules or size dimension ratios (SDRs). SDR relates wall thickness to the diameter of the pipe, using a set ratio of wall thickness to diameter. As the pipe diameter gets bigger, the wall gets thicker. An SDR number, such as SDR 35, designates this ratio. The ratio between a pipe's outside diameter (OD), or the distance between the outer walls of a pipe, and its wall thickness is constant for each pipe size. The larger the SDR number, the thinner the wall versus the diameter. For example, a 2-inch pipe (OD of 2.375 inches) with an SDR of 11 has a wall thickness of 0.216 inches (see *Table 14*). The same pipe with an SDR of 17 has a wall thickness of 0.14 inches. Be aware that pipe can also be measured by its inside diameter (ID), the distance between its inner walls.

In an effort to simplify and standardize the use of plastic pipe and fittings, manufacturers designated two SDR ratios as Schedule 40 (thin wall), which is the most commonly used, and Schedule 80 (thick wall). Most codes designate Schedule 40 as the minimum size for use in or under buildings.

11.4.0 Labeling (Markings)

You must use plastic pipe and fittings that are labeled and approved for use in piping systems. The label on a plastic pipe will include the following:

- Nominal pipe diameter, copper tube size (CTS) or iron pipe size (IPS)
- Schedule or SDR
- Type of material, PVC (polyvinyl chloride), CPVC (chlorinated polyvinyl chloride), and so on
- Pressure rating, usually expressed as pounds per square inch (psi) at a certain temperature
- Relevant standard for the pipe, for example, *American Society for Testing and Materials (ASTM) D-3309*
- Listing body or laboratory, for example, the National Sanitation Foundation
- Name of manufacturer or brand name
- Country of origin

This label ensures that a recognized third party—the listing body or laboratory—has tested the material and approved it for the use intended (see *Figure 46*). For example, the ASTM numbers on cellular core piping are different from those on Schedule 40 solid core piping. Solvent cements must also be listed and labeled for their specific use.

Table 14 Minimum Wall Thickness of 2-Inch Pipe

Plastic Pipe and Fittings Association Minimum Wall Thicknesses of 2-Inch Pipe Based on SDR/SIDR			
IPS-OD SDR (OD = 2.375 in) SDR (D 3035)	Wall	IPS-ID SIDR (ID = 2.067 in) SIDR (D 2239)	Wall
7.0	0.339	5.3	0.390
9.0	0.264	7.0	0.295
11.0	0.216	9.0	0.230
13.5	0.176	11.5	0.180
17.0	0.140	15.0	0.138
21.0	0.113	19.0	0.109
26.0	0.091	24.0	0.089
32.5	0.073	30.5	0.071

209T14.EPS

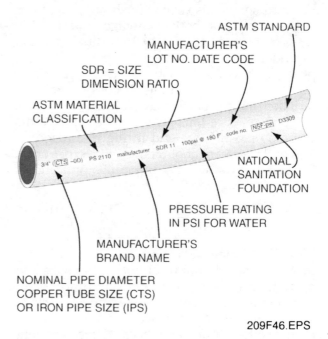

Figure 46 ◆ Typical pipe label.

209F46.EPS

11.5.0 Manufacturers of Plastic Pipe

There are hundreds of plastic pipe manufacturers, and different manufacturers make different pipes for different applications. For example, some specialize in pipes for hot or cold water systems, DWV systems, gas, or radiant heating. Manufacturers of plastic pipe include Charlotte Pipe, Vanguard, Bristol Pipe, J-M Manufacturing, and Cresline Plastic Pipe Co. Manufacturers that specialize in fittings include Nibco Manufacturing Co., Spears Plastics, and Lasco Plastics.

When you specify pipe and fittings, keep in mind that fit is critical to the success of the system installation. When purchasing pipe and fittings, ensure that they all meet the same standard for fit and materials. One manufacturer's pipe may not be compatible with another's fittings, even though both meet code standards.

11.6.0 Types of Plastic Pipe

There are primarily six types of plastic pipe, each with different applications and installation requirements (see *Table 15*). This section reviews commonly used types of plastic pipe.

11.6.1 ABS

ABS pipe and fittings (*Figure 47*) are made from a thermoplastic resin. ABS is the standard material for many types of DWV systems. ABS pipe is available in diameters ranging from 1½ inches to 6 inches. ABS pipe wall comes in both solid wall and cellular core wall construction, which are interchangeable in piping applications.

Because ABS is light, it is easy to handle and install. A 3-inch-diameter, 10-foot-long section weighs less than 10 pounds. ABS performs well at extreme temperatures because it absorbs heat and cold slowly, an important feature for a system that handles both hot and cold wastes.

ABS is highly resistant to household chemicals. In tests, it showed no effect from such common products as detergent, bleach, and household drain cleaners. Sewage treatment plants use ABS because it stands up to the highly corrosive and abrasive liquids commonly found in such systems.

ABS is strong and long-lasting. In 1959, ABS pipe was used in an experimental residence. Twenty-five years later, an independent research firm dug up and analyzed a section of the pipe and found no evidence of rot, rust, or corrosion. The pipe also withstands earth loads, slab foundations, and high surface loads without collapsing.

11.6.2 PVC

PVC is a rigid pipe with high-impact strength that is manufactured from a thermoplastic material (see *Figure 48*). The material has an indefinite life span under most conditions. PVC is frequently used in cold water systems. PVC is interchangeable with ABS pipe for DWV systems. It is also used to transport many chemicals because of its chemical-resistant properties. PVC is available in solid wall and cellular core wall construction.

PVC pipe is normally available in one of the size-dimension ratios (SDRs)—Schedule 40, Schedule 80, and Schedule 120. The wall thickness increase as the nominal size increases. Pipe sizes in Schedule 40 do not approach nominal sizes until 2-inch nominal, running from ⅛ inch pipe, which is approximately ¼ inch ID, to 2-inch pipe,

Table 15 Types of Plastic Pipe and Their Applications

Pipe	Applications
ABS	DWV sanitary systems, corrosive waste
PVC	DWV sanitary systems, cold water service
CPVC	Hot and cold water distribution
PE	Cold water service, corrosive waste, gas service
PEX	Hot and cold water distribution
PB	Hot and cold water distribution

209T15.EPS

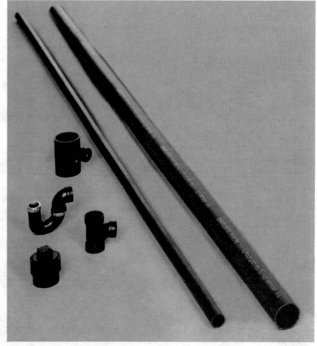

Figure 47 ◆ ABS pipe and fittings.

209F47.EPS

which is about 2.049 inches ID. The markings on the pipe will tell you the nominal size and either the schedule or SDR.

Solid core PVC pipe can be used in high-pressure systems, but only to carry low-temperature water. You must protect it from sunlight because ultraviolet light degrades the thermoplastic materials. It is lightweight, easy to handle and install, has joint flexibility that handles ground movement without leaking, and lasts a long time with no maintenance as long as it is protected from sunlight.

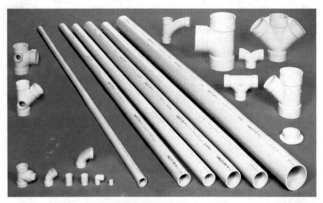

209F48.EPS

Figure 48 ◆ PVC pipe and fittings.

11.6.3 CPVC

CPVC pipe and fittings (*Figure 49*) are made from an engineered vinyl polymer. CPVC is used in hot and cold water distribution systems. Improvements made to its parent polymer, polyvinyl chloride, added high-temperature performance and improved impact resistance to this material. CPVC is produced in standard CTSs from ½ inch to 2 inches, with a full line of fittings.

CPVC is acceptable under many model codes for indoor use. Its molecular structure practically eliminates condensation in the summer and heat loss in the winter, decreasing the likelihood of costly drip damage to walls or structures. CPVC pipe's smooth, friction-free interior surfaces result in lower pressure loss and higher flow rates and provide less opportunity for bacteria growth. CPVC does not break down in the presence of aggressive, or chemically reactive, water. Like other types of plastic, it does not rust, pit, or scale.

CPVC is lightweight and easy to install. Recent improvements to CPVC have made the pipe stronger and more durable during installation. The strength of CPVC is a clear advantage when working in cold-weather states. In laboratory tests down to 20°F, CPVC pipe withstood a water hammer

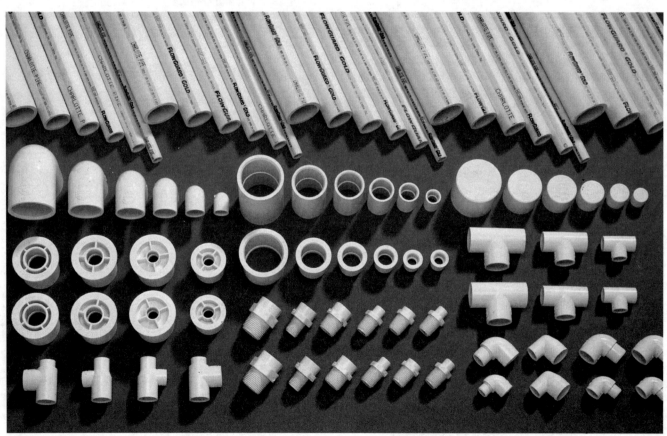

209F49.EPS

Figure 49 ◆ CPVC pipe and fittings.

drop that substantially damaged copper piping under the same conditions. Water hammer is an extreme change in water pressure within a pipe that can cause a loud, banging sound.

11.6.4 PE

Polyethylene (PE) is a thermosetting plastic. PE is commonly used as tubing because of its strength, flexibility, and chemical-resistant properties (see *Figure 50*). PE is also corrosion-resistant, which makes it ideal for transporting chemical compounds. It will not deteriorate when exposed to ultraviolet light, so you can install it outdoors without a protective coating. PE is used for cold water and underground gas service lines (outside buildings). High-density polyethylene (HDPE) is used frequently in contexts where it is possible to use a trenchless pipe-laying method, such as pipe-bursting, because the flexible HDPE pipe can be pre-chlorinated and pulled through the earth behind the bursting head, then fused to the fittings or to the previously laid pipe. Pipe-bursting is becoming a common method of replacing old and weakening plastic or clay pipe. A powerful hydraulic device forces the bursting head through the old pipe, expanding and breaking the pipe as it goes. The pipe is pulled back through the space where the old pipe was. By this procedure, hundreds of feet of pipe can be replaced in an hour.

11.6.5 PEX

Cross-linked polyethylene (PEX) tubing (*Figure 51*) is formed when high-density polyethylene is subjected to heat and high pressure. Because it resists high temperatures, pressure, and chemicals, it is ideal for potable hot and cold water systems, hydronic radiant floor systems, baseboard and radiator connections, and snow-melt systems. PEX is commonly used for manifold plumbing distribution systems because of its flexibility.

11.6.6 PB

Polybutylene (PB) is a thermoplastic pipe used extensively for water supply piping from the late 1970s to the mid-1990s. In many cases, PB piping has become weak and failed without warning. Experts believe this may be because of exposure to oxidants, such as chlorine, in public water systems. For this reason, PB is no longer available, so it is very unlikely that you will install PB piping. Many buildings still contain PB piping, though, and repair fittings are available, so you must be able to identify it (see *Figure 52*).

209F50.EPS

Figure 50 ◆ PE tubing.

209F51.EPS

Figure 51 ◆ PEX tubing.

11.7.0 Material Storage and Handling

To ensure maximum productivity on the job, use common sense when storing and handling plastic pipe. Before starting to work, always plan ahead.

Organize pipe and fittings into groups by pipe size and type. In addition, store pipe and fittings close to where you will be working so they are easy to access. Remember that you must protect many types of plastic pipe from ultraviolet light. When handling pipe and fittings, be careful not to bend or damage them. Pipe that is bent or otherwise damaged not only costs money, but also makes you less productive.

11.8.0 Fittings

A pipe's use determines what kinds of fittings and joints are needed. The following sections discuss the fittings used in water supply and DWV systems. It is important to be familiar with the most common fittings on the market.

11.8.1 Water Supply Fittings

Pressure-type fittings for use with water supply are short turn, or radius, with ledges or shelves. The radius describes the curve or bend of a fitting that changes direction. As the radius increases, the change in flow direction becomes smoother. This is particularly important for waste drain systems, where the smoother flow keeps organic waste from being deposited along the pipeline.

The most common types of water supply fittings include those listed in *Table 16* and shown in *Figure 53*.

11.8.2 DWV (Drain, Waste, and Vent) Fittings

DWV fittings have smooth interior passages with no ledges. They also have a longer radius that makes directional changes smoother and less likely to collect solids (see *Table 17* and *Figure 54*). When selecting DWV fittings, always consult all applicable codes and manufacturers' installation recommendations.

Techniques for measuring, cutting, and joining vary depending on the materials you are using and the function of the pipe (DWV or gas, for instance). There are many tools you will need to become familiar with in order to measure, cut, and join materials properly. The following sections explain methods for working with ABS, PVC, CPVC, PEX, and PE pipe used in DWV and water supply systems.

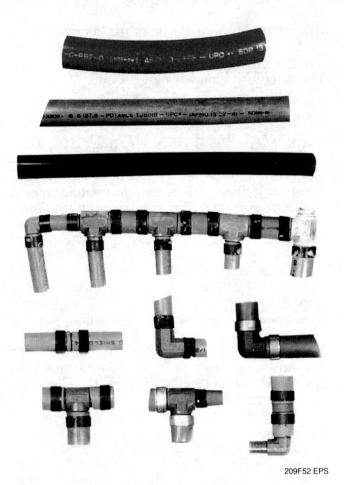

Figure 52 ◆ PB piping and fittings.

209F52.EPS

Table 16 Water Supply Fittings and Their Uses

Fitting	Use
Union	Mechanically connects two pipes, usually at a termination downstream of a service valve
Reducer	Connects pipes of different sizes
Elbow	Changes the direction of rigid pipe by either 90 degrees or 45 degrees
Tee	Provides an opening to connect a branch pipe at 90 degrees to the main pipe run
Coupling	Joins two lengths of the same pipe size when making a straight run
Cap	Plugs water outlets when testing the system or creates an air chamber to eliminate water hammer
Plug	Closes openings in other fittings or seals the end of a pipe
Manifold	Runs several water supply lines from the main supply to different fixtures

209T16.EPS

Figure 53 ◆ Water supply fittings.

Table 17 DWV Fittings and Their Uses

2003 International Plumbing Code®
Table 706.3, Fittings for Change in Direction

| Type of Fitting Pattern | Change in Direction | | |
	Horizontal to Vertical	Vertical to Horizontal	Horizontal to Horizontal
Sixteenth bend	X	X	X
Eighth bend	X	X	X
Sixth bend	X	X	X
Quarter bend	X	X[a]	X[a]
Short sweep	X	X[a,b]	X[a]
Long sweep	X	X	X
Sanitary tee	X[c]		
Wye	X	X	X
Combination wye and eighth bend	X	X	X

For SI: 1 inch = 25.4 mm

a. The fittings shall only be permitted for a 2-inch or smaller fixture drain.

b. Three inches or larger.

c. For a limitation on double sanitary tees, see Section 706.3.

209T17.EPS

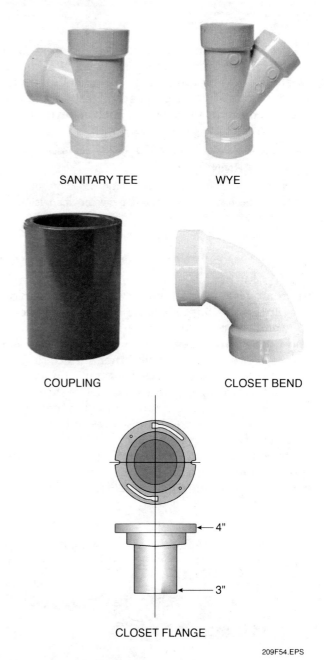

Figure 54 ◆ DWV fittings.

11.9.0 Measuring

It is important to plan ahead when you are installing ABS and PVC pipe and fittings in DWV systems. These systems have a built-in slope, also called a pitch or fall, and you must lay them accurately the first time. You cannot fix mistakes later with heat or hammers.

When you are measuring any pipe, be sure to allow for depth of joints. Take measurements to the full depth of the socket, not with pipe partly inserted into the socket. This is especially important when you use solvent cement to join piping. With this method, you must dry fit the installation, then mark alignment for fittings before you make the joint.

11.10.0 Cutting

Plastic pipe requires a square cut for good joint integrity. Cut tubing as squarely as possible to create the best bonding area within a joint. If you see any indication of damage or cracking at the tubing end, cut off at least 2 inches beyond any visible crack.

 WARNING!

Follow all manufacturer-recommended precautions when cutting or sawing pipe or when using any flame, heat, or power tools. Always wear appropriate personal protective equipment.

After cutting, you should ream the pipe. Reaming removes the small burrs that result when you cut pipe. Burrs left on piping can cause it to corrode and can prevent proper contact between tube and fitting during assembly. A pipe that is reamed correctly provides a smooth inner surface for better flow. You should also remove burrs on the outside of the pipe to ensure a good fit. Create a slight bevel on the end of the tube to make it easier for the tube to fit into the fitting socket and lessen the chances of pushing solvent cement to the bottom of the joint. Tools used to ream pipe ends include the reaming blade on the tubing cutter, files, pocketknives, and deburring tools (*Figure 55*).

DEBURRING TOOLS

INSIDE OUTSIDE REAMER
FOR SMALLER PIPE

209F55.EPS

Figure 55 ◆ Deburring tools.

11.10.1 Cutting PVC and ABS Pipe

You can cut PVC and ABS pipe with appropriate pipe cutters, a handsaw, or a power saw equipped with a carbide tip or abrasive blade. Plastic pipe cutters have one to four cutting wheels that can be rotated around a pipe to cut it. Wheels are available to fit standard cutters. You can also use ratchet shears or lightweight, quick-adjusting cutters designed exclusively for plastic piping (*Figure 56*).

To make sure you get a square cut, use a power saw on large jobs and a miter box on small jobs (see *Figure 57*). Ensure that you use the proper plastic saw for cutting PVC pipe. If these are not available, scribe the pipe and cut to the mark. After cutting the pipe, ream it inside and chamfer the edge to remove burrs, shoulders, and ragged spots. Chamfering involves beveling the edge of the pipe to a 45-degree angle. Chamfering is good practice because it provides for a more secure joint. When the socket and pipe fit tightly, you reduce the possibility of leakage from the pipes.

 CAUTION

Mark carefully! Proper alignment in the final assembly is critical. To ensure alignment, carefully mark the positions of any fittings that will be rolled or otherwise aligned.

11.11.0 Joining

Typically, water supply and DWV fittings are joined using the same installation techniques. Among the installation methods are mechanical fittings, such as ring-tight gaskets, bell-and-spigot, threaded joints, heat fusion, and solvent weld.

Ring-tight gasket fittings have a rubber O-ring or gasket in the socket. You must bevel the pipe at the leading edge to allow it to pass the gasket that forms the seal (see *Figure 58*). The bell-and-spigot pipe has a bell on one end with an internal elastomeric seal. The spigot (straight end) of the next pipe is fitted into the bell end to form a fluid-tight joint. Be sure to lubricate the gasket before inserting the pipe.

You can also connect some plastic pipe to dissimilar pipe with a transition fitting. Refer to your local applicable code and manufacturer's specifications to ensure that you are using the proper transition fittings.

There are many different types of fusion fittings, so always consult your local applicable code and the manufacturer's specifications for proper joining and installation methods (see *Figure 59*). You will learn more about fusion fittings and how to create a fusion weld elsewhere in this curriculum.

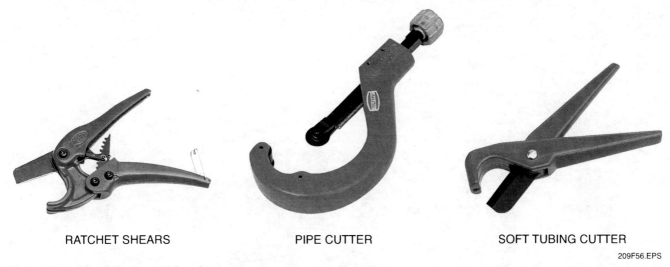

RATCHET SHEARS   PIPE CUTTER   SOFT TUBING CUTTER

209F56.EPS

Figure 56 ◆ Cutting tools used for plastic pipe.

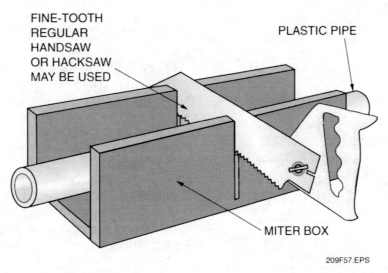

FINE-TOOTH
REGULAR
HANDSAW
OR HACKSAW
MAY BE USED

PLASTIC PIPE

MITER BOX

209F57.EPS

Figure 57 ◆ Using a miter box on small jobs.

11.11.1 Solvent Welding (Solvent Cementing)

ABS, PVC, and CPVC plastic pipe and fittings are often joined with solvent cement to form a solvent weld. HDPE pipe is not glued. The process of solvent welding is also referred to as solvent cementing. Solvent-weld fittings have sockets that the pipe fits into (see *Figure 60*). Before applying solvent, clean the parts to be joined, first by wiping and then by applying a plastic primer or cleaner to all surfaces that will be glued. The solvent cement is applied to the pipe end and the inside of the fitting end, which temporarily softens the joining surfaces. This brief softening period lets you seat the pipe into the socket's interference fit—that is, the fit gets tighter as the pipe is pushed into the socket. The softened surfaces then fuse together,

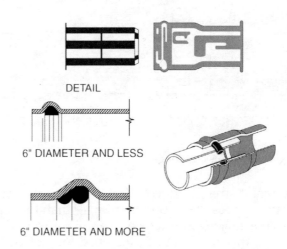

DETAIL

6" DIAMETER AND LESS

6" DIAMETER AND MORE

209F58.EPS

Figure 58 ◆ Ring-tight gasket fitting.

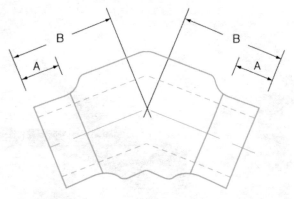

45-DEGREE ELBOW — BUTT FUSION — MOLDED

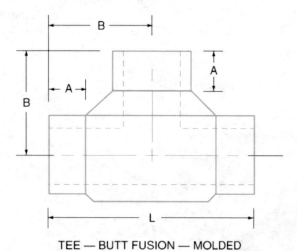

TEE — BUTT FUSION — MOLDED

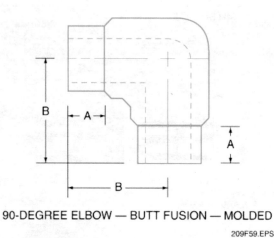

90-DEGREE ELBOW — BUTT FUSION — MOLDED

209F59.EPS

Figure 59 ◆ Fusion fittings.

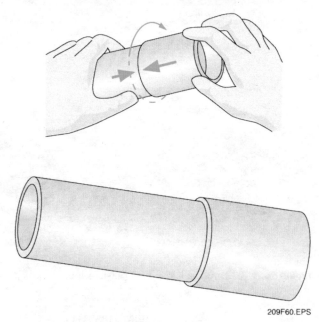

209F60.EPS

Figure 60 ◆ Solvent-welded fittings.

vary based on the size of the pipe and fittings. Be careful to choose the right cement for the job. Most codes require that you use primer before solvent cementing pipe. Always refer to all applicable codes and manufacturer's instructions, and ask your supervisor if you are unsure which cement to choose.

Because the cement hardens fast, you must move quickly and efficiently when joining pipe. The manufacturer's instructions will show minimum cure times for different size tubes at different temperatures before you can pressure-test the joint. Solvent cement and cure times also depend on relative humidity. Cure time is shorter for drier environments, smaller sizes, and higher temperatures.

WARNING!

Avoid unnecessary skin or eye contact with primers and cements. If contact occurs, wash immediately. Use protective eyewear and gloves during any solvent-weld procedure and consult the MSDS for the cement you are using. An MSDS should be available to anyone on the job site.

11.11.2 *Joining CPVC or PVC Pipe and Fittings*

When you join CPVC or PVC pipe and fittings, use only CPVC or PVC cement or all-purpose cement conforming to *ASTM F-493* standards, or the joint may fail. It is also possible to fusion weld PVC or CPVC using special welding equipment.

PVC and CPVC can be assembled for low-pressure applications with threaded joints. The assem-

and joint strength develops as the solvents evaporate. The resulting joint is stronger than the pipe itself. A major advantage of the solvent-weld process is that it eliminates the need for torches or lead pots, which helps prevent the risk of fire during piping installations.

The solvent cements you use will depend on the materials you are solvent welding. Each type of plastic has its own cement, and the cement may bly is the same as carbon steel threaded joints, but Teflon® tape should be used rather than pipe dope, and the tightening is done with a strap wrench.

Use special care when assembling CPVC and PVC systems in extremely low temperatures (below 40°F) or extremely high temperatures (above 100°F). In extremely hot environments, make sure both surfaces to be joined are still wet with cement when you put them together. If the cement has dried, the two surfaces may not adhere. Adapters are available to connect CTS CPVC pipe to CPVC Schedule 40 and 80 pipe for systems requiring piping diameters larger than 2 inches.

To join CPVC or PVC pipe and fittings with solvent cement, follow these steps (see *Figure 61*). Note that these steps illustrate typical instructions. When joining pipe, always follow the cement manufacturer's instructions.

Step 1　Cut the pipes to the desired lengths, then test fit the pipes and fittings. The pipes should fit tightly against the bottom of the hub sockets on the fittings.

Step 2　Clean the surfaces you are joining by lightly scouring the ends of the pipe with emery paper, then wiping with a clean cloth. This ensures that the primer you apply later will soften the plastic before you apply solvent cement.

Step 3　Clean the socket interior and the spigot area of all dirt with a rag or brush.

Step 4　Mark the pipe and fitting with a felt-tipped pen to show the proper position for alignment. Also mark the depth of the fitting sockets on the pipes to make sure that you fit them back in completely when joining.

Step 5　Apply the primer to the surfaces you are joining. The primer will soften the plastic in preparation for the solvent-weld process.

Step 6　While the primer is still wet, apply a heavy, even coat of cement to the pipe end. Use the same applicator without additional cement to apply a thin coat inside the fitting socket. Too much cement can clog waterways. Do not allow excess cement to puddle in the fitting and pipe assembly.

Step 7　Immediately insert the tubing into the fitting socket, rotating the tube one-quarter to one-half turn while inserting. This motion ensures an even distribution of cement in the joint. Properly align the fitting.

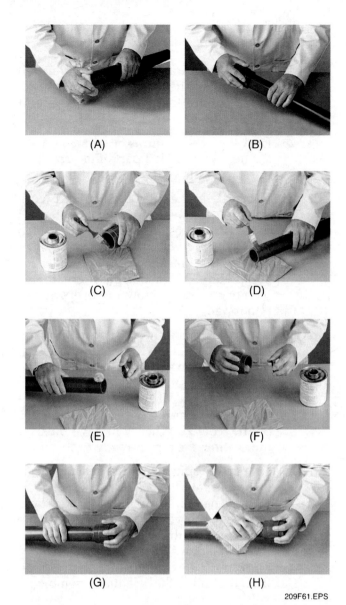

(A)　　　　　　　(B)

(C)　　　　　　　(D)

(E)　　　　　　　(F)

(G)　　　　　　　(H)

209F61.EPS

Figure 61 ◆ Joining CPVC or PVC pipe and fittings with solvent cement.

Step 8　Hold the assembly firmly until the joint sets up. An even bead of cement should be visible around the joint. If this bead does not appear all the way around the socket edge, you may not have applied enough cement. In this case, remake the joint to avoid the possibility of leaks. Wipe excess cement from the tubing and fitting surfaces. Always follow the cement manufacturer's instructions.

11.11.3 Installing PVC Bell-and-Spigot Pipe

PVC bell-and-spigot pipe is generally used outdoors for gravity sewers. These outdoor systems are typically installed to connect with municipal utilities. To install PVC bell-and-spigot pipe, follow these steps (see *Figure 62*):

Step 1   Prepare the inner surface of the bell according to the manufacturer's instructions. Ensure that the groove is free of dirt and other particles.

Step 2   Fold the gasket into a heart shape with the nose or rounded part of its cross section facing out of the mouth of the bell.

Step 3   Insert the gasket into the bell and work it into its groove until it is smooth and free from waves. You may have to snap the gasket, or wet it with clean water or a wet rag, to make sure it goes in place completely. Mark the pipe, creating a memory mark to show the proper position for alignment.

Step 4   After you place the gasket in the bell, thoroughly coat its exposed surface with lubricant. Then apply the lubricant to the entire surface of the spigot end up to the memory mark. Make especially sure that the tapered portion of the spigot is thoroughly coated. When you have finished lubricating, the pipe is ready to be joined.

Step 5   Line up the spigot with the bell and insert it straight into the bell. The spigot end of the pipe has a mark to indicate the proper depth of insertion. This mark will be about flush with the end of the bell when the joint is fully assembled. The memory mark must never be more than ⅜ inch from the end of the bell after assembly.

CAUTION

When you are installing a ring-tight PVC gasket, you must assemble the pipe either by hand or by using a bar or block. Never swing or stab the pipe to join it.

11.11.4 Joining PEX Tubing

Because PEX tubing resists high temperature and chemicals, you cannot join it with solvent cement or heat fusion. The most common method is to use an insert and a crimp-ring system. The other tools used to join PEX tubing include the tubing cutter, hand-crimping tool, and go/no-go gauge (see *Figure 63*).

To join PEX tubing, follow these steps (see *Figure 64*):

Step 1   Square cut the tubing perpendicular to the length of the tubing, using a cutter designed for plastic tubing. Remove all excess material or burrs that might affect the fitting connection.

Step 2   Slide a PEX ring over the end of the tube, and extend it no more than ¹⁄₁₆ inch.

Step 3   Open the handles of an expander tool and insert the tool's expansion head into the end of the tubing until it stops. Be sure you have the correct size expander head in the tool. Place the free handle of the tool against your hip, or place one hand on each handle when necessary.

209F62.EPS

Figure 62 ◆ Installing PVC gravity sewer pipe.

CRIMPSERT INSERT FITTINGS
AND CRIMP RINGS

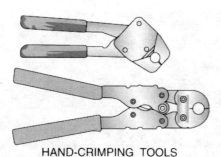

HAND-CRIMPING TOOLS

TUBING CUTTERS

GO NO-GO CRIMP
MEASURING GAUGE

209F63.EPS

Figure 63 ◆ Tools for joining PEX tubing.

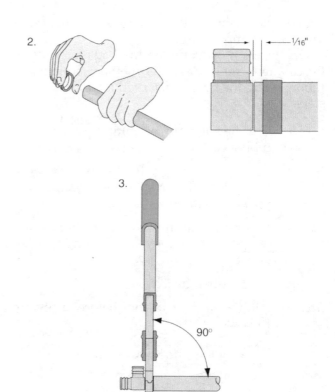

209F64.EPS

Figure 64 ◆ Joining PEX tubing.

Step 4 Fully separate the handles and bring them together. Repeat this process until the tubing and ring are snug against the shoulder on the expansion head. Before the final expansion, withdraw the tubing from the tool and rotate the tool one-eighth of a turn. This prevents the tool from forming ridges in the tubing.

Step 5 Expand the tubing one final time. Immediately remove the tool and slide the expanded tubing over the fitting until the tubing reaches the stop on the fitting. Hold the fitting in place for two or three seconds until the tubing shrinks onto the fitting so that it holds the fitting firmly. For a proper connection, the tubing and PEX ring must be snug against the stop of the shoulder fitting. If there is more than $\frac{1}{16}$ inch between the ring and the fitting, square cut the tubing 2 inches away from the fitting and make another connection using a new PEX ring.

11.11.5 Joining PE Tubing

Because PE tubing is resistant to chemicals, you must join it by heat fusion or with mechanical joints and clamps. PE joined by fusion is similar to a weld on steel—the materials of the joined parts merge so they are indistinguishable from each other. This process gives the joint the same positive characteristics as the pipe itself.

PE fusion often requires special training and certification. Manufacturers of PE products and joining equipment often provide this training and certification free of charge. New techniques that involve compression collars for joining PE are becoming popular because they require less training.

HDPE pipe is also used for sliplining existing pipe to seal corrosion damage or leaks. The polyethylene tube is pulled through the existing pipeline.

Some of the tools that are used in the fusion process include a temperature indicator stick, a heating tool, a fusion timer, a socket face, and a cold ring (see *Figure 65*). The temperature indicator stick is used to make sure that piping has reached the required temperature for successful fusion. The stick is used to mark a particular area on the pipe. When the pipe reaches the desired temperature, the mark will melt. During heat fusion, the surface of the socket face comes in direct contact with the pipe or fitting. The latest machines for HDPE fusion are mobile and automated, and many no longer require the operator to monitor temperature.

One of the most common methods for joining PE tubing is the butt-fusion method. To join PE tubing using the butt-fusion method, follow these steps:

Step 1 Cut the ends of the tubing square with a tubing cutter.

Step 2 Mark the tubing with the proper temperature indicator stick and heat the tubing ends with a heating tool.

Step 3 When the tubing reaches the required temperature, remove the heating tool.

Step 4 Press the tubing ends together to form a tight seal at the joint.

Step 5 Allow the joint to cool before applying force.

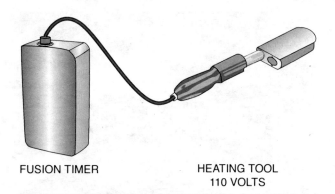

FUSION TIMER · HEATING TOOL 110 VOLTS

HDPE PIPE FUSION MACHINE

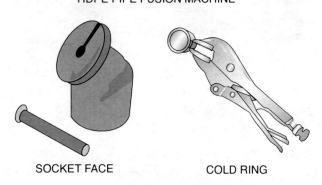

SOCKET FACE · COLD RING

TEMPERATURE INDICATOR STICK

209F65.EPS

Figure 65 ◆ PE fusion tools.

Review Questions

1. The individual responsible for personal safety in and around a trench is the _____.
   a. OSHA inspector
   b. foreman
   c. worker
   d. authority having jurisdiction

2. Toppling, boiling, and sliding are indications that the trench is _____.
   a. too deep
   b. unstable
   c. too wide
   d. too narrow

3. If the width of a 2-foot-wide trench is increased by 6 inches, the load on the pipe in that trench increases by ____ percent.
   a. 5
   b. 25
   c. 50
   d. 150

4. A thrust block is used when the pipeline _____.
   a. is very heavy
   b. is buried below 10 feet
   c. changes directions
   d. runs above ground

5. When a shipment of pipe arrives on the job site, it should be _____ immediately.
   a. inspected
   b. used
   c. protected
   d. covered

6. The laying length of cast iron bell-and-spigot pipe refers to the total _____.
   a. end-to-end length plus the bell length
   b. end-to-end length minus the bell length
   c. end-to-end length
   d. end-to-end length minus the spigot length

7. A compression joint on cast iron pipe requires a bead on the spigot end of the pipe.
   a. True
   b. False

8. The lever tool is typically used to install cast and ductile iron pipe in sizes ranging from _____ inches.
   a. 3 to 8
   b. 3 to 18
   c. 8 to 14
   d. 18 to 24

9. The most common wall thickness classes for ductile iron pipe are _____.
   a. 50 and 52
   b. 51 and 53
   c. 51 and 56
   d. 53 and 56

10. Which of the following is *not* a common type of ductile iron pipe connection?
    a. Compression joint
    b. Mechanical joint
    c. Welded joint
    d. Flange joint

11. To withstand forces within ductile iron pipe, pipefitters must use restraint methods when installing _____ joints.
    a. mechanical or compression
    b. angled or turning
    c. welded
    d. extra heavy

12. The span and rise are equal on which type of concrete pipe?
    a. Vertical elliptical
    b. Circular
    c. Horizontal elliptical
    d. Arch

13. Rubber gaskets used with concrete pipe require _____.
    a. shaping
    b. powder coating
    c. pre-stretching
    d. lubrication

14. The easiest, quickest, and cleanest way to install concrete pipe is with _____.
    a. mastic sealants
    b. rubber gaskets
    c. mortar sealants
    d. external bands

15. Schedule numbers for carbon steel pipe are 40, 80, and _____.
    a. 90
    b. 100
    c. 160
    d. 180

16. When installing dresser couplings, ensure that the ends of the mating pipes are _____.
    a. beveled adequately
    b. cut square
    c. ground down
    d. lubricated fully

17. Duriron bell-and-spigot pipe and fittings are intended for use in applications that are _____.
    a. under pressure
    b. aboveground
    c. gravity-fed
    d. low pressure

18. Duriron is cut in the field with a _____.
    a. chop saw
    b. masonry blade
    c. chain cutter
    d. demolition saw

19. Duriron contains _____ as alloys.
    a. lead and carbon
    b. aluminum and chromium
    c. oxygen, lead, and magnesium
    d. silicon, carbon, and magnesium

20. The standard temperature given in adhesive mixing charts to mix the adhesive for assembling fiberglass pipe is _____.
    a. 70°F
    b. 75°F
    c. 80°F
    d. 85°F

21. An advantage of a Redi-Thread joint is that the joint is _____.
    a. easily glued
    b. permanent
    c. deflectable
    d. lubricated

22. Which of the following is *not* an advantage of thermoplastic pipe?
    a. Inexpensive
    b. Corrosion resistant
    c. Lightweight
    d. High resistance to heat and pressure

23. It is not necessary to clean plastic pipe before solvent welding.
    a. True
    b. False

24. CPVC pipe can withstand a water hammer drop better than copper pipe.
    a. True
    b. False

25. The thermoplastic pipe joined by butt fusion is _____.
    a. CPVC
    b. PE
    c. PEX
    d. PVC

Summary

Underground piping represents only a part of the pipefitters job, but it is an important part. Installing underground piping properly is a complex process requiring attention to detail and specific knowledge of the installation process.

Bad piping installations are expensive and time-consuming to repair. Bad installations can cause system failure leading to loss of life as well as property damage. Following established installation procedures is essential in limiting potential problems.

A key part of underground piping installations is safe trenching. You learned some fundamental guidelines for working in and near a trench, particularly the fact that you are the person responsible for not only your own safety but the safety of the workers around you.

Next you saw how to prepare a trench based on OSHA standards and learned essential facts about pipe bedding, pipe laying, and installing thrust blocks. You saw that there are specific procedures and methods to follow when storing and handling the various types of underground pipe.

Next was a discussion of the types of underground piping materials. Covered were cast iron soil pipe, ductile iron pipe, concrete pipe, carbon steel pipe, iron alloy pipe (Duriron), fiberglass pipe, and finally the various types of thermoplastic pipe. In general, for each type of pipe, you learned about its sizing, fittings, how to join that pipe, and how to cut and install that type of pipe.

Each of the pipe types has particular characteristics that require the fitter to have specialized knowledge, to use tools specific to that pipe, and to follow installation procedures designed for that particular type of pipe. Following the guidelines in this module will help you install underground piping safely and professionally. And this module gave you the established industry guidelines necessary for installing various kinds of pipe, guidelines that are a solid foundation for on-the-job experience and a base for further learning.

Notes

Acrylonitrile-butadiene-styrene (ABS): Plastic pipe and fittings used extensively in drain, waste, and vent (DWV) systems.

Bell-and-spigot pipe: Pipe that has a bell, or enlargement, also called a hub, at one end of the pipe and a spigot, or smooth end, at the other end. The bell and spigot of two different pipes slide together to form a joint. Also called hub-and-spigot pipe.

Bituminous coating: A black coating that consists almost entirely of carbon and hydrogen. It is used as a surfacing for roads, as a water-repellent barrier in buildings, and as a corrosion-resistant barrier for ductile iron pipes.

Boring: Drilling through the ground, either to access resources or to place pipe.

Branch: The outlet or inlet of a fitting, which is not in line with the pipe run but may make any angle off of the pipe run.

Cellular core wall: Plastic pipe wall that is low-density, lightweight plastic containing entrained (trapped) air.

Chemically inert: Does not react with other chemicals.

Chlorinated polyvinyl chloride (CPVC): Plastic pipe and fittings used extensively in hot and cold water distribution systems.

Compression collar: A piece of hardware that uses compression force to connect sections of polyethylene piping.

Cross-linked polyethylene (PEX): Tubing and fittings made with heat and high pressure that resist high temperatures, pressure, and chemicals.

Culvert: A covered channel or a large-diameter pipe that takes a watercourse below ground level.

Elastomeric: Rubberized. Made of an elastic substance such as a polyvinyl elastomer.

Elliptical: Oval shaped.

Exfiltration: The exiting of a fluid or substance from a closed system to the atmosphere or surroundings.

Fusion fitting: A fitting with a butt that has the same outside diameter and inside diameter as the pipe. It is usually joined to a pipe by heat.

Gasket: A packing used to make a pressure-tight joint between two stationary parts.

High-density polyethylene (HDPE): A type of pipe.

Hydronic: A system that heats and cools by circulating water or steam through a closed piping system.

Infiltration: The entrance of a fluid or substance into a closed system from the atmosphere or surroundings.

Inside diameter (ID): The distance between the inner walls of a pipe; the standard measure of piping used in heating and plumbing.

Interference fit: Fit that tightens as the pipe is pushed into the socket.

Invert elevation: A reference elevation point that is the bottom inside diameter of the underground pipe. Plan coordinates of underground piping are referenced to the center of the pipe, whereas elevation is referenced from invert elevation (INV).

Lead: A soft, malleable, ductile, bluish-white metallic element used as a sealant for bell-and-spigot joints.

Mastic: A paintable mixture of finely powdered rock and asphalt commonly used for coating flanges and bolts to prevent corrosion.

Neoprene: A synthetic rubber with very strong resistance to weathering, various chemicals, and oil.

Outside diameter (OD): The distance between the outer walls of a pipe.

Polybutylene (PB): Plastic piping that was formerly used for plumbing pipe; it is no longer used but is still found in some residences.

Polyethylene (PE): Flexible plastic pipe, tubing, and fittings, usually used for water distribution, that do not deteriorate when exposed to sunlight.

Polyvinyl chloride (PVC): Plastic pipe and fittings used for cold water distribution and for industrial water and chemicals, as well as for drain, waste, and vent (DWV) systems.

Portland cement: A hydraulic cement that resembles stone when it hardens. This is the proper name for ordinary cement.

Pounds per square inch (psi): A measurement of pressure.

Pressure rating: The maximum pressure at which a component or system may be operated continuously.

Resin: Any of numerous synthetic materials that are used with fillers, stabilizers, pigments, or other substances to form plastics.

Ring-tight gasket fitting: Fitting with a rubber O-ring or gasket in the socket.

Size dimension ratio (SDR): A measurement of pipe size that relates pipe wall thickness to pipe diameter.

Solid wall: Plastic pipe wall that does not contain trapped air.

Solvent: A liquid capable of dissolving another substance.

Solvent weld: A joint created by joining two pipes using solvent cement that softens the material's surface.

Tap: A threaded hole drilled into a pipe or process vessel.

Thermoplastic pipe: Pipe that can be repeatedly softened by heating and hardened by cooling. When softened, thermoplastic pipe can be molded into desired shapes.

Thermosetting pipe: Pipe that changes chemically when heated, so that once hardened by heat or chemicals, it is hardened permanently.

Thrust block: A poured concrete retaining support that restrains pressurized pipe at directional changes.

Transition fitting: A special fitting used to connect plastic pipe to pipe of a dissimilar material, as specified by applicable code.

Resources & Acknowledgments

Additional Resources

This module is intended to be a thorough resource for task training. The following reference works are suggested for further study. These are optional materials for continued education rather than for task training.

IPT's Pipe Trades Handbook. Robert A. Lee. Clinton, NC: Construction Trades Press.

Figure Credits

Trench Shoring Services, Inc., 209F01, 209F02

Topaz Publications, Inc., 209F10

Vermeer Manufacturing Company, 209F11, 209F12

Ditch Witch®, 209F14

Ridge Tool Co.(Ridgid®), 209F24B, 209F55

Wheeler Manufacturing A Division of Rex International USA, Inc., 209F26

Flowserve Corporation, Table 11, Table 12 Provided by Flowserve Corporation, a global leader in the fluid motion and control business. More information on Flowserve and its products can be found at www.flowserve.com

Plastic Pipe and Fitting Association, Table 14

Charlotte Pipe and Foundry Co., 209F48, 209F49, 209F54 (sanitary tee, wye), 209F61

Hudson Extrusions, Inc., 209F50

Zum Plumbing Products, 209F51

Consumer Plumbing Recovery Center, Inc, 209F52

LASCO Fittings, Inc., 209F53, 209F54 (coupling)

International Code Council, Inc., Table 17 International Plumbing Code ©2003, Falls Church, VA: International Code Council, Inc. Reproduced with permission.

Genova Products, Inc., 209F54 (closet bend)

Reed Manufacturing Company, 209F56

Vanguard Piping Systems, Inc., 209F63

Tempil, Inc. An Illinois Tool Works, Inc., Company, 209F65 (art)

Fast Fusion, 209F65 (photo) Copyright © 2006 www.fast-fusion.com

Cianbro Corporation, Module Opener

NCCER CURRICULA — USER UPDATE

NCCER makes every effort to keep its textbooks up-to-date and free of technical errors. We appreciate your help in this process. If you find an error, a typographical mistake, or an inaccuracy in NCCER's curricula, please fill out this form (or a photocopy), or complete the online form at **www.nccer.org/olf**. Be sure to include the exact module ID number, page number, a detailed description, and your recommended correction. Your input will be brought to the attention of the Authoring Team. Thank you for your assistance.

Instructors – If you have an idea for improving this textbook, or have found that additional materials were necessary to teach this module effectively, please let us know so that we may present your suggestions to the Authoring Team.

NCCER Product Development and Revision

13614 Progress Blvd., Alachua, FL 32615

Email: curriculum@nccer.org
Online: www.nccer.org/olf

❏ Trainee Guide   ❏ Lesson Plans   ❏ Exam   ❏ PowerPoints   Other _____

Craft / Level: _____   Copyright Date: _____

Module ID Number / Title: _____

Section Number(s): _____

Description: _____

_____

_____

_____

_____

Recommended Correction: _____

_____

_____

_____

_____

Your Name: _____

Address: _____

_____

Email: _____   Phone: _____

Glossary of Trade Terms

Acid: A chemical compound that reacts with and dissolves certain metals to form salts.

Acrylonitrile-butadiene-styrene (ABS): Plastic pipe and fittings used extensively in drain, waste, and vent (DWV) systems.

Actuator: The part of a regulating valve that converts electrical or fluid energy to mechanical energy to position the valve.

Adjacent side: The side of a right triangle that is next to the reference angle.

Align: To make straight or to line up evenly.

Alloy: Two or more metals combined to make a new metal.

Ambient temperature: The same temperature as the surrounding atmosphere; room temperature.

American Society for Testing Materials International (ASTM International): Founded in 1898, a scientific and technical organization formed for the development of standards on the characteristics and performance of materials, products, systems, and services. Formerly known as the American Society for Testing and Materials.

Angle of repose: The greatest angle above the horizontal place at which a material will adjust without sliding.

Angle valve: A type of globe valve in which the piping connections are at right angles.

Apex: The point at which the lines of a figure converge.

Arithmetic numbers: Numbers that have definite numerical values, such as 4, 6.3, and ⅝.

Ball valve: A type of plug valve with a spherical disc.

Battery limits: The outside perimeter of a project.

Bedding: The surface or foundation on which a pipe rests in a trench.

Bell-and-spigot pipe: Pipe that has a bell, or enlargement, also called a hub, at one end of the pipe and a spigot, or smooth end, at the other end. The bell and spigot of two different pipes slide together to form a joint. Also called hub-and-spigot pipe.

Bench mark: A point of known elevation from which the surveyors establish all of the grades.

Benching: A method of protecting workers from cave-ins by excavating the sides of an excavation to form one or a series of horizontal levels or steps, usually with vertical or near-vertical surfaces between levels.

Bevel: An angle cut or ground on the end of a piece of solid material.

Bituminous coating: A black coating that consists almost entirely of carbon and hydrogen. It is used as a surfacing for roads, as a water-repellent barrier in buildings, and as a corrosion-resistant barrier for ductile iron pipes.

Body: The main part of the valve. It contains the disc, seat, and valve ports. The body of the valve is directly connected to the piping by threaded, welded, or flanged ends.

Bonnet: The part of a valve containing the valve stem and packing.

Boring: Drilling through the ground, either to access resources or to place pipe.

Branch: The outlet or inlet of a fitting not in line with the run but which may make any angle.

Branch: The outlet or inlet of a fitting, which is not in line with the pipe run but may make any angle off of the pipe run.

Burn-through: A hole that is formed in a weld due to improper grinding or welding.

Bushing: A pipe fitting that connects a pipe with a fitting of a larger nominal size. A bushing is a hollow plug with internal and external threads to suit the different diameters.

Butterfly valve: A quarter-turn valve with a plate-like disc that stops flow when the outside area of the disc seals against the inside of the valve body.

Carbonation: A chemical process in which carbon accumulates and causes a breakdown of surrounding minerals, such as dirt.

Catchment basin: A bowl area constructed to capture and retain runoff water.

Caustic: A material that is capable of burning or corroding by chemical action.

Cellular core wall: Plastic pipe wall that is low-density, lightweight plastic containing entrained (trapped) air.

Chamfer: An angle cut or ground only on the edge of a piece of material.

Glossary of Trade Terms

Chase: A recess on the inside of a wall, used for piping or electrical lines.

Check valve: A valve that allows flow in one direction only.

Chemically inert: Does not react with other chemicals.

Chlorinated polyvinyl chloride (CPVC): Plastic pipe and fittings used extensively in hot and cold water distribution systems.

Circle: A continuous curved line that encloses a space, with every point on the line the same distance from the center of the circle.

Circumference: The distance around a circle.

Close nipple: A nipple that is about twice the length of a standard thread and threaded from end to end with no shoulder.

Coffer dam: A temporary dam constructed to divert or contain water.

Compaction: The action of forcing small soil particles between larger ones until most of the air space or voids and moisture have been squeezed out. The result is a hard, solid, rock-like mass.

Competent person: A person who is capable of identifying existing and predictable hazards in the area or working conditions that are unsanitary, hazardous, or dangerous to employees, and who has the authority to take prompt corrective measures to fix the problem.

Compression collar: A piece of hardware that uses compression force to connect sections of polyethylene piping.

Compression strength: The ability of soil to hold a heavy weight.

Concentric reducer: A reducer that maintains the same center line between the two pipes that it joins.

Condensate: The liquid product of steam, caused by a loss in temperature or pressure.

Control valve: A globe valve automatically controlled to regulate flow through the valve.

Corrosive: Causing the gradual destruction of a substance by chemical action.

Cross braces: The horizontal members of a shoring system installed perpendicular to the sides of the excavation, the ends of which bear against either uprights or wales.

Cross: A fitting with four branches all at right angles to each other.

Cross-linked polyethylene (PEX): Tubing and fittings made with heat and high pressure that resist high temperatures, pressure, and chemicals.

Cubic: The designation of a given unit representing volume.

Culvert: A covered channel or a large-diameter pipe that takes a watercourse below ground level.

Cylinder: A shape created by a circle moving in a straight line through space perpendicular to the surface of the circle.

Deformation: A change in the shape of a material or component due to an applied force or temperature.

Disc: Part of a valve used to control the flow of system fluid.

Disturbed soil: Soil that has been previously backfilled or removed from the earth and replaced by any means.

Eccentric reducer: A reducer that displaces the center line of the smaller of the two joining pipes to one side.

Elastomeric: Rubberized. Made of an elastic substance such as a polyvinyl elastomer.

Elastomeric: Elastic or rubberlike. Flexible, pliable.

Elbow: A fitting that makes an angle between adjacent pipes. An elbow is always 90 degrees unless another angle is stated; also known as an ell.

Elliptical: Oval shaped.

Excavation: Any man-made cut, cavity, trench, or depression in the earth's surface, formed by earth removal.

Exfiltration: The exiting of a fluid or substance from a closed system to the atmosphere or surroundings.

Exponent: A number or symbol placed to the right and above another number, symbol, or expression, denoting the power to which the latter is to be raised.

Factors: The numbers that can be multiplied together to produce a given product.

Fillet weld: A weld with a triangular cross section joining two surfaces at right angles to each other.

Finished grade: Any surface that has been cut or built to the elevation indicated for that point.

Fit up: To put piping material in position to be welded together.

Glossary of Trade Terms

Flange: A fitting on the end of a pipe that is shaped like a rim and allows the pipe to be bolted to another flanged pipe.

Flange: A form of attachment between pipes consisting of a ring perpendicular to the line of pipe, with holes drilled through the ring parallel to the pipe line, through which bolts may be used to attach the pipe or fitting to another flanged pipe or fitting.

Formula: An equation that states a rule.

Full-penetration weld: Complete joint penetration for a joint welded from one side only.

Fusion fitting: A fitting with a butt that has the same outside diameter and inside diameter as the pipe. It is usually joined to a pipe by heat.

Galling: Deformity of the threads in which the threads are stripped.

Galling: An uneven wear pattern between trim and seat that causes friction between the moving parts.

Gasket: A packing used to make a pressure-tight joint between two stationary parts.

Gate valve: A valve with a straight-through flow design that exhibits very little resistance to flow. It is normally used for open/shut applications.

Globe valve: A valve in which flow is always parallel to the stem as it goes past the seat.

Grade pin: Steel rod driven into the ground at each surveyor's hub. A string is attached between them at the grade indicated on the information stakes.

Grade rod: A small length of round or rectangular wood or metal used in place of a rule for checking grades.

Grade: The surface level required by the plans or specifications.

Haunch: The area below the middle of the pipe in a bench.

Head loss: The loss of pressure due to friction and flow disturbances within a system.

Heat-tracing: The addition of heat-producing wire to the pipe to prevent the material inside from being frozen.

High-density polyethylene (HDPE): A type of pipe.

Hub: A point-of-origin stake that identifies a point on the ground. The top of the hub establishes the point from which soil elevations and distances are computed.

Hydration: A chemical process in which water or liquids accumulate and cause a breakdown of surrounding minerals, such as dirt.

Hydronic: A system that heats and cools by circulating water or steam through a closed piping system.

Hypotenuse: The longest side of a right triangle. It is always located opposite the right angle.

Infiltration: The entrance of a fluid or substance into a closed system from the atmosphere or surroundings.

Information stake: A marker that shows elevations and grades that must be established and the distances to them.

Insert: A type of reducer that fits into the socket of a fitting to reduce the line size.

Inside diameter (ID): The distance between the inner walls of a pipe; the standard measure of piping used in heating and plumbing.

Interference fit: Fit that tightens as the pipe is pushed into the socket.

Invert elevation: A reference elevation point that is the bottom inside diameter of the underground pipe. Plan coordinates of underground piping are referenced to the center of the pipe, whereas elevation is referenced from invert elevation (INV).

Isometric drawing (ISO): A three-dimensional drawing of a piping system that is not drawn to scale to provide clarity.

Kinetic energy: Energy of motion.

Lateral: A fitting or branch connection that has a side outlet that is any angle other than 90 degrees to the run.

Lead: A soft, malleable, ductile, bluish-white metallic element used as a sealant for bell-and-spigot joints.

Lift: A layer of fill that can be either loose or compacted.

Line number: A group of abbreviations that specify size, service, material class/specification, insulation thickness, and tracing requirements of a given piping segment.

Literal numbers: Letters that represent arithmetic numbers, such as x, y, and h. Also known as algebraic numbers.

Glossary of Trade Terms

Malleable iron: Metallic fitting material that is generally used for air and water and has a pressure rating of 125 to 150 psi.

Mastic: A paintable mixture of finely powdered rock and asphalt commonly used for coating flanges and bolts to prevent corrosion.

Material safety data sheet (MSDS): A document that describes the composition, characteristics, health hazards, and physical hazards of a specific chemical. It also contains specific information about how to safely handle and store the chemical and lists any special procedures or protective equipment required.

Monitoring well: A well dug to test the effect of construction on groundwater.

National Pipe Thread (NPT): The United States standard for pipe threads. This thread has a 1/16-inch taper per inch from back to front.

Neoprene: A synthetic rubber with very strong resistance to weathering, various chemicals, and oil.

Nipple: A short piece of pipe threaded on both ends and less than 12 inches long. Any pipe over 12 inches is referred to as cut pipe.

Opposite side: The side of a right triangle that is directly across from the reference angle.

Orthographic projection: The projection of a single view of an object.

Outside diameter (OD): The distance between the outer walls of a pipe.

Oxidation: A chemical process in which oxygen accumulates and causes a breakdown of surrounding minerals, such as dirt.

Oxide: A type of corrosion that is formed when oxygen combines with a base metal.

Packing: Material used to make a dynamic seal, preventing system fluid leakage around a valve stem.

Perpendicular: At a right angle to the plane of a line or surface.

Phonographic: When referring to the facing of a pipe flange, serrated grooves cut into the facing, resembling those on a phonograph record.

Pi: A number that represents the ratio of the circumference to the diameter of a circle (π). Pi is approximately 3.1416.

Piping and instrumentation drawing (P&ID): A schematic flow diagram of a complete system or systems that shows function, instrument, valving, and equipment sequence.

Plug valve: A quarter-turn valve with a ported disc.

Plug: The moving part of a valve trim (plug and seat) that either opens or restricts the flow through a valve in accordance with its position relative to the valve seat, which is the stationary part of a valve trim.

Polybutylene (PB): Plastic piping that was formerly used for plumbing pipe; it is no longer used but is still found in some residences.

Polyethylene (PE): Flexible plastic pipe, tubing, and fittings, usually used for water distribution, that do not deteriorate when exposed to sunlight.

Polyvinyl chloride (PVC): Plastic pipe and fittings used for cold water distribution and for industrial water and chemicals, as well as for drain, waste, and vent (DWV) systems.

Portland cement: A hydraulic cement that resembles stone when it hardens. This is the proper name for ordinary cement.

Positioner: A field-based device that takes a signal from a control system and ensures that the control device is at the setting required by the control system.

Pounds per square inch (psi): A measurement of pressure.

Pounds per square inch gauge (psig): Amount of pressure in excess of the atmospheric pressure level.

Pressure rating: The maximum pressure at which a component or system may be operated continuously.

Protective system: A method of protecting employees from cave-ins, from material that could fall or roll from an excavation face or into an excavation, or from the collapse of adjacent structures. Protective systems include support systems, sloping and benching systems, and shielding systems.

Pyramid: A shape with a multi-sided base, and sides that converge at a point.

Radius: A straight line from the center of a circle to a point on the edge of the circle.

Glossary of Trade Terms

Raveling: The process of small particles breaking away from excavation walls.

Rectangular: Description of a shape having parallel sides and four right angles.

Reducer: A pipe fitting with female threads on each end, with one end being one or more sizes smaller than the other end.

Relief valve: A valve that automatically opens when a preset amount of pressure is exerted on the valve disc.

Resin: Any of numerous synthetic materials that are used with fillers, stabilizers, pigments, or other substances to form plastics.

Ring-tight gasket fitting: Fitting with a rubber O-ring or gasket in the socket.

Root opening: The space between the pipes at the beginning of a weld; the gap is usually 1/16 to 1/8 of an inch.

Run: A length of pipe made up of more than one length of pipe.

Run: The horizontal distance from one pipe to another.

Saturated steam: Steam in contact with water.

Screw jack: A screw- or hydraulic-type jack that is used as cross bracing in a trench.

Seat: The part of a valve against which the disc presses to stop flow through the valve.

Set: The vertical distance from the line of flow of a pipe and the line of flow of the pipe to which it is attached.

Sheeting: Plywood or sheet metal that is placed against and braced between the walls of an excavation.

Shield: A structure that is able to withstand the forces imposed on it by a cave-in and thereby protect employees within the structure. Shields can be permanent structures or can be designed to be portable and moved along as work progresses. Additionally, shields can be either pre-manufactured or job-built in accordance with *29 CFR 1926.652 (c)(3) or (c)(4).*

Shore: Timber or other material used as a temporary prop for excavations. It may be sloping, vertical, or horizontal.

Shoring: A structure such as a metal hydraulic, mechanical, or timber shoring system that supports the sides of an excavation and is designed to prevent cave-ins.

Size dimension ratio (SDR): A measurement of pipe size that relates pipe wall thickness to pipe diameter.

Skeleton: A condition that occurs when individual timber uprights or individual hydraulic shores are not placed in contact with the adjacent member.

Sketch: A drawing representing the primary features of an object, usually a rough draft or freehand drawing.

Sloping: A method of protecting employees working in excavations from cave-ins by cutting the excavation walls to the angle of repose of the soil being excavated.

Slurry: A mixture of liquid and suspended solids, like muddy water or paper pulp.

Solid wall: Plastic pipe wall that does not contain trapped air.

Solid: A figure enclosing a volume.

Solvent weld: A joint created by joining two pipes using solvent cement that softens the material's surface.

Solvent: A liquid capable of dissolving another substance.

Sphere: A shape whose surface is everywhere the same distance from a central point.

Spoils: Excavated soil that is removed from the hole and piled next to the excavation. The distance between the spoils and the excavation is regulated by OSHA.

Spool: A piping segment of a pipe system or an ISO.

Straight tee: A fitting that has one side outlet 90 degrees to the run.

String line: A nylon line usually strung tightly between supports to indicate both direction and elevation.

Subsidence: A depression in the earth that is caused by unbalanced stresses in the soil surrounding an excavation.

Superheated steam: Steam used in heavy industrial applications.

Support system: A structure, such as an underpinning, bracing, or shoring, that provides support to an adjacent structure, underground installation, or the walls of an excavation.

Swage: A type of reducer.

Glossary of Trade Terms

Tack-weld: A weld made to hold parts together in proper alignment until the final weld is made.

Takeout: The dimension that the fittings take out of a center-to-center measurement. Also known as takeup, takeoff, or makeup.

Tap: A threaded hole drilled into a pipe or process vessel.

T-bars: T-shaped wood frames used in place of steel pins to support a string line over trenches.

Tee: A fitting that has one side outlet 90 degrees to the run.

Teflon® tape: Tape made of Teflon® that is wrapped around the male threads of a pipe before it is screwed into a fitting to serve as a lubricant and sealant.

Thermal transients: Short-lived temperature spikes.

Thermoplastic pipe: Pipe that can be repeatedly softened by heating and hardened by cooling. When softened, thermoplastic pipe can be molded into desired shapes.

Thermosetting pipe: Pipe that changes chemically when heated, so that once hardened by heat or chemicals, it is hardened permanently.

Throttling: The regulation of flow through a valve.

Thrust block: A poured concrete retaining support that restrains pressurized pipe at directional changes.

Tight sheeting: The use of specially edged timber planks, such as tongue-and-groove planks, that are at least 3" thick. Steel sheet piling or similar construction that resists the lateral pressure of water and prevents loss of backfill is called tight sheeting.

Torque: A twisting force used to apply a clamping force to a mechanical joint.

Transition fitting: A special fitting used to connect plastic pipe to pipe of a dissimilar material, as specified by applicable code.

Travel: The diagonal distance from one pipe to another.

Trench box: A premanufactured steel box that provides a safe means of shoring by allowing workers to lay pipe within the box while it is inside the excavation.

Trench: A narrow excavation made below the surface of the ground. In general, a trench is no wider than it is deep and should never be more than 15 feet wide.

Trim: Functional parts of a pump or valve, such as seats, stem, and seals, that are inside the flow area.

Union: A fitting used to connect pipes.

Uprights: The vertical members of a trench shoring system placed in contact with the earth and usually positioned so that individual members do not contact each other. Uprights placed so that individual members are closely spaced, in contact with, or interconnected to each other, are often called sheeting.

Valve body: The part of a valve containing the passages for fluid flow, valve seat, and inlet and outlet connections.

Valve stem: The part of a valve that raises, lowers, or turns the valve disc.

Valve trim: The combination of the valve plug and the valve seat.

Void: A space in a soil mass not occupied by solid material.

Volume: The amount of space occupied by an object.

Wales: Horizontal members of a shoring system placed parallel to the excavation face whose sides bear against the vertical members of the shoring system or the earth.

Water hammer: An increase in pressure in a pipeline caused by a sudden change in the flow rate. In a steam line, water hammer is caused by condensate blocking the flow of steam at a pipe bend.

Wedge: The disc in a gate valve.

Wire drawing: The erosion of a valve seat under high velocity flow through which thin, wire-like gullies are eroded away.

Wye: A fitting that has a side outlet of 45 degrees to the run; also referred to as a lateral.

Yoke bushing: The bearing between the valve stem and the valve yoke.

Index

Index

Subsidence, 8.7, 8.25, 9.3, 9.4
Sulfuric acid, 1.2
Sunlight, 9.45, 9.48, 9.49
Support for piping, 1.5, 2.14, 2.16, 2.18–2.19, 3.28, 9.11, 9.45
Support systems within trench, 8.2, 8.25
Surveyor, 8.15, 8.16
Swage
    definition, 6.21
    symbols, 2.4, 5.11, 6.9, 7.10
    types, 5.5, 5.7, 6.3, 6.5
SWP. *See* Steam working pressure
Symbols
    mathematical, 4.5
    overview, 2.3
    pipe support, 2.16, 2.18
    piping and piping fittings
        butt weld, 7.8, 7.10
        overview, 2.4, 5.9
        socket weld, 6.6, 6.9
        threaded, 3.33, 5.11
    valves and valve connections, 3.33–3.34, 5.11, 6.9, 7.10

T
Tables, how to use, 4.4–4.5
Takeout of takeoff, 5.12, 5.14, 5.28, 5.29, 6.10, 7.13–7.15
Tank(s), 1.3, 3.20, 3.22, 4.12–4.13, 4.14, 7.13
Tap, 9.24, 9.63
Tape, Teflon®, 1.3, 5.23, 5.29, 9.55
Taper, of a thread, 5.10, 5.12, 5.13, 5.24
Tar and felt paper, 9.36
T-bar, 8.19, 8.20, 8.25
Tee
    alignment of pipe to, 7.27
    cast iron soil pipe fittings, 9.16
    definition, 5.29
    ductile iron pipe fittings, 9.24
    plastic, 9.50, 9.51, 9.54
    sewer main service, 8.16, 8.17
    straight, 6.3, 6.21
    symbols, 2.4, 5.11, 6.9, 7.10
    threaded, 5.5
    types, 6.3, 6.4, 6.5, 7.4, 7.5
Teflon®
    liquid, 5.23
    piping for acids, 1.2
    tape, 1.3, 5.23, 5.29, 9.55
    valve seat, 6.17
Tell-tale hole, 7.4
Temperature
    ambient, 1.12
    bituminous coating for extremes, 9.22
    effects on plastic pipe, 9.45
    freezing conditions and valves, 3.28, 3.30
    indicator stick for PE tubing, 9.58
    iron alloy piping to withstand high, 9.40
    valve parts to withstand high, 3.8, 3.15
    and valve selection, 3.30
Temperature changes. *See* Expansion, thermal; Insulation
Tensile strength, 7.2
Tetrafluoroethylene (TFE), 3.8, 3.9, 3.11
TFE. *See* Tetrafluoroethylene
Thermal transients, 3.4, 3.38
Thermoplastic *vs.* thermosetting, 9.45, 9.63
Thread(s)
    definition, 5.10
    left-hand, 3.33
    overview, 5.10
    procedure, 5.20–5.23, 5.28
    symbols, 3.33

types, 5.10, 5.12, 5.13
valve, 3.30, 3.33, 5.24–5.25
Threaders, 5.20–5.23
3-4-5-foot method, 7.25–7.26
Throttling ability
    and ball valves, 3.7–3.8
    definition, 3.2, 3.38
    and diaphragm valves, 3.18
    and globe valves, 3.13, 3.14
    and plug valves, 3.9
    and valve selection, 3.30
    and Y-type valves, 3.15
Timber, 8.9, 8.10, 9.10, 9.26, 9.36
Timer, fusion, 9.58
Tin, 1.3
Titanium, 5.2
Title block, 2.2, 2.3
Tools. *See also* Equipment
    alignment, 3.28, 7.17–7.21
    beveling, 7.8, 7.11, 7.13
    clamps, 2.18, 3.4, 3.20, 7.12, 7.17–7.21
    for cutting pipe. *See* Cutter, pipe; Saw
    deburring, 9.52
    Hi-Lo gauge, 7.22–7.23
    to install cast iron piping, 9.18–9.21
    level. *See* Level
    for PE tubing, 9.58
    for PEX tubing, 9.56, 9.57
    scale, 4.2–4.4
    square, 6.12, 6.13, 6.15–6.17, 7.23–7.25, 7.27
    storage during trench work, 8.6, 9.2
Torch, oxyacetylene cutting, 7.8, 7.11
Torque
    actuation, 3.10
    for ball valve seal, 3.7
    for carbon steel pipe joint, 9.39
    for cast iron pipe joint, 9.17, 9.22
    definition, 3.38
    for ductile iron pipe joint, 9.27, 9.29
Traffic, 8.2, 8.6, 8.9, 8.14, 9.2
Trap, cast iron soil pipe fittings, 9.15, 9.16
Travel, 4.19, 4.23, 5.18, 5.20
Trench(es)
    definition, 8.2, 8.25
    effects of shape on load, 9.3, 9.5
    failure (cave-in), 8.2, 8.4, 8.6, 8.7, 8.8–8.9, 9.3, 9.4
    for fuel oil piping, 1.3
    hazards, 8.3–8.4, 8.5, 8.6
    indications of an unstable, 8.7–8.8, 9.3
    insulation in, 1.10
    protective systems. *See* Shielding; Shoring; Sloping
    safety issues, 8.2, 8.3–8.4, 8.6–8.7, 8.9–8.15, 8.16, 8.23, 9.2–9.3
Trenchless pipelaying, 9.6–9.8, 9.9
Triangles, 4.10, 4.19–4.20, 7.25–7.26
Trim, valve, 3.8, 3.14, 3.32–3.33, 3.38
Tripping, 3.28
Tubing
    brass, 5.2
    cutting, 9.52, 9.56
    polyethylene, 9.49, 9.57–9.58
    polyethylene cross-linked, 9.49, 9.56–9.57
Turbine, steam, 1.5
Two-hole method, 6.5, 6.7, 6.13, 6.14

U
U-loop, 1.8
Ultraviolet radiation. *See* Sunlight
U.N. (unless noted otherwise), 2.2

Underground systems. *See* Piping, underground; Trench
Union(s)
    definition, 5.29
    plastic, 9.50, 9.51
    socket weld, 6.3, 6.4
    symbols, 6.9
    threaded, 5.5, 5.6
Units, mathematical, 4.6
Upright, 8.9, 8.10, 8.11, 8.25
U.S. Department of Transportation (DOT), 8.10
Utilities, underground, 8.2

V
Valuable materials, 1.7
Valve(s)
    for acid piping systems, 1.2
    angle, 3.14, 3.38, 6.9
    ball, 3.7–3.9, 3.38, 5.11
    butterfly, 3.15–3.17, 3.18, 3.38
    bypass, 3.4
    check, 3.10, 3.22–3.24, 3.34, 3.38, 5.11
    clamping or binding, 3.4
    clogged, 1.2
    cone, 3.8
    control, 3.14, 3.18, 3.20–3.22, 3.38, 5.11
    diaphragm, 3.18, 3.19, 5.11
    end protector, 3.28
    end types, 3.34
    flap, 3.22
    foot, 3.23–3.24
    gate, 3.2–3.6, 3.34, 3.38, 5.11
    globe, 3.2, 3.13–3.15, 3.34, 3.38, 5.11
    installation, 3.29–3.30, 5.24–5.25, 6.16–6.17, 7.27–7.28
    knife gate, 3.6–3.7
    markings and nameplate information, 3.31–3.33
    needle, 3.15, 3.18, 3.20
    overview, 3.2, 3.37
    pinch, 3.18
    in piping and instrumentation drawings, 2.12
    plug, 3.9–3.11, 3.34, 3.38, 5.11, 7.10
    rating designation, 3.32
    relief (pressure-relief), 3.21–3.22, 3.34, 3.38
    safety (pop-off), 3.18, 3.20–3.21
    selection, 3.30–3.31
    size, 3.30, 3.33
    stop, 3.34
    storage and handling, 3.28–3.29
    symbols, 3.33–3.34, 6.9, 7.10
    that regulate direction of flow. *See* Valve, check
    that regulate flow, 3.13–3.18, 3.19, 3.20
    that relieve pressure, 3.18, 3.20–3.22
    that start and stop flow, 3.2–3.12
    three-way, 3.5, 3.11–3.12
    types and applications, 3.31
    wafer, 3.15–3.16, 3.17
    wafer lug, 3.16, 3.17
    warm-up procedure, 3.4
    Y-type, 3.15, 3.16
Valve failure, 3.27
Vapors, 1.2, 1.3, 1.4, 1.7, 3.28, 6.2
Vehicles. *See* Traffic
Vendors, 2.12, 2.14
Vent, three-position, 3.4, 3.5
Venturi pattern, 3.8
Vibration, 5.2, 8.10, 8.14, 9.24, 9.27
Vise, 5.24, 5.25, 6.12, 7.12
Void, 8.21, 8.25
Volume, 4.12–4.17, 4.23
VS. *See* Equipment, vendor-supplied

W
Wachs pipe beveller, 7.11
Wale, 8.9, 8.10, 8.11, 8.25
Washer, 3.14, 5.6
Wastewater
    acids in neutralization process, 1.2
    drain, waste, and vent applications, 9.46, 9.47, 9.50–9.51
    ductile iron piping for, 9.22
    iron alloy iron piping for, 9.39–9.40
    sewer line projects, 8.16–8.17, 8.18
    valves for, 3.6, 3.11, 3.22
Wastewater treatment plants, 3.11, 9.24, 9.27, 9.47
Water
    cooling, 1.5, 1.6, 9.11
    demineralized, 1.5, 1.6
    distilled, 1.5
    for fire control, 9.11
    groundwater, 8.4, 8.7, 8.9, 8.14, 9.2, 9.3, 9.30
    piping systems, 1.5, 1.6, 5.9, 9.22, 9.47, 9.48, 9.49
    potable, 1.5, 1.6, 9.49
    recycled for cooling or irrigation, 1.7
    storm, 8.4, 9.13, 9.29, 9.30
    testing, 8.4
    underground lines, 8.2
    utility, 1.5
Water hammer, 1.5, 1.12, 9.48–9.49
Waterways, 8.10
Wedge(s), gate valve, 3.2, 3.3, 3.5, 3.38
Weep hole, 7.4
Weld
    butt, 6.6, 7.32, 9.37, 9.54, 9.58
    failure, 7.4
    fillet, 6.2, 6.3, 6.21
    full-penetration, 7.15, 7.33
    solvent, 9.53–9.55, 9.63
    tack, 6.11, 6.13–6.17, 6.21, 7.19
Weld draw, 7.22
Welding
    to align pipe and fittings, 5.11, 6.13–6.16, 6.20, 7.21–7.27
    burn-through during, 7.15, 7.33
    clamps, 7.21
    codes, 7.18, 7.19
    safety issues, 5.2, 6.2, 7.21, 7.23
    solvent, 9.53–9.55, 9.63
    use of backing rings, 7.16–7.17, 7.18
    valve installation, 6.16–6.17
Weldolet, 7.4, 7.5
Well, 8.4, 8.24
Wheel, grinder, 7.13
Whitworth threads, 5.10
Wire drawing, 3.14, 3.15, 3.38
Working level, definition, 1.8
Wrench, 3.28, 5.24, 5.25, 9.22, 9.27–9.29
Wrench makeup, 5.12, 5.14, 5.24
Wye, 5.4, 5.29, 9.51

Y
Y pipe fitting, 9.15, 9.16, 9.24
Y sewer service, 8.16, 8.17, 8.18
Yard, side, 2.9
Yield strength, 7.2
Yoke, 3.20, 3.30

Z
Zinc, 1.3, 5.2